Assessing Water Quality by Statistical Methods

Assessing Water Quality by Statistical Methods

Editor

Alina Barbulescu

MDPI • Basel • Beijing • Wuhan • Barcelona • Belgrade • Manchester • Tokyo • Cluj • Tianjin

Editor
Alina Barbulescu
Civil Engineering
Transilvania University of Brasov
Brasov
Romania

Editorial Office
MDPI
St. Alban-Anlage 66
4052 Basel, Switzerland

This is a reprint of articles from the Special Issue published online in the open access journal *Water* (ISSN 2073-4441) (available at: www.mdpi.com/journal/water/special_issues/Water_Quality_Statitcal_Analysis).

For citation purposes, cite each article independently as indicated on the article page online and as indicated below:

LastName, A.A.; LastName, B.B.; LastName, C.C. Article Title. *Journal Name* **Year**, *Volume Number*, Page Range.

ISBN 978-3-0365-6359-6 (Hbk)
ISBN 978-3-0365-6358-9 (PDF)

Contents

About the Editor

Alina Barbulescu

Alina Bărbulescu is a Full Professor at the Transilvania University of Brașov, Department of Civil Engineering. She graduated from the University of Craiova (Romania), Faculty of Mathematics, and from the Petre Andrei University of Iași (Romania), Faculty of Law. After a MSc in Mathematics at Bucharest University (Romania), she obtained a Ph.D. in Mathematics from Al. I. Cuza University of Iasi (Romania), a Ph.D. in Cybernetics and Economic Statistics from the Academy of Economic Studies Bucharest (Romania), and a Ph.D. in Civil Engineering, with Magna cum Laude, at the Technical University of Civil Engineering, Bucharest (Romania). She obtained her habilitation in Civil Engineering in 2014, with the thesis "Modeling the spatially distributed precipitations, and in Cybernetics and Economics Statistics" in 2019, with the thesis "Contribution to statistical analysis and modeling in economics and finance."

Before her current position, she taught at the Ovidius University of Constanța, Romania, Higher Colleges of Technology, U.A.E, and was a visiting professor at Abu Dhabi University. Her research interests span applied statistics, econometrics, hydrological modeling, water, and air pollution. She has authored over 200 articles, 30 books, and book chapters, was a coeditor of 20 Special Issues of scientific journals and conference proceedings and presented invited lectures at international conferences. She is a member of IAHS, IAENG, and SSMR and a topic editor at different journals, including "Water" and "Atmosphere".

Preface to "Assessing Water Quality by Statistical Methods"

Water, indispensable for life, has became scarcer in the last period due to overexploitation and pollution. Deforestation, urbanization, and, generally, agricultural and economic development lead to the water quality decreasing not only of the surface water, but also of the groundwater, endangering the water reserve. Toxic substances reaching water bodies without satisfactory treatment and climate change negatively impact water quality. Moreover, billions of inhabitants worldwide suffer from water scarcity, having limited or challenging access to drinking water. Therefore, evaluating the groundwater vulnerability and the threats to the surface waters, like detecting the pollution sources, their seasonal and spatial extent, and assessing the water quality are essential for making insight decisions on reducing and eliminating, if possible, the pollution impact.

Recent studies on water quality increasingly used statistical methods to select the water parameters that significantly impact the water quality or address pollution extent. They highlighted the advantages of the proposed approach in terms of time and cost savings. The advantage of using statistical methods also relies on validating the result that may be used in the next phase of forecasting. Therefore, the Special Issue "Assessing the Water Quality by Statistical Methods" of "Water" the book is based on contains articles on the following topics:

- Spatial and temporal analysis of the pollutants dissipation in water;
- Assessing the water quality using different water quality indicators;
- Assessing the groundwater vulnerability by parametric and nonparametric methods;
- Multivariate statistical approaches for evaluation of complex water quality datasets;
- Integrated frameworks for detecting the water quality in different catchments;
- Assessing the pollutants' transport in water;
- Quantitative and qualitative analysis of spatial and spatio-temporal hydrological data;
- Artificial intelligence methods for modeling the water parameters series.

The topics may interest water resources scientists and the larger public, treating actual environmental problems. Some methodologies proposed in the articles can be used or adapted to analyze other complex databases or solve spatiotemporal environmental problems.

The editor thanks the authors for sharing their research, and the reviewers whose valuable suggestions significantly improved the submitted articles.

We also thank the editorial staff, who ensured an excellent editorial process.

Alina Barbulescu
Editor

water

MDPI

Editorial

Assessing Water Quality by Statistical Methods

Alina Bărbulescu [1,*] and Cristian Ștefan Dumitriu [2,*]

1 Department of Civil Engineering, Transilvania University of Brașov, 5 Turnului Street, 900152 Brașov, Romania
2 S.C. Utilnavorep S.A., 55 Aurel Vlaicu Avenue, 900055 Constanța, Romania
* Correspondence: alina.barbulescu@unitbv.ro (A.B.); cris.dum.stef@gmail.com (C.Ș.D.)

Citation: Bărbulescu, A.; Dumitriu, C.Ș. Assessing Water Quality by Statistical Methods. *Water* **2021**, *13*, 1026. https://doi.org/10.3390/w13081026

Received: 5 April 2021
Accepted: 7 April 2021
Published: 9 April 2021

Water is one of the natural resources most affected by anthropogenic activities, like industry, agriculture, and traffic. Moreover, pollutants resulting from different sources are reaching water bodies with insufficient or without proper treatment. Climate change impacts the water quality as well, reducing the quantities of drinking water. Apart from human health, profoundly affected by water pollution, organisms living in the water and many ecosystems are affected more and more. Considering the difficult access to the drinking water of the population from different parts of the world, preserving the water resources, assessing their quality, and taking measures for the remediation of the water bodies' natural properties is a must.

In this context, the Special Issue with the title "Assessing Water Quality by Statistical Methods" addressed topics related to groundwater vulnerability [1–4], estimating water quality parameters in natural and artificial lakes and gulfs [5–7], investigating the causes of Rivers pollution [8–10], and proposing new tools for estimating the regional extent [11] and mitigating the effects of high precipitations on the water quality [12].

The single review [1] published in this volume addresses the groundwater vulnerability evaluation by the DRASTIC method and DRASTIC-like approaches, without including the particular case of the karst aquifer. Since its proposal, DRASTIC has become an essential tool for assessing the groundwater vulnerability, being adapted to different cases, like lithology, regions, land-uses, pollutants, land cover, and types of human settlements (rural and urban). The review is based on 128 articles, giving a detailed image of the previous work and opening possible study directions.

The article of Baena-Ruiz and Pulido-Velazquez [2] adapts the DRASTIC index to obtain reliable assessments in carbonate aquifers while maintaining its original conceptual formulation. The authors combine spatial statistics and decision trees to establish the DRASTIC parameters' domains and their associated weights. Comparisons of the new method's results with those obtained using the COP for the karst aquifer [13] show a concordance on 75% of the study area, the Upper Guadiana Basin. Given its good output, this approach should be validated in future studies.

In the same idea of evaluating the groundwater quality, in another part of the world—the United Arab Emirates—the article [3] proposes a combined approach involving a multivariate statistical analysis and water quality indices' use. The study relies on a database containing 42 concentration series of 19 water parameters collected in the Liwa area. Since the hypothesis that the series of water parameters recorded at different locations are similar was rejected, the samples were grouped in clusters and the main parameters that determined the differences between the clusters were determined by Principal Component Analysis (PCA). Finally, a quality index for assessing the water suitability for drinking was computed in two scenarios. The authors emphasized the necessity of using more than one technique to evaluate water quality for different purposes and to cross-validate the results [3].

Du et al. [4] employed the Factorial Analysis (FA) and Correspondence Analysis (CA), combined with geospatial tools, to identify the sources of inorganic nitrogen compounds in shallow groundwater, to determine the governing influencing factors and analyze the formation process in an agricultural area in Northeast China [4].

In the article [5], the authors investigated the long-term time series of salinity and water level of Techirghiol Lake (Romania), renowned in Europe for its sapropelic mud's therapeutic properties. These variables trends have been studied as functions of the precipitation, overland flow, groundwater recharge, and the hydraulic works from the lake's neighborhood. The study shows that the Techirghiol Lake is a heavily modified water body.

Mamun et al. [6] realized a comprehensive spatial and temporal analysis of the water quality of the Paldang reservoir (Korea), which is an important source of freshwater. The tools utilized to carry out the research were the PCA, FA, and stepwise spatial discriminant analysis (DA), together with the Trophic State Index (TSI) and Trophic State Index deviation (TSID). They show that anthropogenic activity is mainly responsible for the significant spatial variations of most water parameters. The water parameters that significantly vary during different seasons are the temperature, the biological oxygen demand, and total suspended solids. The water parameters that significantly vary at different sites are the water temperature, the electrical conductivity, the total nitrogen, the ratio biological oxygen demand: chemical oxygen demand, and total nitrogen: total phosphorus.

In their research, Al-Taani et al. [7] investigated the water quality of the Aqaba Gulf in Jordan, a zone with high touristic potential. The investigation of 19 parameters of water samples collected in different parts of the Littoral did not present specific patterns, except for the metal contents that showed increasing values close to an industrial complex by comparison to other sampling places.

The article *Characteristics and Causes of Long-Term Water Quality Variation in Lixiahe Abdominal Area, China* [8] presents the results of a set of statistical tests (the Mann–Kendall trend test and Sen's slope estimator) and the analysis of the correlation between the water quality variation, the water level, and the water diversion employing the cross wavelet transform and wavelet coherence. The results show that the comprehensive water quality index (CWQI) included periodic fluctuations on multiple scales from 0.25 to 5 years [8].

Yu et al. [9] analyzed twenty water parameters from samples collected during March, May, July, September, and November 2010 in Chaobai River (China). They found that the eutrophication level was severe in most locations where the water was collected, and the water was not appropriate for irrigation water (indicated by the sodium adsorption ratio). Cluster analysis was used to determine the significant spatial and temporal variation among different data series. The Gibbs plot indicated that the water–rock interaction mainly controls the water chemistry, whereas the PCA showed that river water had undergone the minerals dissolution.

Kim et al. [10] study the land-cover influence on the streams water quality in urban zones from South Korea. After classifying the watersheds in three clusters, the factorial analysis is used to select the water parameters that participate in the computation of a newly defined water quality indicator (WQImin), further employed to estimate the degree of water contamination. WQImin appears as a competitor for the other water quality indexes already known.

Costache et al. [11] introduced an integrated framework for detecting the areas prone to flooding generated by flash-floods in small river catchments as an indicator for water quality. Three models were generated in the first stage, by a combination of frequency ratio (FR), weights of evidence (WOE), and statistical index (SI), with fuzzy analytical hierarchy process (FAHP), and the best one (FAHP-WOE) was selected. In the next stage, the first step's output was weighted using the flow accumulation method to determine the valleys with different levels of susceptibility to flood. Finally, ten flood conditioning factors were used to determine flood susceptibility through the analytical hierarchy process model. Given that the higher the susceptibility to flooding is, the higher the pollution with different sediments and other materials carried by the water is, this method provides an interesting tool for pollution warning in small river catchments.

The article [12] proposes a new evolutionary approach for determining an optimal parameter in the Inverse Distance Weighting Interpolation (IDW) method that has a large applicability in the spatial interpolation of regional data series. Proposed for the scenario, the algorithm was tested on 41 series collected at ten locations, in Dobrogea, Romania. In terms of the mean absolute error (MAE) and the mean standard error (MSE) the new algorithm outperformed the classical IDW, the ordinary kriging, and the Particle Swarm Optimization of IDW. The algorithm should be validated on other data sets.

Author Contributions: Conceptualization, A.B. and C.Ș.D.; methodology, A.B. and C.Ș.D.; formal analysis, C.Ș.D.; writing—A.B. and C.Ș.D.; writing—review and editing, A.B.; supervision, A.B.; project administration, A.B. All authors have read and agreed to the published version of the manuscript.

Funding: This research received no external funding.

Conflicts of Interest: The authors declare no conflict of interest.

References

1. Bărbulescu, A. Assessing the groundwater vulnerability: DRASTIC method and its versions: A review. *Water* **2020**, *12*, 1356. [CrossRef]
2. Baena-Ruiz, L.; Pulido-Velazquez, D. A Novel Approach to Harmonize Vulnerability Assessment in Carbonate and Detrital Aquifers at Basin Scale. *Water* **2020**, *12*, 2971. [CrossRef]
3. Bărbulescu, A.; Nazzal, Y.; Howari, F. Assessing the groundwater quality in the Liwa area, the United Arab Emirates. *Water* **2020**, *12*, 2816. [CrossRef]
4. Du, X.; Feng, J.; Fang, M.; Ye, X. Sources, Influencing Factors, and Pollution Process of Inorganic Nitrogen in Shallow Groundwater of a Typical Agricultural Area in Northeast China. *Water* **2020**, *12*, 3292. [CrossRef]
5. Maftei, C.; Buta, C.; Carazeanu Popovici, I. The Impact of Human Interventions and Changes in Climate on the Hydro-Chemical Composition of Techirghiol Lake (Romania). *Water* **2020**, *12*, 2261. [CrossRef]
6. Mamun, M.; Kim, J.Y.; An, K.-G. Multivariate Statistical Analysis of Water Quality and Trophic State in an Artificial Dam Reservoir. *Water* **2021**, *13*, 186. [CrossRef]
7. Al-Taani, A.A.; Rashdan, M.; Nazzal, Y.; Howari, F.; Iqbal, J.; Al-Rawabdeh, A.; Al Bsoul, A.; Khashashneh, S. Evaluation of the Gulf of Aqaba Coastal Water, Jordan. *Water* **2020**, *12*, 2125. [CrossRef]
8. Jiang, C.; Zhou, J.; Wang, J.; Fu, G.; Zhou, J. Characteristics and Causes of Long-Term Water Quality Variation in Lixiahe Abdominal Area, China. *Water* **2020**, *12*, 1694. [CrossRef]
9. Yu, Y.; Song, X.; Zhang, Y.; Zheng, F. Assessment of Water Quality Using Chemometrics and Multivariate Statistics: A Case Study in Chaobai River Replenished by Reclaimed Water, North China. *Water* **2020**, *12*, 2551. [CrossRef]
10. Kim, T.; Kim, Y.; Shin, J.; Go, B.; Cha, Y. Assessing Land-Cover Effects on Stream Water Quality in Metropolitan Areas Using the Water Quality Index. *Water* **2020**, *12*, 3294. [CrossRef]
11. Costache, R.; Bărbulescu, A.; Pham, Q.B. Integrated framework for detecting the areas prone to flooding generated by flash-floods in small river catchments—A useful indicator for water quality. *Water* **2021**, *13*, 758. [CrossRef]
12. Bărbulescu, C.; Șerban, M.-L. Indrecan, Improving spatial interpolation quality. IDW versus a genetic algorithm. *Water* **2021**, *13*, 863. [CrossRef]
13. Ravbar, N.; Goldscheider, N. Comparative application of four methods of groundwater vulnerability mapping in a Slovene karst catchment. *Hydrogeol. J.* **2008**, *17*, 725–733. [CrossRef]

water

Review

Assessing Groundwater Vulnerability: DRASTIC and DRASTIC-Like Methods: A Review

Alina Barbulescu

Department of Mathematics and Computer Science, Ovidius University of Constanta, 900527 Constanta, Romania; alinadumitriu@yahoo.com

Received: 30 March 2020; Accepted: 8 May 2020; Published: 11 May 2020

Abstract: Groundwater vulnerability studies are sources of essential information for the management of water resources, aiming at the water quality preservation. Different methodologies for estimating the groundwater vulnerability, in general, or of the karst aquifer, in particular, are known. Among them, DRASTIC is one of the most popular due to its performance and easy-to-use applicability. In this article, we review DRASTIC and some DRASTIC-like methods introduced by different scientists, emphasizing their applications, advantages, and drawbacks.

Keywords: aquifer; DRASTIC; index; groundwater; vulnerability

---

1. Introduction

In recent decades, water scarcity and its pollution became a major issue all over the world. Preserving the groundwater quality is very important for assuring the drinking water resources, given that billions of people all over the world do not have access to water or suffer from water scarcity [1].

Since 1968, when Margat [2] introduced the concept of groundwater vulnerability, many definitions were proposed for this concept. For example, Hirata and Bertolo [3] defined the groundwater vulnerability as "the property of a groundwater system that depends on the sensitivity of the material in permitting the degradation of the saturated zone by pollutant substances originating from human activities", while the National Research Council [4] defined this term as "the relative ease with which a contaminant (in this case a pesticide) applied on or near the land surface can migrate to the aquifer of interest under a given set of agronomic management practices, pesticide characteristics, and hydrogeological sensitivity conditions".

The intrinsic vulnerability describes the water vulnerability to different pollutants (independent of their nature) resulted from human activities and is related to the hydrological, geological, and hydrogeological aquifer's characteristics. Given that the aquifers have different reactions to the same contaminant due to their physicochemical characteristics, the specific vulnerability shows the groundwater vulnerability to a pollutant (or a group of pollutants), determined by the pollutant's properties, taking into account the time of impact, its intensity, and the interaction between the intrinsic vulnerability components and the contaminant [5,6].

Adams and Foster [7] emphasized that the aquifer vulnerability depends on the properties of the layers situated above the saturated zone to attenuate the pollutants' effect, by retention or neutralization by chemical reactions.

Gogu and Dassargues [6] divided the approaches of assessing the groundwater vulnerability in three groups, as a function of the groundwater protection. The first group takes into account only the soil and unsaturated zone, the second one takes into consideration the groundwater flow and the contaminant transfer to some extent [8], whereas the third focuses on the soil, the unsaturated medium, and the aquifer.

Different approaches are used for estimating groundwater vulnerability. They can be grouped into three categories. The first group is formed by the index-based methods, which take into consideration only the characteristics of soil and unsaturated zone. They are divided into Hydrogeological Complex and Settings methods (HCS) [9]; Matrix Systems [10], approaches based on the combination of two parameters, and Rating Systems [11–13]. They work by building water vulnerability groups using different ratings associated with the physical characteristics of the study media. The second group contains the statistical approaches that assess the groundwater vulnerability through statistical analysis or regression models [14–16]. The third one contains the methods based on simulation, which uses simulation techniques for forecasting the processes related to contaminant transport [17–20]. The index-based techniques have the advantage that they do not depend on data availability or similarities [21].

The procedures that belong to the first and second categories are used for studying the intrinsic vulnerability of large areas [22].

The most used index methods for studying the groundwater vulnerability are DRASTIC [23], GOD [12], AVI rating system [13], DIVERSITY [24], ISIS [11], PRAST [25], SEEPAGE, SINTACS [26–29]. For the karst aquifer, EPIK [5], REKS [30], RISKE [31], RISKE 2 [32], COP and COP + K [33,34], PaPRIKa [35], PI and the Slovene approach [36,37] have been proposed.

Introduced in 1985, DRASTIC is among the most popular approaches used in groundwater vulnerability estimation due to its capability and easy-to-use. In the following, we shall focus on reviewing this method, and some of the DRASTIC-like procedures that aim to improve the performance of the groundwater vulnerability estimation, emphasizing the differences between them. We shall not focus on the methods assessing the groundwater vulnerability for the karst aquifer because of the extensive literature for the general case and the lack of space.

Some classifications of the methods that will be presented in next sections are:

1. Based on the extent of their use:
   a. With general applicability—DRASTIC, GOD
   b. For specific regions—SINTACS, DRAMIC, DRIST, DRAV
   c. That considers the land use—DRASTIC-LU, DRASIC-LU, SINTACS-LU
   d. For urban area—DRAMIC, DRASTICA
2. Based on the specific vulnerabilities assessed:
   a. Lithological-oriented—methods assessing the kart aquifer vulnerability [5,29–36] and for the fractured environment (referred in the following by Modified DRASTIC)
   b. Pollutants' oriented—Pesticide-DRASTIC, Modified Pesticide–DRASTIC, SI DRARCH.

We shall indicate the references to the articles treating these methods in the next sections, together with a description of approaches.

The methods (and corresponding parameters) for groundwater vulnerability assessment discussed in this article are summarized in Table 1.

2. DRASTIC

DRASTIC is a model that considers the main hydrological and geological factors with a potential impact on aquifer pollution. Its acronym stands for D—depth to groundwater, R—recharge rate, A—aquifer, S—soil, T—topography, I—vadose zone's impact, and C—aquifer's hydraulic conductivity [38].

The depth to water table (D) [m] is the thickness of the layer crossed by the pollutant before reaching the aquifer. The aquifer vulnerability is inverse proportional to the depth to the water table.

Table 1. Methods and corresponding parameters for groundwater vulnerability assessment.

Parameter/ Method	Depth to the Water Table	Net Recharge	Hydrogeological Features	Soil Characteristics	Topographic Slope	Characteristics of Unsaturated Zone	Aquifer Hydraulic Conductivity	Liniment Density	Stream Network	Aquifer Tickness	Landuse [#]	Anthropogenic [##] Impact (LU)	Pesticides	Specific Region
DRASTIC	x	x	x	x	x	x	x							
DRASTICM	x	x	x	x	x	x	x	x						
DRIST	x	x		x	x	x								x
DRAV	x	x	x	*										
DRAMIC	x	x	x			x	**			x		x		
DRASTICA	x	x	x	x	x		x					x		
DRASTIC-LU	x	x	x	x	x	x	x					x		
DRASIC-LU	x	x	x	x		x	x					x		
SI	x	x	x		x							x		x
DRARCH	x	x	***			x	x			x				
SINTACS	x	x	x		x	x	x		x	x	x			x
SINTACS-LU	x	x	x		x	x	x		x	x	x	x		x
Pesticide DRASTIC	x	x	x	x	x	x	x						x	
Pesticide DRASTIC LU	x	x	x	x	x	x	x					x	x	

[#] The land use parameter characterize the human activity as effect on the runoff coefficient, not as the contaminants' nature. [##] Refers to the impact of the human activity as impact of the built environment or the nature of pollutant. * replaced by the vadose zone lithology. ** replaced by the contaminant impact. *** replaced by the ratio of the clay layers' thickness to the vadose zone thickness.

The net recharge (R) [mm/year] represents the volume of infiltrated water that reaches the aquifer. The contamination possibility increases if the net recharge increases. Three types of recharges can be distinguished: direct, indirect, and localized [39,40]. The aquifer media (A) consists of different types of rocks serving as an aquifer.

The upper part of the vadose zone, with intense biological activity, is defined to be the soil media (S).

The topography (T) (%) is defined by the terrain slope, together with its variation. A low slope will determine a small surface flow and a high pollution risk.

The vadose zone's impact (I)—The unsaturated or discontinuously saturated layer situated above the water table is called vadose. The pollutant's transfer is influenced by the vadose zone's lithology.

The aquifer hydraulic conductivity (C) is the aquifer materials' capacity to leave the water to pass through it. The aquifer vulnerability is low for reduced hydraulic conductivities.

The hypotheses of the DRASTIC models are:

- The pollutants are produced at the surface of the Earth;
- The pollutants are transported into the soil by precipitation;
- The pollutants' travel velocity is that of the water;
- The affected area must be big enough.

Firstly, a rate from 1 to 10 is assigned to each parameter, 1 being the least important [38]. Then, the DRASTIC index score is built, using the weights fixed for each parameter. The formula for DRASTIC index is:

$$\text{DRASTIC index} = D_R D_W + R_R R_W + A_R A_W + S_R S_W + T_R T_W + I_R I_W + C_R C_W \quad (1)$$

where R is the rate and w is the parameter weight.

The weights have been set up by EPA (the United States Environmental Protection Agency) based on the experts' knowledge after studying different regions. In the original DRASTIC algorithm the weights range from 1 to 5 (1 being the least important), the smallest possible index score is 23 and the highest, 230. Tables 2 and 3 contain the weights and ratings of the components, firstly provided in [38]. Lower groundwater vulnerability is described by a lower index score.

Table 2. DRASTIC D, R, T, and C rating and weighting [38].

Depth to Water (mm) – weight = 5							
range	0–1.5	1.5–4.6	4.6–9.1	9.1–15.2	15.2–22.8	22.8–30.4	>30.4
rating	10	9	7	5	3	2	1
Net Recharge (mm) – weight = 4							
range	0–50.8	50.8–101.6	101.6–177.8	177.8–254	>254		
rating	1	3	7	8	9		
Hydraulic Conductivity of the Aquifer (m/day) – weight = 3							
range	0.04–4.1	4.1–12.3	12.3–28.7	28.7–41	41–82	>82	
rating	1	2	4	6	8	10	
Topography (slope %) – weight = 1							
range	0–2	2–6	6–12	12–18	>18		
rating	10	9	5	3	1		

Table 3. DRASTIC A, I, S rating and weighting

Aquifer Media		Vadose Zone Material		Soil Media	
weight = 3	**rating**	**weight = 5**	**rating**	**weight = 2**	**rating**
Massive shale	2	Silt/clay	1	Non-srinking and non-aggregated clay	1
Metamorphic/igneous	3	Shale	3	Muck	2
Weathered metamorphic/igneous	4	Metamorphic/igneous	4	Clay loam	3
Thin-bedded sandstone, limestone, shale sequences	6	Limestone	6	Silty loam	4
Massive sandstone	6	Sandstone	6	Loam	5
Massive limestone	8	Bedded limestone, Sandstone, shale	6	Sandy loam	6
Sand and gravel	8	Sand and gravel with significant silt and clay	6	Shrinking and/or aggregated clay	7
Basalt	9	Sand and gravel	8	Peat	8
Karst limestone	10	Basalt	9	Sand	9
		Karst limestone	10	Gravel	10
				Thin or absent	10

Different authors [41,42] pointed out the DRASTIC results' accuracy, the small amount of input data, its application's low cost [38,43], reduced computational time, and simple computational procedure [44]. DRASTIC proved to be useful for evaluating the aquifer vulnerability in priority monitoring areas and as a valuable indicator where detailed hydrogeological evaluation is necessary. Other authors emphasized the limited validation procedure of the DRASTIC methodology [45,46] and a low correlation between the experimental data and the model's output [47,48]. Wang et al. [49] remarked on the necessity of procedure adaption for urban areas, while the parameters' weight choice in the DRASTIC index was criticized by Merchant [50]. Therefore, several approaches were proposed for improving the groundwater vulnerability estimation accuracy, each of them involving a different number of parameters. In the following, we shall present some of these methods and the rationale for their use.

Although the DRASTIC model was intended to be used in mapping applications, it was not expressly designed for use in a GIS, its initial applications employing a manual map overlay and computation procedure [50]. The main importance of vulnerability maps is that their analysis can provide effective information for making informed decisions for water management [51].

Merchant et al. [52] were the first that used GIS for DRASTIC implementation. Since then, due to their capability of retrieving, storing, organizing, analyzing, and presenting geographically referenced spatial data, GIS methods have been successfully employed for assessing the groundwater vulnerability [53–60].

The main GIS advantage is its efficiency of combining data layers and changing the parameters used for the vulnerability classification [49]. For producing a groundwater pollution risk map (Figure 1), it is necessary to prepare the seven individual maps (one for each component in the model). Therefore, all data should be available, accurate enough [50], and introduced in a GIS database.

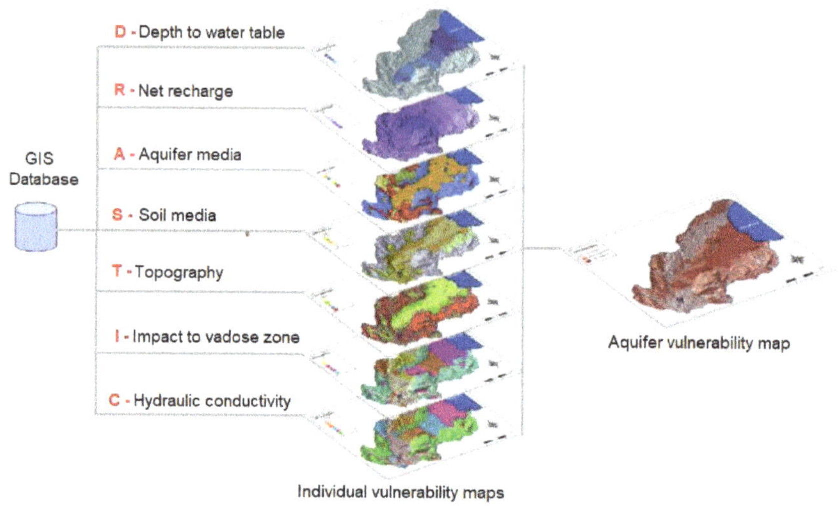

Figure 1. Flowchart for building a vulnerability map in DRASTIC (adapted from [61]).

The D parameter layer is generated using topographic maps. Then, the IDW method is applied for interpolating the water level data and obtaining the depth to the water layer. The general water balance equations are used for generating the recharge layer, R. The lithology, and the type of aquifer media are considered for the estimation of the A factor. The soil media, S, is determined by using the soil textural classification chart. The topographic slope, T, is determined by using a digital elevation model and the data extraction from the topography layer. Then the slope layer (%) is generated, and the range is reclassified taking into account the DRASTIC ranges. The impact of the V factor is determined using the depth to the water layer and the well logs report. The hydraulic conductivity data is retrieved by experimental measurements. Finally, the pollution risk layer is produced using the seven layers previously built by GIS, and all the DRASTIC thematic layers are combined [62].

3. Modified DRASTIC (DRASTICM)

Scientific studies pointed out that geologic structures have a significant impact on highly fractured environments' vulnerability. Therefore, in a study performed for a region form Nicaragua, Mendoza and Barmen [63] modified the DRASTIC index by including the influence of the length, connectivity, and lineament density. They introduced the lineament influence, denoted by M, in the new model, called Modified DRASTIC, whose index, MDI, is defined by

$$\mathrm{MDI} = \mathrm{DRASTIC\ index} + 5\mathrm{M_R}, \tag{2}$$

where R is the rating, M is the lineament factor.

A rate between 0 and 3 was assigned to the influence of the lineament.

Data collected from the field and photographic interpretation were normalized and combined in a map to assess the lineament influence. This map and the other seven (from DRASTIC) contributed to building the Modified DRASTIC map.

Mendoza and Barmen [63] also proposed the classification of groundwater vulnerability degree as very high (MDI > 199), high (MDI between 160 and 199), moderate (MDI in the range 120–159), low (MDI between 80 and 119), and very low (MDI < 79).

The results show that D and T are the factors with a significant influence on vulnerability prediction. Compared with DRASTIC, the modified DRASTIC gave a better estimation of the contamination risk in zones with high fractured structures.

4. DRIST and Modified DRASTIC

Introduced for investigating the underground water vulnerability in Grombalia, the DRIST model was adapted to the hydrogeological system properties from this region. DRIST considers only parameters related to the unsaturated aquifer zone, while DRASTIC works with the aquifer saturated zone characteristics [40]. The calculation of the DRIST vulnerability index is similar to that for DRASTIC (but ignoring A and C parameters).

In the same article, Chenini et al. [40] proposed a Modified DRASTIC method. The difference between these approaches resides in the estimation of the factors A and I. In the new one the lithology is substituted by the permeability, as suggested in [64]. The other maps are created by the same procedure as in DRASTIC.

The permeability map of the vertical vadose zone is realized based on the vertical permeability formula:

$$K_1 = H \Big/ \sum_{i=1}^{p} (h_i / k_i), \tag{3}$$

where K_1 is the vertical average permeability (m/s), H—the unsaturated zone total thickness (m), h_i—the thickness of the ith layer (m), k_i—the permeability of the ith layer (m/s), and p—the number of layers [64].

The saturated zone's permeability map is determined using the formula of the horizontal permeability [65]:

$$K_2 = \left(\sum\nolimits_{i=1}^{p} (h_i k_i)\right) / \left(\sum\nolimits_{i=1}^{p} h_i\right), \tag{4}$$

where K_2 is the average horizontal permeability (m/s), h_i, k_i and p have the same significance as in formula (3), while $\sum_{i=1}^{p} h_i$ at the denominator of formula (10) is the saturated zone total thickness (m).

Comparing the two vulnerability maps, Chenini et al. [40] showed that there are differences between them. The area with medium vulnerability is more significant in the Modified DRASTIC due to the minimization of the saturated zone effect, as an effect of the permeability replacement by lithology in the process of parameters' estimation. DRIST map reflects the effect of removing the factors related to the saturated zone.

Sakala et al. [66] used the same model and a neural network approach to generate a groundwater vulnerability model. The network used as input the DRIST parameters, and as the training dataset, the sulfate and Total Dissolved Solids (TDS) concentrations retrieved from five groundwater samples. The groundwater vulnerability model was finally obtained by applying a fuzzy operator for combining the training and classification results. The model's results are well correlated with the available data and the output of the DRIST model.

5. DRAV

DRAV is a model designed by modifying DRASTIC for taking into account the groundwater characteristics from the arid zones [67]. Since generally, in those areas, there is no horizontal runoff, the DRASTIC T term was removed, and S was replaced by V (vadose zone's lithology). The factors D, R, and A were kept in the new model.

The DRAV index is a linear combination of the factors D, R, A, and V with the normalized weights 0.20, 0.15, 0.31, and 0.34, respectively.

The scores for the D factor are 1, 2, 3, 5, 7, 10, for groundwater depths (m) greater than 30, in the interval (10, 30], between 6 and 10, in the interval (3, 6], in the range 1–3, less than 1, respectively [66].

The scores for the R factor are 1, 2, 4, 6, 8, 10, for recharge modules (×104 m^3/km^2/area) less than 5, in the interval [5, 10), between 10 and 20, from 20 to 30, in the interval [30, 50), greater than 50, respectively [67].

The scores for the A factor are 1, 3, 5, 7, and 10, for a storativity (m^3/day/m) smaller than 2, between 2 and 20, in the interval [20, 200), from 200 to 1000, and greater than 1000, respectively [66].

The scores (for the V factor) 10, 4, 2, and 1 were associated with Sandy gravel, Sandy loam, Sandy clay, and Silty and fine sand, respectively [67].

Five classes of phreatic water vulnerability (extremely high, high, medium, low, and extremely low) corresponds to vulnerability indices above 8, the interval (6, 8], between 4 and 6, the interval (2, 4], ≤ 2, respectively.

DRAV was used to analyze the pore groundwater in the northwestern part of China, but no comparison with other methods is provided. Therefore more studies are necessary to validate this approach.

6. DRAMIC

Many scientists emphasized the limitations of DRASTIC's application for urban areas [49,68], as follows. (1) The terrain where the cities are situated is mostly flat, so the T factor in the DRASTIC model is not relevant. (2) The values of the soil media can be hardly obtained because the ground surface is mostly covered by concrete. (3) The hydraulic conductivity is not relevant. Therefore, they built the DRAMIC index, by replacing in DRASTIC the S factor by the thickness of the aquifer (M), and the C factor by the contaminant impact (denoted by C as well). It must be noticed that DRAMIC does not consider the pollutants' properties, but its stability and infiltration capacity into the aquifer. The parameters (and ratings) in DRAMIC are [49]:

- Aquifer thickness (m): 0–6 (9), 6–15 (7), 15–25 (5), 25–32 (4), 32–40 (3), 40–50 (2), >50 (1);
- Contaminant's characteristics:
    - Stability, infiltration easiness (9)
    - Stability, infiltration relative easiness (7)
    - Stability, infiltration uneasiness, and Relative stability, infiltration easiness (5)
    - Relative stability, infiltration relative easiness (4)
    - Relative stability, infiltration uneasiness, and Instability, infiltration easiness (3)
    - Instability, infiltration relative easiness (2)
    - Instability, infiltration uneasiness (1)

The DRAMIC Index is computed by the relation

$$\text{DRAMIC index} = 2D_R + 3R_R + 4A_R + 2M_R + 5I_R + 1C_R \tag{5}$$

where R is the rating.

The main factors considered in DRAMIC are the stability of the pollutant and the easiness of the pollutant infiltration. The results of this model applied in a study from China (Wuhan region) were compared with the field data, showing a good correlation. Despite promising results, other studies are needed to validate this method for other urban areas.

7. DRASTICA

DRASTICA is a modified DRASTIC model, which includes the anthropogenic influence in urbanized environments [68,69]. A new factor (A-anthropic factor) was introduced, with the weight equal to 5. The index is computed as in DRASTIC, adding the new term, $A_R A_w$, where A_R is the rating and A_w the weight. The anthropic factors and the rating assigned are the following [68]: Effluents/sewage/industrial waste (untreated), Oil spillage/gas flaring and E-wastes – 9, Open dumpsites (non-sanitary landfill) and Emissions from automobiles/generators – 8, Cementary/soakaway/pit latrine (unlined) and Fertilizer/agrochemicals–7, Domestic waste (organic/degradable) – 6, Effluents/sewage/industrial waste (treated) and Sanitary landfill – 5, Cementary/soakaway/pit-latrine (lined) and Bush burning – 4. The rating and weighing of the other parameters were kept as in DRASTIC.

Four vulnerability categories were built (low, moderate, high, and very high), corresponding to values of vulnerability indexes in the intervals 140–159, 160–179, 180–199, 200–215.

In a study of the water pollution impact in Lucknow, India, DRASTICA better performs by comparison to DRASTIC, when the models were validated using field data. The sensitivity study emphasized that the less sensitive factors were A (aquifer), followed by S and T. The parameters with the highest impact are D, followed by A (anthropogenic factor) and C.

Another research concerning the groundwater vulnerability in the Niger Delta [69] concluded that the anthropic activity (incorporated in the A factor) had a consistent impact on the groundwater contamination.

8. DRASTIC-LU

Studies concerning the groundwater vulnerability showed an increasing impact of land use on water contamination [51,70,71]. Alam et al. [51] indicated that industrial and sewage pollution, pesticides, and fertilizers alter groundwater quality. They proposed a new index, DRASTIC-LU, adding "the land use pattern" (LU) parameter. The land use categories considered (and the rating) are respectively: urban and industrial (10), rural and industrial (9), rural and agriculture (8), with a weight of 5.

The DRASTIC-LU index is computed by:

$$\text{DRASTIC-LU} = D_R D_w + R_R R_w + A_R A_w + S_R S_w + T_R T_w + I_R I_w + C_R C_w + L_R L_w, \quad (6)$$

where the land use rating and weight are L_R and L_w, respectively.

The other acronyms have the same significance as in the DRASTIC index.

The parameter of the vadose zone impact (I_R) is computed by [70]:

$$I_R - T\Big/\left(\sum_{i=1}^{n} \frac{T_i}{I_{r_i}}\right), \quad (7)$$

where T is the vadose zone total thickness, T_i is the ith layer thickness and I_{r_i} is the ith layer rating.

Since this approach considers many layers of the vadose zone, it is expected to provide more accurate results.

The values of the DRASTIC-LU index are situated in the interval [158, 190], divided into subintervals as follows: less than 160 (corresponding to low vulnerability zone), 160–170 (medium vulnerability zone), 170–180 (corresponding to high vulnerability zone), greater than 180 (very high vulnerability).

Some research [51,70,72,73] investigated the groundwater vulnerability in different regions of India. In a study related to a zone of Central India, Alam [50] showed that the most significant parameters in the model DRASTIC-LU model are D, I, C, and LU.

In a vulnerability analysis in the Basin of Damodar River, Kumar and Khrisma [72] compared the performance of DRASTIC and DRASTIC-LU, emphasizing the significant impact of the LU component. The sequence of impact intensities I > D > C > LU > S > T > R > A resulted after investigating the map sensitivity. At the models' validation stage, a better correlation between the field data and the estimated ones resulted in the DRASTIC-LU model (0.893 against 0.781 for DRASTIC). Therefore one can conclude that the essential factors that should be taken into account for assessing the vulnerability in the study zone are A, T, I, and LU.

In the sensitivity analysis by map removal in a DRASTIC-LU approach for Karun Basin, Sinha et al. [73] found a different sequence of impact intensities by comparison with [72] (LU > S > T > D > I > A > R). Therefore, the LU and S factors have the main effect on the DRASTIC-LU index. This result is concordant with the field reality (the aquifer' shallow waters). Sensitivity analysis revealed that depth of water table, land use, and topography produce large variations of vulnerability index by comparison with other parameters.

9. DRASIC-LU

DRASIC-LU is a version of DRASTIC, initially used for assessing the groundwater pollution risk in some sub-regions of India (Ganga Plain) [71]. Due to the topographic small variation, the parameter T was removed from the DRASTIC index and was replaced by the parameter L (land use), which reflects the land use impact on the water quality [74]. The land use categories are the same as in the DRASTIC-LU, and the vadose zone impact parameter is computed by the relation (7). Qinghai et al. [75] introduced the hydraulic conductivity values in concordance with the experimental data. They are respectively:

- The ratings for D (depth to the water table) are 2, 3 and 5, while the weighting factor is 5;
- The rating for net recharge (R) is 9, and the weight scale is 4;
- The rating for aquifer media (A) is 8, and the weight scale is 3;
- The ratings for soil media (S) are 5 and 6, and the weight scale is 2;
- The ratings for vadose zone impact (I) are 1 and 2, while the weight scale is 5;
- The ratings for hydraulic conductivity (C) are 4, 8 and 10, while the weight scale is 3;
- The ratings for land use (L) are 8, 9, and 10, and the weight scale is 5.

The new index is defined by:

$$\text{DRASIC-LU} = D_R D_W + R_R R_W + A_R A_W + S_R S_W + I_R I_W + C_R C_W + L_R L_W, \tag{8}$$

The terms have the same significance as in Equations (1) and (6). The index varies in the interval [140, 180], which is divided in four sub-intervals: [140, 150]—for low vulnerability zones, [150, 160]—for moderate vulnerability zones, [160, 170] for high vulnerability zones and [170, 180]—for very high, respectively.

Studying an aquifer from the Ganga Plain, Umar et al. [74] concluded that D, C, I, and LU are the main factors to be considered for vulnerability mapping.

10. SI Index

Ribeiro [76] introduced the SI method for the estimation of the groundwater vulnerability to pollutants generated in areas at medium and large in Portugal. SI is obtained by removing S, I, and C from DRASTIC an including the land use parameter (LU) that incorporates the agricultural activities' impact (especially nitrates) on the water quality [77]. Therefore this method assesses the specific vulnerability of groundwater.

The SI index is computed by:

$$SI = D_R D_W + R_R R_W + A_R A_W + T_R T_W + LU_R LU_W, \tag{9}$$

where the parameters' weights are [77]:

$$D_W = 0.186, R_W = 0.212, A_W = 0.259, T_W = 0.121, LU_W = 0.222.$$

The essential land use activities classes and the corresponding rating values (displayed inside the brackets) [77] varies between 0 (for semi-natural zones and forest) and 100 (for mines, landfill, and industrial discharge), with intermediate values as follows:

- 90—Paddy fields, Irrigated perimeters irrigated,
- 80—Shipyard and quarry,
- 75—Green and continuous urban zones, and artificially covered zones
- 70—Discontinuous urban zones, and Permanent cultures
- 50—Aquatic media, agro-forest zones, pastures.

From 2000, when it was introduced, the SI index was applied for vulnerability studies in Columbia, Malaysia, Morocco, Portugal, and Tunisia [59,76–84]. Hamza and Added [82] show that DRASTIC does not consider the contaminant's nature and gives great weight to the hydrogeological factors. The case study supports the idea that the SI method was designed for taking into account the nitrates properties and the relations between them and the intrinsic vulnerability. LU factor integrates the land use types, allowing the integration of different particular characteristics.

The results of Stigter et al. [77] and Hamza et al. [83] show that permeable aquifer and high recharge are responsible for the pollution vulnerability increase. For chloride or nitrate contaminants in specific conditions, the dilution potential may have a significant role in the determination of contamination degree [81]. Validating the vulnerability maps using the measured nitrites concentration, Stigter et al. [77] emphasized the groundwater vulnerability underestimation when DRASTIC was used instead of SI. Another comparative analysis of these methods validated in the field showed a better concordance when using the SI approach [83].

11. DRARCH

This model was introduced for studying the water vulnerability at arsenic in the Taiyuan basin and is based on simulation of the solute transport. The procedure can be summarized as follows [75]:

(1) Build a series of contaminant transport models employing Hydrus1D and use each model index in the simulations of the contaminant transport.
(2) Increase the accepted index value and compute the associated migration distance of the contaminant.
(3) Analyze the relationship between the index values and the pollutant' simulated migration distances and determine the indexes' ratings.
(4) Use the factorial analysis to determine the weighting of each index.
(5) Apply the ordinary kriging for estimating the vulnerability spatial variation over the basin.

The D and R indices from DRASTIC are kept in the DRARCH model, while the other indices were replaced by:

- Aquifer thickness (A);
- The ratio of the clay layers' thickness to the vadose zone thickness (R), introduced for emphasizing that the clay has a specific surface area and an adsorption capacity greater than other sediments;
- The coefficient of pollutant's adsorption by the sediment in the vadose zone (C);
- Aquifer hydraulic conductivity (H).

The indices weights are 2, 1, 7, 9, 7, and 5, respectively. The rating values associated with these indices are given in [49]. The range, in meters, and the rating associated with the depth to the water table are respectively: 0–2 (10), 2–5 (9), 5–7 (7), 7–10 (5), 10–12 (3), 12–15 (2) and >15 (1). For the net recharge, the range, in millimeters, and the rating are respectively: 0–50 (1), 50–70 (2), 70–80 (3), 80–100 (4), 100–150 (6), 150–200 (9) and >200 (10). For the aquifer thickness, the range, in meters, and the rating are respectively: 0–5 (10), 5–15 (9), 15–25 (8), 25–30 (4), 30–50 (2) and >50 (1).

For the ratio of the cumulative thickness of clay layers to the total thickness of vadose zone, the range, in %, and the rating are respectively: 0–5 (10), 5–10 (8), 10–20 (5), 20–30 (3), 60–100 (1). For the contaminant adsorption coefficient of sediment in the vadose zone, the range, in L/kg, and the rating are respectively: 0–1 (10), 1–2 (9), 2–5 (7), 5–15 (5), 15–30 (3), 30–50 (2) and >50 (1). For the hydraulic conductivity of the aquifer, the range, in m/d, and the rating are respectively: 0–5 (1), 5–10 (2), 10–15 (4), 15–20 (7), 20–25 (8), >25 (10).

The total vulnerability score V is computed by:

$$V = D_R D_w + R_R R_w + A_R A_w + R_R R_w + C_R C_w + H_R H_w \quad (10)$$

where V is the DRARCH score, R—the rating value, w—the parameter weight.

The vulnerability index values are between 31 and 310 and five vulnerability classes were adopted: very low (31–86), low (87–142), moderate (143–198), high (199–254), and very high (255–310).

Other approaches of the aquifer vulnerability to arsenic used a GIS-based DRASTIC [85], with the vulnerability classifications and the indices values given in [86].

12. SINTACS

SINTACS was proposed and developed by Civita [26] and Civita and De Maio [27,28] for improving and adapting the DRASTIC model to the particularities of Italy. The letters in SINTACS are the first letters of the Italian words that define the models' factors. They are the depth to the water table (Soggicenza), effective infiltration (Infiltrazione), attenuation capacity of the unsaturated zone (Nonsaturo), type of the soil media (Tipologia della copertura), characteristics of the saturated zone (Acquifero), hydraulic conductivity (Conducibilità), and topographic slope (Superficie topografica) [25,29].

Civita [29] remarked that for using one or another method for assessing the groundwater vulnerability, one should consider the density of the observation points, the data availability, its completeness, and reliability, the homogeneity of the study region. In a critical review of some methods he presents the reasons for searching a better approach for the evaluation of groundwater vulnerability:

- The soil action is isolated from the action of the embedding system.
- The climatic factors and their influence on the water system is not considered
- Most methods have only a local application
- The use of vulnerability maps for the prevention of the groundwater quality deterioration should be supported by a deep insight into the mechanism of the contaminant production and its risk level [21].

Based on the use of the same parameters, the SINTACS structure has a higher complexity than the DRASTIC one.

For a complete and reliable database, the SINTACS procedure is the following [25,29].

- Select the factors used in the study
- Divide the factors into types or subintervals containing the factors' values
- Assign a rating, P, between 1 and 10, to each subinterval, in concordance with its importance in the last step of the algorithm (Figure 2)

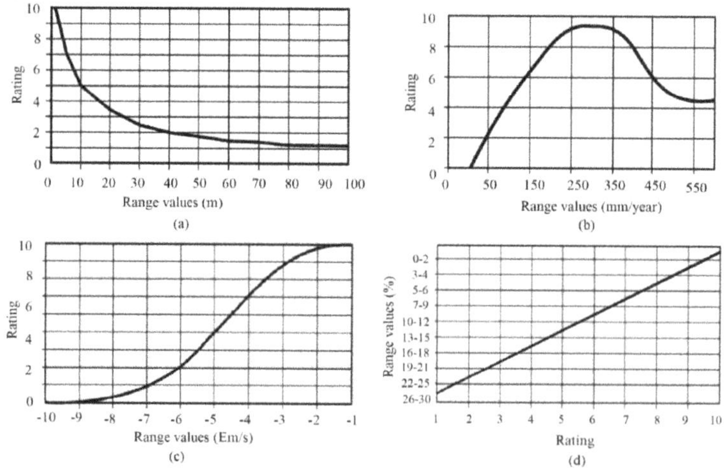

Figure 2. Ratings in SINTACS for (**a**) S, (**b**) I, (**c**) C, (**d**) S parameters (adapted from [29]).

- Choose the strings of weights, W, and multiply the factor ratings (Table 4).

Table 4. Strings of weights in SINTACS (adapted from [29]).

Parameter	S	I	N	T	A	C	S
Normal	5	4	5	3	3	3	3
Severe	5	5	4	5	3	2	2
Seepage	4	4	4	2	5	5	2
Karst	2	5	1	3	5	5	5
Fissured	3	3	3	4	4	5	4
Nitrates	5	5	4	5	2	2	3

The vulnerability index is computed as a weighted average, by the formula:

$$I_S = \sum_{i=1}^{7} w_i \times P_i, \tag{11}$$

P_i being the rating value and w_i is the corresponding weight.

One of the SINTACS advantages is the possibility of simultaneous use in different zones since each situation has assigned a specific weighting rate. Notice the differences between the DRASTIC and SINTACS procedures of weighting and rating, the last one operating in parallel with different strings of weights (Table 4) to describe the environmental conditions [6]. The most difficult task remains the range selection and the assignation of weight and ratings.

Ratings in SINTACS for the net recharge (R) are Coarse alluvial deposit 6–9, karstified limestone 8–10, Fractured limestone 4–8, Fissured dolomite 2–5, Medium-fine alluvial deposit 3–6, sandstone complex 4–7, Sandstone and conglomerate 5–8, Fissured plutonic rock 3–5, Turbidic sequence 2–5, Fissured volcanic rock 5–10, Marl and claystone 1–2, Coarse moraine 4–6, Clay, silt and peat 2–4, Medium –fine moraine 1–2, Pyrolastic rock 2–5, Fissured metamorphic rock 2–6 [27].

The six vulnerability classes (and the corresponding intervals of the vulnerability index) are very high ($I_S > 210$), high ($186 < I_S < 210$), moderately high ($140 < I_S < 186$), medium ($105 < I_S < 140$), low ($80 < I_S < 105$), and very low ($I_S < 80$).

For extending the applicability of SINTACS to the entire Italian territory, a new approach was introduced, by combining the SINTACS Release 5 [28] with the GNDCI_CNR Basic Method [29]. Since its release, SINTACS became one of the most used methods for the assessment of groundwater vulnerability in countries as Algeria, Italy, Jordan, Morocco, Thailand [87–93].

Corniello et al. [89] remarked that lithological and morphological settings play an important role in the process of generating SINTACS vulnerability maps. In a comparative study of three methods [90] on sites situated in a Mediterranean region, it is shown that the climatic conditions have a significant influence on the methods' performance, DRASTIC providing better results than SINTACS and AVI. A comparative study of the vulnerability maps produced by DRASTIC and SINTACS for an aquifer situated in Algeria [91] shows that the results are statistically concordant. Luoma et al. [92] emphasize in their research on a coastal aquifer that the SINTACS vulnerability maps are concordant with the field reality.

From the comparative analysis of the results provided by performing DRASTIC, SINTACS, and GOD methods on a database from Central Romania [93], one can remark on the similarity of the maps generated by the first two methods, there are few differences in the extent of the class of low vulnerability. In zones with small vulnerability variations, GOD performed worst. Therefore this method should be used only for regions with big vulnerability variations.

Aiming at detecting the capabilities of five groundwater vulnerability approaches, Civita and De Regibus [11] developed their research in three zones (mountains, hills, and flat). SINTACS and DRASTIC could adapt to the various situations, by comparison to the other competitors (GOD being among them) due to their flexibility.

Secunda et al. [94] and Noori et al. [95] used a SINTACS-LU approach in their research. The new factor, LU, was introduced by analogy to DRASTIC-LU, for considering the land use effect on the groundwater vulnerability. The new approaches better performed than SINTACS in case studies from Israel and Iran. Both SINTACS-LU and DRASTIC-LU vulnerability maps delimited the zones highly affected by human activity [94]. The sensitivity analysis for SINTACS-LU [95] showed that the parameter with the highest impact was the vadose zone, followed by the land use. The analysis of the correlation between the vulnerability index and the nitrate values (recorded on-field) was the highest for SINTACS-LU (0.75), by comparison to those of DRASTIC-LU (0.68) and SI (0.64).

13. Groundwater Vulnerability Assessment to Specific Pollutants

Even if all the described methods could be applied to assess the groundwater vulnerability to contamination, new approaches have been proposed to account for the specific properties of some pollutants. These include Pesticide DRASTIC, Pesticide DRASTIC-LU [38,96], Modified Pesticide DRASTIC [80,81,96], Modified DRASTIC for nitrate [20,46,96–99]. The factors weights in Pesticide DRASTIC and Pesticide DRASTIC LU differs from those of DRASTIC, the rating being preserved. They are presented in Table 5. The ratings of the land use in Pesticide DRASTIC-LU are 1, 5, 7, or 8.

Table 5. Weights assigned to the parameters in Pesticide DRASTIC, Pesticide DRASTIC-LU.

Parameter	D	R	A	S	T	I	C	LU
Pesticide DRASTIC	5	4	3	5	3	4	2	-
Pesticide DRASTIC-LU	5	4	3	5	3	4	2	5

DRASTIC, Pesticide DRASTIC, and Pesticide DRASTIC-LU were used for a study in a part of the Gangetic Plain with intense agricultural activities. Statistical analysis of the average values revealed that the most significant contribution to calculation the vulnerability indexes were I, T (in DRASTIC and Pesticide DRASTIC), and D, followed by R (in Pesticide DRASTIC LU). The sensitivity analysis found that A and R factors had the highest impact on all the models. The less significant parameters were S, and T—in DRASTIC, I and C—in Pesticide DRASTIC. Pesticide DRASTIC was the best model point of view of the correlation between the field data and prediction [96].

DRASTIC, Pesticide DRASTIC, and SI were applied in a case study of an aquifer from Tunisia. SI and Pesticide DRASTIC better detected the pollution risk. The concordance between the categories of vulnerability determined by these approaches was 64%. The authors [80] recommend the use of these two approaches for different purposes; the first one for monitoring, whereas the second one as a part of a multicriteria decision tool for allocating different zones to specific anthropic activities.

The performance of the same three models, together with Modified DRASTIC were compared on an aquifer in India (Telangana) [81]. The D factor has a considerable impact, followed by soil, the smaller one being that of R. The vulnerability classes are almost the same in SI, modified DRASTIC, and modified Pesticide DRASTIC because of the effect of LU inclusion. The Modified Pesticide DRASTIC map contains a higher area with high vulnerability, compared to Pesticide DRASTIC. The scientists remarked [81] the DRASTIC vulnerability underestimation and SI overestimation. All the models (but SI) are in concordance by at least 60%. It seems that the modified Pesticide DRASTIC provided the best predictions.

The nitrate is not a natural compound of soil, being the result of human activities, like the fertilizers used in agriculture or defecation [100]. While some authors used the Pesticide DRASTIC or Pesticide DRASTIC-LU to study the soil contamination with nitrates [46], other scientists [97–99] developed new approaches for improving the weights assigning in DRASTIC. Antonakos and Lambrakis [98] proposed DRASTIC-based hybrid methods, Huan et al. [99] adjusted the DRASTIC rating and weighting system. They validated their models on study cases from Greece and China, respectively.

Kazakis and Voudouris [97] replaced the A, S, and I factors of DRASTIC by the thickness of the aquifer, losses of nitrogen from the soil, and hydraulic resistance. The second factor was estimated by the GLEAMS model [101]. The parameters' range and weights were also modified. The two new methods, named DRASTIC-PA and DRASTIC-PAN, were compared to DRASTIC and LOSN-PN. Their performance were validated by the sensitivity analysis.

14. Other Approaches

In addition to the discussed approaches and those listed in Introduction [5,12,13,23–37], Modified versions of DRASTIC have also been developed. Some of them, for carbonate aquifers, are presented in the articles of Davies et al. [102] (that introduced KARST), Witkowski et al. [103], Różkowski [104], Denny et al. [105] (that introduced DRASTIC-Fm), Jiménez-Madrid et al. [106] (that introduced DRISTPI). We shall not insist here on them, due to lack of space.

As already mentioned, one of the main criticism of DRASTIC was that it does not take into account the study particular characteristics of each study region, and does not adapt the ratings and weights [46]. To surpass this inconvenience, other techniques have been proposed, as follows:

- Approaches for improving the weighting techniques [107]
- Approaches that use Analytic Hierarchy Process (AHP) and AHP-fuzzy [46,60,108,109]
- Approaches that use the fuzzy logic [110–114]
- Approaches that use genetic algorithms and neural networks [45,115–117].
- Correspondence Analysis [118], aiming at minimizing the redundancy between factors using a multivariate statistical method
- Calibration techniques, proposed in [98,119], used in [99,100].
- For a deep insight into these approaches, the readers can access the cited articles.

15. Models' Validation

Different authors used many models to validate the results of the vulnerability maps [25,90,120]. Kumar et al. [21] emphasize that this comparison is not advisable because various approaches use different parameters, so the vulnerability maps might not be similar. The benefit of such procedures is to offer an insight into the existence and spatial distribution of groundwater pollution. Therefore, other techniques should be used, as the validation of the vulnerability maps on contaminants data sets collected on-site from wells distributed in the study region. This is usually done using the concentrations of nitrates in the collected samples. A method whose results are most contrasting could be considered most sensitive, so it can be used [6].

Napolitano-Fabri [121] proposed the single-parameter sensitivity analysis (SPSA), which is the most frequently used technique for evaluating the significance of the parameters in the vulnerability models [46,57,58,64,99,122–127]. SPSA provides information on the rating and weighting assigned to each parameter, enabling its evaluation by the researcher. SPSA compares the theoretical and effective weights assigned to each parameter in a model.

Lodwick et al. [128] introduced the map removal sensitivity analysis (MRSA) for assessing the uncertainty degree of the models' output. MRSA consists of removing one map from the analysis of vulnerability and computing a variation index.

Higher SPSA (MRSA) is, higher the importance of the parameter is.

Promising methods of validation involve statistical methods, such those proposed by Masetti [15], Neukum [17], Panagopoulos et al. [118], and Pacheco et al. [121].

16. Conclusions

In this article, we reviewed DRASTIC and the main DRASTIC-like approaches proposed by scientists for improving the initial algorithm. The methods for assessment of the groundwater vulnerability in karstic regions were not discussed here. DRASTIC is among the most used tools for groundwater vulnerability evaluation. Generally, it uses readily available geodata with no experimental data. DRASTIC employs numerous parameters, and its outputs are only sometimes compared with field-collected data. Therefore DRASTIC-based forecast should be rigorously checked before making management decisions.

The groundwater vulnerability maps are important tools for assessing the groundwater vulnerability and planning future land use. No method developed for creating vulnerability maps is the most reliable, each of them depending on the aquifer characteristics, the land use, the data availability, the parameters involved in the model, the weightings, and rating assigned to each parameter.

Some aspects should be addressed in the next studies:

- Development of analytical methods for choosing and validating the ratings and weightings attached to each parameter in the models
- Integrating the models of water flow and pollutants' transport in different soils types in the methodology of choosing the weighting values of different parameters
- Detecting the relationships between the parameters used in the models by statistical methods and removing the effect of this correlation by adjustment of the ratings and weightings attached to the corresponding parameters
- Development of unified models that should include the soil and geological characteristics
- Development of hybrid models to reduce the influence of subjectivity in the parameters' settings and use the statistical methods for the results' validation.
- Improvement of the databases containing hydro-chemical elements and their integration into GIS software
- Improvement of GIS software by integrating analytical methods with groundwater vulnerability methods
- Development of spatio-temporal methods for the groundwater vulnerability assessment.

Funding: This research received no external funding.

Conflicts of Interest: The authors declare no conflict of interest.

References

1. World Wildlife Fund. Available online: https://www.worldwildlife.org/initiatives/fresh-water (accessed on 15 March 2019).
2. Margat, J. *Ground Water Vulnerability to Contamination*; Bases de la Cartographie, Doc. 68 SGC198 HYD; BRGM: Orleans, France, 1968. (In French)
3. Hirata, R.; Bertolo, R. Groundwater Vulnerability in Different Climatic Zones. In *Encyclopedia of Life Support Systems (EOLSS), Groundwater—Vol. II*; Available online: https://www.eolss.net/Sample-Chapters/C07/E2-09-04-06.pdf (accessed on 20 March 2020).
4. *Ground Water Vulnerability Assessment: Contamination Potential under Conditions of Uncertainty*; National Research Council National Academy Press: Washington, DC, USA, 1993; Available online: https://www.nap.edu/read/2050/chapter/3#17 (accessed on 4 September 2019).
5. Doerfliger, N.; Jeannin, P.Y.; Zwahlen, F. Water vulnerability assessment in karst environments: A new method of defining protection areas using a multi-attribute approach and GIS tools (EPIK method). *Environ. Geol.* **1999**, *39*, 165–176. [CrossRef]
6. Gogu, R.C.; Dassargues, A. Current trends and future challenges in groundwater vulnerability assessment using overlay and index methods. *Environ. Geol.* **2000**, *39*, 549–559. [CrossRef]
7. Adams, B.; Foster, S.S.D. Land-surface zoning for groundwater protection. *Water Environ. J.* **1992**, *6*, 312–319. [CrossRef]
8. Derouane, J.; Dassargue, A. Delineation of groundwater protection zones based on traces tests and transport modelling in alluvial sediments. *Environ. Geol.* **1998**, *36*, 27–32. [CrossRef]
9. Albinet, M.; Margat, J. Groundwater Pollution Vulnerability Mapping. *Bull. Bur. Res. Geol. Min. Bull BRGM 2nd Ser.* **1970**, *3*, 13–22.

10. Goossens, M.; Van Damme, M. Vulnerability mapping in Flanders, Belgium. In *Vulnerability of Soil and Groundwater to Pollutants International Conference*; van Duijvenbooden, W., van Waegeningh, G.H., Eds.; Proceedings and Information 38; TNO Committee on Hydrological Research: The Hague, The Netherlands, 1987; pp. 355–360. Available online: https://pascal-francis.inist.fr/vibad/index.php?action=getRecordDetail&lang=en&idt=7424922 (accessed on 23 March 2020).
11. Civita, M.; De Regibus, C. Sperimentazione di alcune metodologie per la valutazione della vulnerabilità degli aquiferi. *Q Geol. Appl. Pitagora Bologna* **1995**, *3*, 63–71.
12. Foster, S.S.D. Fundamental concepts in aquifer vulnerability, pollution risk and protection strategy. *Hydrol. Resour. Proc. Inf.* **1987**, *38*, 69–86.
13. Van Stempvoort, D.L.; Evert, L.W. Aquifer vulnerability index: A GIS compatible method for groundwater vulnerability mapping. *Can. Water Resour. J.* **1993**, *18*, 25–37. [CrossRef]
14. Eckhardt, D.A.V.; Stackelberg, P.E. Relation of ground-water quality to land use on Long Island, New York. *Ground Water* **1995**, *33*, 1019–1033. [CrossRef]
15. Masetti, M.; Sterlacchini, S.; Ballabio, C.; Sorichetta, A.; Poli, S. Influence of threshold value in the use of statistical methods for groundwater vulnerability assessment. *Sci. Total Environ.* **2009**, *407*, 3836–3846. [CrossRef]
16. Yen, S.T.; Liu, S.; Kolpin, D.W. Analysis of nitrate in near surface aquifers in the midcontinental United States: An application of the inverse hyperbolic sine tobit model. *Water Resour. Res.* **1996**, *32*, 3003–3011. [CrossRef]
17. Neukum, C.; Hotzl, H.; Himmelsbach, T. Validation of vulnerability mapping methods by field investigations and numerical modelling. *Hydrogeol. J.* **2008**, *16*, 641–658. [CrossRef]
18. Pineros Garcet, J.D.; Ordoñez, A.; Roosen, J.; Vanclooster, M. Metamodelling: Theory, concepts, and application to nitrate leaching modelling. *Ecol. Model.* **2006**, *193*, 629–644. [CrossRef]
19. Singhal, V.; Goyal, R. Development of conceptual groundwater flow model for Pali Area, India. *Afr. J. Environ. Sci. Technol.* **2011**, *5*, 1085–1092. [CrossRef]
20. Fusco, F.; Alloca, V.; Coda, S.; Cusano, D.; Tufano, R.; De Vita, P. Quantitative Assessment of Specific Vulnerability to Nitrate Pollution of Shallow Alluvial Aquifers by Process-Based and Empirical Approaches. *Water* **2020**, *12*, 269. [CrossRef]
21. Kumar, P.; Bansod, K.S.; Debnath, K.S.; Thakur, P.K.; Ghanshyam, C. Index-based groundwater vulnerability mapping models using hydrogeological settings: A critical evaluation. *Environ. Impact Assess. Rev.* **2015**, *51*, 38–49. [CrossRef]
22. Rao, P.S.C.; Alley, W.M. Pesticides. In *Regional Groundwater Quality*; Alley, W.M., Ed.; Van Nostrand: Reinhold, NY, USA, 1993; pp. 345–382.
23. Aller, L.; Bennet, T.; Lehr, J.H.; Petty, R.J. *DRASTIC: Standardized System for Evaluating Ground Water Pollution Potential using Hydrogeologic Settings*; Office of Research Development, US EPA: Ada, OK, USA, 1985.
24. Ray, I.A.; Odell, P.W. DIVERSITY: A new method for evaluating sensitivity of groundwater to contamination. *Environ. Geol.* **1993**, *22*, 344–352. [CrossRef]
25. Civita, M. *Le Carte Della Vulnerabilita Degli Acquiferi all Inquinamento: Teoria e Pratica*; Pitagora Editrice: Bologna, Italy, 1994; pp. 325–333.
26. Civita, M. *Legenda Unificata per le Carte Della Vulnerabilita' dei Corpi Idrici Sotterranei/Unified Legend for the Aquifer Pollution Vulnerability Maps. Studi sulla Vulnerabilita' degli Acquiferi, 1 (Append.)*; Pitagora: Bologna, Italy, 1990; p. 13.
27. Civita, M.; De Maio, M. *SINTACS. Un Sistema Parametrico per la Valutazione e la Cartografia Della Vulnerabilita' Degli Acquiferi All'inquinamento. Metodologia and Automatizzazione*; Pitagora: Bologna, Italy, 1997.
28. Civita, M.; de Maio, M. *SINTACS R5-Valutazione e Cartografia Automatica Della Vulnerabilità Degli Acquiferi All'inquinamento con il Sistema Parametrico*; Pitagora: Bologna, Italy, 2000; pp. 226–232.
29. Civita, M. The combined approach when assessing and mapping groundwater vulnerability to contamination. *J. Water Resour. Prot.* **2010**, *2*, 14–28. [CrossRef]
30. Malik, P.; Svasta, J. REKS: An alternative method of Karst groundwater vulnerability estimation. In Proceedings of the XXIX Congress of the International Association of Hydrogeologists, Bratislava, Slovakia, 6–10 September 1999; pp. 79–85. [CrossRef]
31. Pételet-Giraud, E.; Dörfliger, N.; Crochet, P. RISKE: Méthode d'évaluation multicritère de la vulnérabilité des aquifers karstiques. Application aux systèmes des Fontanilles et Cent-Fonts (Hérault, Sud de la France). *Hydrogéol* **2000**, *4*, 71–88.

32. Plagnes, V.; Théry, S.; Fontaine, L.; Bakalowicz, M.; Dörfliger, N. Karst vulnerability mapping: Improvement of the RISKE method. In Proceedings of the KARST 2005, Water Resources and Environmental Problems in Karst, Belgrade, Serbia, 14–19 September 2005.
33. Andreo, B.; Ravbar, N.; Vias, J.M. Source vulnerability mapping, in carbonate (karst) aquifers by extension of the COP method: Application to pilot sites. *Hydrogeol. J.* **2009**, *17*, 749–758. [CrossRef]
34. Vías, J.M.; Andreo, B.; Perles, M.J.; Carrasco, F.; Vadillo, I.; Jiménez, P. Proposed method for groundwater vulnerability mapping in carbonate (karstic) aquifers: The COP method. *Hydrogeol. J.* **2006**, *14*, 912–927. [CrossRef]
35. Kavouri, K.; Plagnes, V.; Tremoulet, J.; Dörfliger, N.; Rejiba, F.; Marchet, P. PaPRIKa: A method for estimating karst resource and source vulnerability—Application to the Ouysse karst system (southwest France). *Hydrogeol. J.* **2011**, *19*, 339–353. [CrossRef]
36. Goldscheider, N. Karst groundwater vulnerability mapping: Application of a new method in the Swabian Alb, Germany. *Hydrogeol. J.* **2005**, *13*, 555–564. [CrossRef]
37. Ravbar, N.; Goldscheider, N. Proposed methodology of vulnerability and contamination risk mapping for the protection of karst aquifers in Slovenia. *Acta Carsologica* **2007**, *36*, 461–475. [CrossRef]
38. Aller, L.; Bennett, T.; Lehr, J.H.; Petty, R.J.; Hackett, G. *DRASTIC: A Standardized System for Evaluating Groundwater Potential Using Hydrogeologic Settings*; EPA/600/2-85/018; U.S. Environmental Protection Agency: Washington, DC, USA, 1987.
39. Fan, J.; Oestergaard, K.T.; Guyot, A.; Lockington, D.A. Estimating groundwater recharge and evapotranspiration from water table fluctuations under three vegetation covers in a coastal sandy aquifer of subtropical Australia. *J. Hydrol.* **2014**, *519*, 1120–1129. [CrossRef]
40. Chenini, I.; Zghibi, A.; Kouzana, L. Hydrogeological investigations and groundwater vulnerability assessment and mapping for groundwater resource protection and management: State of the art and a case study. *J. Afr. Earth Sci.* **2015**, *109*, 11–26. [CrossRef]
41. Kalinski, R.J.; Kelly, W.E.; Bogardi, I.; Ehrman, R.L.; Yamamoto, P.O. Correlation between DRASTIC vulnerabilities and incidents of VOC contamination of municipal wells in Nebraska. *Groundwater* **1994**, *32*, 31–34. [CrossRef]
42. McLay, C.D.A.; Dragden, R.; Sparling, G.; Selvarajah, N. Predicting groundwater nitrate concentrations in a region of mixed agricultural land use: A comparison of three approaches. *Environ. Pollut.* **2001**, *115*, 191–204. [CrossRef]
43. Akhavan, S.; Mousavi, S.; Abedi-Koupai, J.; Abbaspour, K.C. Conditioning DRASTIC model to simulate nitrate pollution case study: Hamadan-Bahar plain. *Environ. Earth Sci.* **2011**, *63*, 1155–1167. [CrossRef]
44. Barbash, J.E.; Resek, E.A. *Pesticides in Ground Water: Distribution, Trends, and Governing Factors*; Ann Arbor Press Inc.: Ann Arbor, MI, USA, 1996.
45. Jang, W.S.; Engel, B.; Harbor, J.; Theller, L. Aquifer Vulnerability Assessment for Sustainable Groundwater Management Using DRASTIC. *Water* **2017**, *9*, 792. [CrossRef]
46. Neshat, A.; Pradhan, B.; Pirastehm, S.; Shafri, H.Z.M. Estimating groundwater vulnerability to pollution using a modified DRASTIC model in the Kerman agricultural area, Iran. *Environ. Earth Sci.* **2014**, *71*, 3119–3131. [CrossRef]
47. Holden, L.R.; Graham, J.A.; Whitmore, R.W.; Alexander, W.J.; Pratt, R.W.; Liddle, S.F.; Piper, L.L. Results of the national Alachlor well water survey. *Environ. Sci. Technol.* **1992**, *26*, 936–943. [CrossRef]
48. Maas, R.P.; Kucken, D.J.; Patch, S.C.; Peek, B.T.; Van Engelen, D.L. Pesticides in eastern North Caroline rural supply wells: Landuse factors and persistence. *J. Environ. Qual.* **1995**, *24*, 426–431. [CrossRef]
49. Wang, Y.; Merkel, B.J.; Li, Y.; Ye, H.; Fu, S.; Ihm, D. Vulnerability of groundwater in Quaternary aquifers to organic contaminants: A case study in Wuhan City, China. *Environ. Geol.* **2007**, *53*, 479–484. [CrossRef]
50. Merchant, J.W. GIS-Based Groundwater Pollution Hazard Assessment: A Critical Review of the DRASTIC model. *Photogramm. Eng. Remote Sens.* **1994**, *60*, 1117–1127.
51. Alam, F.; Umar, R.; Ahmad, S.; Dar, A.F. A new model (DRASTIC-LU) for evaluating groundwater vulnerability in parts of Central Ganga plain, India. *Arab. J. Geosci.* **2012**, *7*, 927–937. [CrossRef]
52. Merchant, J.W.; Whittemore, D.; Whistler, O.J.L.; McElwee, C.D.; Woods, J.J. Groundwater Pollution Hazard Assessment: A GIS Approach. In Proceedings of the International Geographic Information Systems (IGIS) Symposium, Association of American Geographers, Washington, DC, USA, 15–18 November 1987; Volume 3, pp. 103–115.

53. Hrkal, Z. Vulnerability of groundwater to acid deposition, Jizerske Mountains, northern Czech Republic: Construction and reliability of a GIS-based vulnerability map. *Hydrogeol. J.* **2001**, *9*, 348–357. [CrossRef]
54. Rupert, M.G. Calibration of the DRASTIC ground water vulnerability mapping method. *Ground Water* **2001**, *39*, 625–630. [CrossRef]
55. Lake, I.R.; Lovett, A.A.; Hiscock, K.M. Evaluating factors influencing groundwater vulnerability to nitrate pollution: Developing the potential of GIS. *J. Environ. Manag.* **2003**, *68*, 315–328. [CrossRef]
56. Babiker, I.S.; Mohammed, M.A.A.; Hiyama, T.; Kato, K. A GIS-based DRASTIC model for assessing aquifer vulnerability in Kakamigahara Heights, Gifu Prefecture, central Japan. *Sci. Total Environ.* **2005**, *345*, 127–140. [CrossRef]
57. Dixon, B. Groundwater vulnerability mapping: A GIS and fuzzy rule based integrated tool. *Appl. Geogr.* **2005**, *25*, 327–347. [CrossRef]
58. Rahman, A. A GIS based DRASTIC model for assessing groundwater vulnerability in shallow aquifer in Aligarh, India. *Appl. Geogr.* **2008**, *28*, 32–53. [CrossRef]
59. Shirazi, S.M.; Imran, H.M.; Akib, S.; Yusop, Z.; Harun, Z.B. Groundwater vulnerability assessment in the Melaka State of Malaysia using DRASTIC and GIS techniques. Environ. *Earth Sci.* **2013**, *70*, 2293–2304. [CrossRef]
60. Thirumalaivasan, D.; Karmegam, M.; Venugopal, K. AHP-DRASTIC: Software for specific aquifer vulnerability assessment using DRASTIC model and GIS. *Environ. Model. Softw.* **2003**, *18*, 645–656. [CrossRef]
61. Sener, E.; Sener, S.; Davraz, A. Assessment of aquifer vulnerability based on GIS and DRASTIC methods: A case study of the Senirkent-Uluborlu Basin (Isparta, Turkey). *Hydrogeol. J.* **2009**, *17*, 2023. [CrossRef]
62. Ariffin, S.M.; Zawawi, M.A.M.; Man, H.C. Evaluation of groundwater pollution risk (GPR) from agricultural activities using DRASTIC model and GIS. In *IOP Conference Series: Earth and Environment Science*; IOP Publishing: Bristol, UK, 2016; Volume 37, p. 012078. [CrossRef]
63. Mendoza, J.A.; Barmen, G. Assessment of groundwater vulnerability in the Río Artiguas basin, Nicaragua. *Environ. Geol.* **2006**, *50*, 569–580. [CrossRef]
64. Saidi, S.; Bouri, S.; Dhia, H.B. Groundwater vulnerability and risk mapping of the Hajeb-jelma aquifer (Central Tunisia) using a GIS-based DRASTIC model. *Environ. Earth Sci.* **2010**, *59*, 1579–1588. [CrossRef]
65. Castany, G. *Principe et Methodes de L'hydrogeologie*; Dunod: Paris, France, 1982.
66. Sakala, E.; Fourie, F.; Gomo, M.; Coetzee, H.; Magadaza, L. Specific groundwater vulnerability mapping: Case study of acid mine drainage in the Witbank coalfield, South Africa. In Proceedings of the Sixth IASTED International Conference, Gaborone, Botswana, 5–7 September 2016. [CrossRef]
67. Zhou, J.; Li, G.; Liu, F.; Wang, Y.; Guo, X. DRAV model and its application in assessing groundwater vulnerability in arid area: A case study of pore phreatic water in Tarim Basin, Xinjiang, Northwest China. *Environ. Earth Sci.* **2010**, *60*, 1055–1063. [CrossRef]
68. Singh, A.; Srivastav, S.K.; Kumar, S.; Chakrapani, G.J. A modified-DRASTIC model (DRASTICA) for assessment of groundwater vulnerability to pollution in an urbanized environment in Lucknow, India. Environ. *Earth Sci.* **2015**, *74*, 5475–5490. [CrossRef]
69. Amadi, N.; Olasehinde, P.I.; Nwankwoala, H.O.; Dan-Hassan, M.A.; Okoye, N.O. Aquifer vulnerability studies using DRASTICA Model. *Int. J. Eng. Sci. Invent.* **2014**, *3*, 1–10.
70. Hussain, M.H.; Singhai, D.C.; Joshi, H.; Kumar, S. Assessment of groundwater vulnerability in tropical alluvial interfluves, India. *Bhu-Jal News J.* **2006**, *1–4*, 31–43.
71. Khan, M.M.A.; Umar, R.; Lateh, H. Assessment of aquifer vulnerability in parts of Indo Gangetic plain, India. *Int. J. Phys. Sci.* **2010**, *5*, 1711–1720.
72. Kumar, A.; Khrisma, A.K. Groundwater vulnerability and contamination risk assessment using GIS-based modified DRASTIC-LU model in hard rock aquifer system in India. *Geocarto Int.* **2019**. [CrossRef]
73. Sinha, M.K.; Verma, M.K.; Ahmad, I.; Baier, K.; Jha, R.; Azzam, R. Assessment of groundwater vulnerability using modified DRASTIC model in Kharun Basin, Chhattisgarh, India. *Arab. J. Geosci.* **2016**, *9*, 11–22. [CrossRef]
74. Umar, R.; Ahmed, I.; Alam, F. Mapping Groundwater Vulnerable Zones Using Modified DRASTIC Approach of an Alluvial Aquifer in Parts of Central Ganga Plain, Western Uttar Pradesh. *J. Geol. Soc. India* **2009**, *73*, 193–201. [CrossRef]

75. Qinghai, G.; Yanxin, W.; Xubo, G.; Teng, M. A new model (DRARCH) for assessing groundwater vulnerability to arsenic contamination at basin scale: A case study in Taiyuan basin, northern China. *Environ. Geol.* **2007**, *52*, 923–932. [CrossRef]
76. Ribeiro, L. *SI: A New Index of Aquifer Susceptibility to Agricultural Pollution*; ERSHA/CVRM, Instituto Superior Técnico: Lisboa, Portugal, 2000.
77. Stigter, T.Y.; Riberio, L.; Dill, A.M.M.C. Evaluation of an intrinsic and a specific vulnerability assessment method in comparison with groundwater salinization and nitrate contamination levels in two agricultural regions in the south of Portugal. *Hydrogeol. J.* **2006**, *14*, 79–99. [CrossRef]
78. Ribeiro, L.; Pindo, J.C.; Dominguez-Granda, L. Assessment of groundwater vulnerability in the Daule aquifer, Ecuador, using the susceptibility index method. *Sci. Total Environ.* **2017**, *574*, 1674–1683. [CrossRef]
79. Anane, M.; Abidi, B.; Lachaal, F.; Limam, A.; Jellali, S. GIS-based DRASTIC, Pesticide DRASTIC and the Susceptibility Index (SI): Comparative study for evaluation of pollution. *Hydrogeol. J.* **2013**, *21*, 715–731. [CrossRef]
80. Brindha, K.L.E. Cross comparison of five popular groundwater pollution vulnerability index approaches. *J. Hydrol.* **2015**, *524*, 597–613. [CrossRef]
81. El Himer, H.; Fakir, Y.; Stigter, T.Y.; Lepage, M.; El Mandour, A.; Ribeiro, L. Assessment of the vulnerability to pollution of a wetland watershed. The case study of Oualidia-Sidi Moussa wetland, Morocco. *Aquat. Ecosyst. Health* **2013**, *16*, 205–215. [CrossRef]
82. Hamza, M.H.; Added, A. Validity of DRASTIC and SI vulnerability methods. In *GeoSpatial Visual Analytics: Geographical Information Processing and Visual Analytics for Environmental Security*; De Amicis, R., Stojanovic, R., Conti, G., Eds.; Springer: Dordrecht, The Netherlands, 2009; pp. 395–407.
83. Hamza, M.H.; Added, A.; Francés, A.; Rodríguez, R. Validité de l'application des méthodes de vulnérabilité DRASTIC, SINTACS et SI à l'étude de la pollution par les nitrates dans la nappe phréatique de Metline–Ras Jebel–Raf Raf (Nord-Est tunisien). *Comptes Rendus Geosci.* **2007**, *339*, 493–505. [CrossRef]
84. Teixeira, J.; Chaminé, H.I.; Marques, J.E.; Carvalho, J.M.; Pereira, A.J.S.C.; Carvalho, M.R.; Fonseca, P.E.; Pérez-Alberti, A.; Rocha, F. A comprehensive analysis of groundwater resources using GIS and multicriteria tools (Caldas da Cavaca, Central). *Environ. Earth Sci.* **2014**, *73*. [CrossRef]
85. Rana, S.; Kumar, P.; Puri, S.; Bansod, B.K.; Debnath, S.; Ghanshyam, C.; Kapur, P. Localization of Arsenic Contaminated Zone of Domkal Block in Murshidabad, West Bengal using GIS-based DRASTIC model. In Proceedings of the International Conference on Communication and Computing (ICC-2014), Bangalore, India, 10–14 June 2014. [CrossRef]
86. Narany, T.; Ramli, M.F.; Aris, A.Z.; Sulaiman, W.N.A.; Fakharian, K. Spatial assessment of groundwater quality monitoring wells using indicator kriging and risk mapping, Amol-Babol Plain, Iran. *Water* **2013**, *6*, 68–85. [CrossRef]
87. Boufekane, A.; Saighi, O. Assessment of groundwater pollution by nitrates using intrinsic vulnerability methods: A case study of the Nil valley groundwater (Jijel, North-East Algeria). *Afr. J. Environ. Sci. Technol.* **2013**, *7*, 949–960. [CrossRef]
88. Al Kuisi, M.; El-Naqa, A.; Hammouri, N. Vulnerability mapping of shallow groundwater aquifer using SINTACS model in the Jordan Valley area, Jordan. *Environ. Geol.* **2006**, *50*, 651–667. [CrossRef]
89. Corniello, A.; Ducci, D.; Monti, G.M. Aquifer pollution vulnerability in the Sorrento Peninsula, Southern Italy, evaluated by SINTACS Method. *Geofís. Int.* **2004**, *43*, 575–581.
90. Draoui, M.; Vias, J.; Andreo, B.; Targuisti, K.; El Messari, J.S. A Comparative Study of Four Vulnerability Mapping Methods in a Detritic Aquifer under Mediterranean Climatic Conditions. *Environ. Geol.* **2008**, *54*, 455–463. [CrossRef]
91. Kaddour, K.; Houcine, B.; Smail, M. Assessment of the vulnerability of an aquifer by DRASTIC and SYNTACS methods: Aquifer of Bazer—Geult Zerga area (northeast Algeria). *E3 J. Environ. Res. Manag.* **2014**, *5*, 0169–0179.
92. Majandang, J.; Sarapirome, S. Groundwater vulnerability assessment and **sensitivity analysis** in Nong Rua, Khon Kaen, Thailand, using a GIS-Based SINTACS Model. *Environ. Earth Sci.* **2013**, *68*, 2025–2039. [CrossRef]
93. Gogu, R.C.; Pandele, A.; Ionita, A.; Ionescu, C. Groundwater vulnerability analysis using a low-cost Geographical Information System. In Proceedings of the MIS/UDMS Conference WELL-GIS WORKSHOP's Environmental Information Systems for Regional and Municipal Planning, Prague, Czech Republic, 20 November 1996; pp. 35–49.

94. Secunda, S.; Collin, M.; Melloul, A.J. Groundwater Vulnerability Assessment Using a Composite Model Combining DRASTIC with Extensive Land Use in Israel's Sharon Region. *J. Environ. Manag.* **1998**, *54*, 39–57. [CrossRef]
95. Noori, R.; Ghahremanzadeh, H.; Kløve, B.; Adamowski, F.J.; Baghvand, A. Modified-DRASTIC, modified-SINTACS and SI methods for groundwater vulnerability assessment in the southern Tehran aquifer. *J. Environ. Sci. Health* **2019**, *54*, 89–91. [CrossRef] [PubMed]
96. Saha, D.; Alam, F. Groundwater vulnerability assessment using DRASTIC and Pesticide DRASTIC models in intense agriculture area of the Gangetic plains, India. *Environ. Monit. Assess.* **2014**, *186*, 8741–8763. [CrossRef]
97. Kazakis, N.; Voudouris, K.S. Groundwater vulnerability and pollution risk assessment of porous aquifers to nitrate: Modifying the DRASTIC method using quantitative parameters. *J. Hydrol.* **2015**, *525*, 13–25. [CrossRef]
98. Antonakos, K.; Lambrakis, N.L. Development and testing of three hybrid methods for assessment of aquifer vulnerability to nitrates, based on the DRASTIC model, an example from NE Korinthia, Greece. *J. Hydrol.* **2007**, *333*, 288–304. [CrossRef]
99. Huan, H.; Wang, J.; Teng, Y. Assessment and validation of groundwater vulnerability to nitrate based on a modified DRASTIC model: A case study in Jilin City of northeast China. *Sci. Total Environ.* **2012**, *440*, 14–23. [CrossRef] [PubMed]
100. Javadi, S.; Kavehkar, N.; Mohammadi, K.; Khodadadi, A.; Kahawita, R. Calibrating DRASTIC using field measurements sensitivity analysis and statistical methods to assess groundwater vulnerability. *Water Int.* **2011**, *36*, 719–732. [CrossRef]
101. Harmel, D.; Knisel, W.; Leonard, R.; Davis, F. Groundwater Loading Effects of Agricultural Management Systems. Available online: https://www.ars.usda.gov/plains-area/temple-tx/grassland-soil-and-water-research-laboratory/docs/gleams/ (accessed on 26 April 2020).
102. Davis, A.D.; Long, A.J.; Wireman, M. KARSTIC: A sensitivity method for carbonate aquifers in karst terrain. *Environ. Geol.* **2002**, *42*, 65–72. [CrossRef]
103. Witkowski, A.J.; Rubin, K.; Kowalczyk, A.; Różkowski, A.; Wróbel, J. Groundwater vulnerability map of the Chrzanów karst-fissured Triassic aquifer (Poland). *Environ. Geol.* **2003**, *44*, 59–67. [CrossRef]
104. Różkowski, J. Evaluation of intrinsic vulnerability of an Upper Jurassic karst-fissured aquifer in the Jura Krakowska (southern Poland) to anthropogenic pollution using the DRASTIC method. *Geol. Q.* **2007**, *51*, 17–26. Available online: https://gq.pgi.gov.pl/article/view/7434 (accessed on 13 January 2020).
105. Denny, S.C.; Allen, D.M.; Journeay, J.M. DRASTIC-Fm: A modified vulnerability mapping method for structurally-controlled aquifers. *Hydrogeol. J.* **2007**, *15*, 483–494. [CrossRef]
106. Jiménez-Madrid, A.; Carrasco, F.; Martínez, C.; Gogu, R.C. DRISTPI, A new groundwater vulnerability mapping method for use in karstic and non-karstic aquifers. *Q. J. Eng. Geol. Hydroeol.* **2013**, *46*, 245–255. [CrossRef]
107. Pacheco, F.A.L.; Pires, L.M.G.R.; Santos, R.M.B.; Sanches Fernandes, L.F. Factor weighting in DRASTIC modeling. *Sci. Total Environ.* **2015**, *505*, 474–486. [CrossRef]
108. Hailin, Y.; Ligang, X.; Chang, Y.; Jiaxing, X. Evaluation of groundwater vulnerability with improved DRASTIC method. *Procedia Environ. Sci.* **2011**, *10*, 2690–2695. [CrossRef]
109. Sener, E.; Davraz, A. Assessment of groundwater vulnerability based on a modified DRASTIC model, GIS and an analytic hierarchy process (AHP) method: The case of Egirdir Lake basin (Isparta, Turkey). *Hydrogeol. J.* **2013**, *21*, 701–714. [CrossRef]
110. Shouyu, C.; Guangtao, F. A DRASTIC-based fuzzy pattern recognition methodology for groundwater vulnerability evaluation. *Hydrol. Sci. J.* **2003**, *48*, 211–220. [CrossRef]
111. Afshar, A.I.; Marino, M.A.; Asce, H.M.; Ebtehaj, M.; Moosa Vi, J. Rule-based fuzzy system for assessing groundwater vulnerability. *J. Environ. Eng.* **2007**, *133*, 532–540. [CrossRef]
112. Bojórquez-Tapia, L.A.; Cruz-Bello, G.M.; Luna-González, L.; Juárez, L.; Ortiz-Pérez, M.A. V-DRASTIC: Using visualization to engage policymakers in groundwater vulnerability assessment. *J. Hydrol.* **2009**, *373*, 242–255. [CrossRef]
113. Pathak, D.R.; Hiratsuka, A. An integrated GIS based fuzzy pattern recognition model to compute the groundwater vulnerability index for decision making. *J. Hydro-Environ. Res.* **2011**, *5*, 63–77. [CrossRef]

114. Rezaei, F.; Safavi, H.R.; Ahmadi, A. Groundwater vulnerability assessment using fuzzy logic: A case study in the Zayandehrood Aquifers, Iran. *Environ. Manag.* **2013**, *51*, 267–277. [CrossRef]
115. Fijani, E.; Nadiri, A.; Moghaddam, A.A.; Dixon, B. Optimization of DRASTIC Method by Supervised Committee Machine Artificial Intelligence to Assess Groundwater Vulnerability for Maragheh-Bonab Plain Aquifer, Iran. *J. Hydrol.* **2013**, *503*, 89–101. [CrossRef]
116. Barzegar, R.; Moghaddam, A.A. Combining the advantages of neural networks using the concept of committee machine in the groundwater salinity prediction. *Model. Earth Sys. Environ.* **2016**, *2*, 26. [CrossRef]
117. Barzegar, R.; Moghaddam, A.A.; Baghban, H. A supervised committee machine artificial intelligent for improving DRASTIC method to assess groundwater contamination risk: A case study from Tabriz plain aquifer, Iran. *Stoch. Environ. Res. Risk Assess.* **2016**, *30*, 883–899. [CrossRef]
118. Panagopoulos, G.P.; Antonakos, A.K.; Lambrakis, N.J. Optimization of the DRASTIC method for groundwater vulnerability assessment via the use of simple statistical methods and GIS. *Hydrogeol. J.* **2006**, *14*, 894–911. [CrossRef]
119. Neshat, A.; Pradhan, B.; Shafri, H.Z.M. An integrated GIS based statistical model to compute groundwater vulnerability index for decision maker in agricultural area. *J. Indian Soc. Remote Sens.* **2014**, *42*, 777–788. [CrossRef]
120. Ghazavi, R.; Ebrahimi, Z. Assessing groundwater vulnerability to contamination in an arid environment using DRASTIC and GOD models. *Int. J. Environ. Sci. Technol.* **2015**, *12*, 2909–2918. [CrossRef]
121. Pacheco, F.A.L.; Sanches Fernandes, L.F. The multivariate statistical structure of DRASTIC model. *J. Hydrol.* **2013**, *476*, 442–459. [CrossRef]
122. Napolitano, P.; Fabbri, A.G. Single parameter sensitivity analysis for aquifer vulnerability assessment using DRASTIC and SINTACS. In *Proceedings of the HydroGIS: Application of Geographical Information Systems in Hydrology and Water Resources Management*; Kovar, K., Nachtnebel, H.P., Eds.; IAHS Publication no. 235; IAHS: Wallingford, UK, 1996; pp. 559–566.
123. Neshat, A.; Pradhan, B.; Dadras, M. Groundwater vulnerability assessment using an improved DRASTIC method in GIS. *Resour. Conserv. Recycl.* **2014**, *86*, 74–86. [CrossRef]
124. Al-Hanbali, A.; Kondoh, A. Groundwater vulnerability assessment and evaluation of human activity impact (HAI) within the Dead Sea groundwater basin, Jordan. *Hydrogeol. J.* **2008**, *16*, 499–510. [CrossRef]
125. Almasri, M.N. Assessment of intrinsic vulnerability to contamination for Gaza coastal aquifer, Palestine. *J. Environ. Manag.* **2008**, *88*, 577–593. [CrossRef]
126. Hasiniaina, F.; Zhou, J.; Guoyi, L. Regional assessment of groundwater vulnerability in Tamtsag basin, Mongolia using drastic model. *J. Am. Sci.* **2010**, *6*, 65–78.
127. Samake, M.; Tang, Z.; Hlaing, W.; M'Bue, I.; Kasereka, K. Assessment of groundwater pollution potential of the Datong Basin, Northern China. *J. Sustain. Dev.* **2010**, *3*, 140–152. [CrossRef]
128. Lodwik, W.A.; Monson, W.; Svoboda, L. Attribute error and sensitivity analysis of maps operation in geographical information systems-sustainability analysis. *Int. J. Geogr. Inf. Syst.* **1990**, *4*, 413–428. [CrossRef]

water

MDPI

Article

A Novel Approach to Harmonize Vulnerability Assessment in Carbonate and Detrital Aquifers at Basin Scale

Leticia Baena-Ruiz * and David Pulido-Velazquez

Spanish Geological Survey (IGME), Urb. Alcázar del Genil, 4. Edificio Zulema, bajo, 18006 Granada, Spain; d.pulido@igme.es

* Correspondence: l.baena@igme.es

Received: 23 September 2020; Accepted: 21 October 2020; Published: 23 October 2020

Abstract: The DRASTIC (D: Depth to water; R: Net recharge; A: Aquifer media; S: Soil media; T: Topography; I: Impact of vadose zone; C: Hydraulic conductivity) index is usually applied to assess intrinsic vulnerability in detrital and carbonate aquifers, although it does not take into account the particularities of karst systems as the COP (C: Concentration of flow; O: Overlying layers above water table; P: precipitation) method does. In this paper we aim to find a reasonable correspondence between the vulnerability maps obtained using these two methods. We adapt the DRASTIC index in order to obtain reliable assessments in carbonate aquifers while maintaining its original conceptual formulation. This approach is analogous to the hypothesis of "equivalent porous medium", which applies to karstic aquifers the numerical solution developed for detrital aquifers. We applied our novel method to the Upper Guadiana Basin, which contains both carbonate and detrital aquifers. Validation analysis demonstrated a higher confidence in the vulnerability assessment provided by the COP method in the carbonate aquifers. The proposed method solves an optimization problem to minimize the differences between the assessments provided by the modified DRASTIC and COP methods. Decision trees and spatial statistics analyses were combined to identify the ranges and weights of DRASTIC parameters to produce an optimal solution that matches the COP vulnerability classification for carbonate aquifers in 75% of the area, while maintaining a reliable assessment of the detrital aquifers in the Basin.

Keywords: groundwater vulnerability; carbonate aquifers; optimized DRASTIC; COP; decision trees; nitrate validation

1. Introduction

Groundwater pollution is a widespread problem affecting most aquifers all over the world due to the increasing agricultural and industrial human activity [1,2]. It is the result of the interaction between the anthroposphere and the hydrosphere, where substances from the different land uses penetrate the groundwater, leading to impacts on environmental water quality and human health. The protection of the groundwater resources has become a priority due to the importance of groundwater for human supply, irrigation and dependent ecosystems, especially in semi-arid regions [3–5]. The degree of protection depends on intrinsic groundwater vulnerability, which is defined as the susceptibility of aquifers to pollution arising from anthropogenic activity [6]. Several methods have been proposed to assess intrinsic vulnerability by using conceptual groundwater flow models and taking transport processes into account [7,8]. The most frequently employed methods are the index-based approaches [9,10] of which DRASTIC (D: Depth to water; R: Net recharge; A: Aquifer media; S: Soil media; T: Topography; I: Impact of vadose zone; C: Hydraulic conductivity) [9] is the most common. DRASTIC has been applied to various types of aquifer, though some authors have revealed that it is not suitable for assessing vulnerability in some karstic aquifers [10–13].

Karstic aquifers present physical particularities that, in general, make them more sensitive to contamination [14,15]. Specific approaches, such as the COP (C: Concentration of flow; O: Overlying layers above water table; P: precipitation) method [10], have been developed specifically to assess groundwater vulnerability in carbonate (karstic) aquifers. The COP method has been extensively applied in different research studies to this type of aquifer [16–21].

However, despite significant physical differences between detrital and karstic aquifers, the same approaches are often applied to simulate groundwater processes in both types of aquifers. In numerical groundwater flow models, the "equivalent porous medium" assumes the validity of the Darcy's law in karstic aquifers, which is the most frequent numerical approach deduced for detrital aquifers. We find many examples of application of this approach both for detrital aquifers [22–24] and karstic aquifers [25,26]. In the same way, index methods such as DRASTIC could be suited to assessing vulnerability in both detrital and karstic aquifers.

In this context, recent research studies have attempted to adapt the DRASTIC method to karst aquifers in recent years [12,27–30]. Some of these [12,28,29] modified DRASTIC by including and/or removing parameters to account for karst characteristics. Other studies [27,30] adapted DRASTIC by modifying the classic assessment according to the distribution of nitrate concentration, although the most contaminated areas do not always imply higher vulnerability [31,32]. A detailed review of different approaches proposed to modify the DRASTIC method is given in [13].

These modifications of the DRASTIC method were performed using different methodologies/approaches: single-parameter sensitivity analysis [3,5,33], calibration by correlation analysis with pollutants [4,34,35], analytic hierarchy process [36–38] amongst others. Different data mining algorithms have also been applied to improve DRASTIC performance [39–42], predict areas vulnerable to groundwater contamination or to identify hydrogeological factors influencing groundwater contamination [39,43,44]. Data mining algorithms aim to extract knowledge from previously unknown and indistinguishable data, and are used as operational tools to find optimal solutions in high-dimension problems. Although many authors have employed these techniques for different purposes, more research is needed to explore the potential of the statistical techniques—including data mining—to deal with the uncertainties in the weights and ratings of index-based methods in the groundwater vulnerability assessments [45].

In general, the aforementioned index-based methods have been proven to provide satisfactory vulnerability assessment in a variety of regions. However, different index methods often provide dissimilar results in a single study area, making it difficult to compare and validate results [46–48]. In complex groundwater systems containing various aquifers types, an integrated vulnerability assessment is needed to homogenize criteria and compare results at basin scale and between different case studies. In general, a harmonized method that is applicable to all aquifer types would be more likely to be implemented worldwide [12,49]. Previous work [12] has aimed at developing a new method (DRISTPI; Depth to water, Recharge, Impact of vadose zone, Soil, Topography, Preferential Infiltration) by adding and removing certain parameters from the classic DRASTIC method that can be applied to any type of aquifer. It was tested in karstic aquifers, but not in detrital aquifers.

In this paper we propose a novel approach to standardize the DRASTIC method to assess vulnerability in a basin composed by both, detrital and carbonate aquifers. We proposed an adaptation of the most commonly applied method for detrital aquifers—the DRASTIC method—to obtain a reliable vulnerability assessment for carbonate aquifers. The DRASTIC method can be applied in both detrital and carbonate aquifers, though it does not always provide appropriate vulnerability assessment in carbonate aquifers. In contrast, the COP method is specific to carbonate aquifers and it yielded better results in the validation analysis in our case study. Therefore, we proposed to adapt DRASTIC to harmonize the vulnerability assessment for different aquifer typologies, in order to make the results comparable in a basin that contains a wide variety of geological formations. Harmonization of criteria in the vulnerability assessment would allow the dissimilarities provided by different vulnerability methods to be dealt with. Although in the literature we can find several attempts to adapt the DRASTIC method to the particularities of each case study, none of them have proved the applicability of the modified DRASTIC in different aquifer typologies.

In this study, an optimization problem is solved to establish a correspondence between DRASTIC and COP by maintaining the original formulation of the DRASTIC method. It aims to find the weights and ranges of the DRASTIC parameters that maximize the correspondence between the qualitative vulnerability classes obtained using DRASTIC and COP methods. The optimization problem is solved through a new approach that combines spatial statistics analysis and data mining (decision trees). Although decision trees have been applied to predict the sensitivity to contaminants based on groundwater vulnerability [50,51], they have not been used to adapt/improve the DRASTIC method.

The optimization methodology was applied to five carbonate aquifers in the Upper Guadiana Basin, where the validation analysis demonstrated that COP method produces better vulnerability assessment than the original DRASTIC one. The optimal solution obtained for carbonate aquifers (O-DRASTIC) was also tested in the three detrital aquifers within the basin and the validation analysis also shows significant improvement in the results comparing with the original DRASTIC. A sensitivity analysis of the changes introduced to define the optimum DRASTIC reveals the influence of the various intrinsic characteristics on the severity of vulnerability for different aquifer typologies.

2. Materials and Methods

2.1. Study Area and Data

The Upper Guadiana Basin (Figure 1) is located in the central part of Spain. It is composed of eight groundwater bodies including five unconfined mixed (carbonate and detrital) and three unconfined detrital aquifers extending over approximately 14,000 km^2.

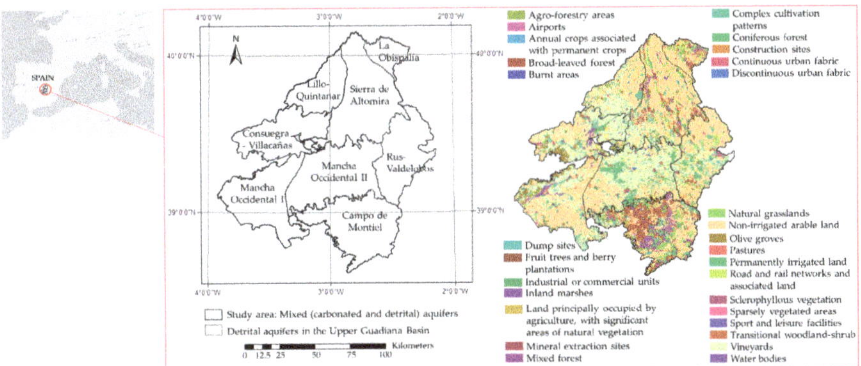

Figure 1. Location of study area within the Upper Guadiana Basin (Spain).

The climate in this area is typically continental and semiarid. Mean annual precipitation over the period 1974–2015 was 445 mm and the mean annual temperature is 14.7 °C. Mean potential evapotranspiration is 700 mm/year.

The area is predominantly flat, bounded by mountain landscapes to the north (Sierra de Altomira) and south (Campo de Montiel).

Connectivity between groundwater bodies is structurally complex, with strong natural interaction between groundwater and surface water. Under natural conditions groundwater flow discharges into the central aquifer of the system (Mancha Occidental Aquifer) [52] which flows eastwards.

The predominantly dry climate and the prevalence of irrigated agriculture means that the Upper Guadiana Basin has been intensively pumped, and this has led to some of its aquifers being declared overexploited [53]. The water table is highly variable, lying at less than 1 m to more than 250 m, though over most of the basin it lies below 15 m.

Rainfall is the main source of aquifers' recharge. Mean annual recharge varies between 45 and 70 mm/year, although there is disagreement about these values [52–54].

The soils in the basin mainly belong to the cambisol group according to the FAO (Food and Agriculture Organization of the United Nations) classification [54,55]. Regosol and others, such as luvisol and podzol, can be found in the southeast [54]. Soil texture in the northern part of the basin is predominantly silty-loam, whereas the soil in the southern part it is peaty.

The geology is complex, including mixed carbonate–detrital aquifers. In the southern half of the Upper Guadiana Basin, the aquifers are predominantly composed of limestones, with many karstified zones [53]. In general, the karst is not very developed and there are no swallow holes in these aquifers. Many parts of the central aquifer are formed by Tertiary detrital deposits [53]. The northern aquifers are more heterogeneous. There are no large karstified areas and other formations of metamorphic materials can be found. The detrital aquifers are mainly composed of Tertiary and Quaternary alluvial materials.

The unsaturated zone is formed by poorly permeable lithologies in the northern part of the basin, with higher permeabilities in the southern part [54].

Conductivity in the Upper Guadiana Basin varies widely. To the north, conductivity is low (below 1.5×10^{-4} m/s); in the central part there are zones with values higher than 5×10^{-4} m/s, while conductivity to the southern is mainly in the range 3.5×10^{-4}–5×10^{-4} m/s [56].

2.2. DRASTIC and COP Vulnerability Maps

DRASTIC and COP vulnerability maps were calculated in the five mixed aquifers following the proposal made in [9,10], respectively (Figure 2). A detailed explanation of these methods can be found in the Appendix A. Tables A4 and A5 (Appendix B) summarizes the data sources and the methodology applied to calculate the parameters of DRASTIC (hereinafter, we will call it "original DRASTIC") and COP, respectively.

The "original DRASTIC" values vary between 41 and 171 within the study area. The final index was reclassified into five qualitative classes (Very low: <100; Low: 100–120; Moderate: 120–160; High: 160–180; Very high: ≥180) by grouping the original categories proposed in [9] to obtain the same number of classes as the COP vulnerability map. The values of parameters Aquifer media, Soil media, Impact of vadose zone and Conductivity (Figure 2(a3,a4,a6,a7)) show significant heterogeneity, with clear differences between the northern and central–southern areas. These differences are finally reflected in the vulnerability map (Figure 2(a8)).

Figure 2b shows the results of applying the COP method as proposed by [10]. The protection conferred by Overlaying the layers (Figure 2(b2)) generally decreases from north to south. It is very low in the southern part due to the presence of limestones and dolomites outcrops. The surface features related to the Concentration of flow (Figure 3(b1)) produce only a slight reduction of protection over 80% of the area, except in the southern Campo de Montiel aquifer, where the protection from Overlying layers is greatly reduced.

While the highest values of the DRASTIC index are located in the center and southern part of the basin (classified as "Moderate vulnerability"), the highest values of the COP index appear in the southern zone (Campo de Montiel aquifer). Both indices give their lowest values in the north area (Sierra de Altomira and Lillo Quintanar aquifers).

The percentage overlap between the vulnerability classes obtained using COP and DRASTIC is 55.75%. There is a significant coincidence in the "Very low" and "Low" vulnerability classes, which cover nearly 80% of the area in the COP map. However, the coincidence in "High" and "Very high" classes is almost null. A misclassification of the highest vulnerability areas of groundwater dependent ecosystems such as the Upper Guadiana Basin would lead to erroneous planning and management decisions, possibly leading to significant environmental impacts.

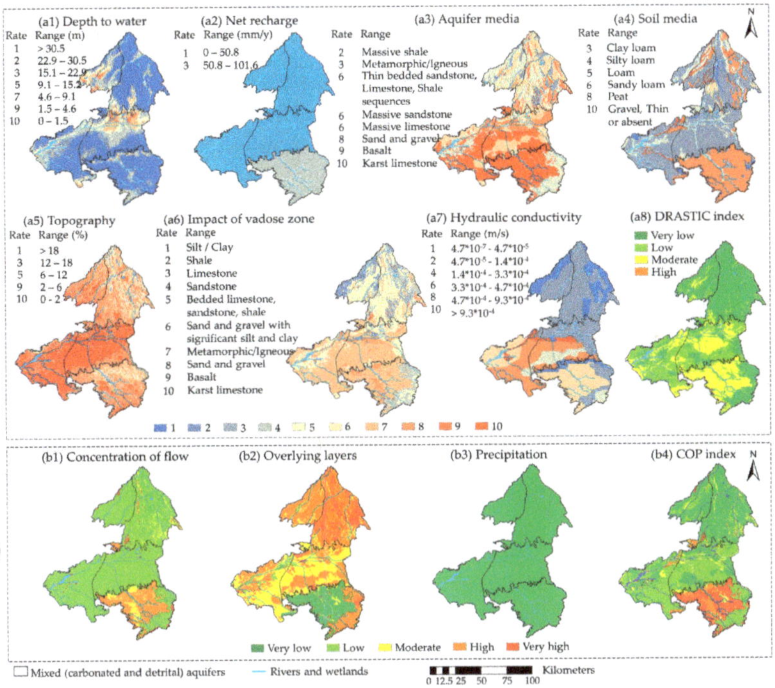

Figure 2. (**a**) DRASTIC (D: Depth to water; R: Net recharge; A: Aquifer media; S: Soil media; T: Topography; I: Impact of vadose zone; C: Hydraulic conductivity) maps: (**a1**) Depth to water; (**a2**) Net recharge; (**a3**) Aquifer media; (**a4**) Soil media; (**a5**) Topography; (**a6**) Impact of vadose zone; (**a7**) Hydraulic conductivity; (**a8**) Vulnerability map; (**b**) COP(C: Concentration of flow; O: Overlying layers above water table; P: precipitation) maps: (**b1**) Reduction of protection due to Concentration of flow; (**b2**) Degree of protection from Overlying layers; (**b3**) Reduction of protection due to the precipitation factor; (**b4**) Vulnerability map.

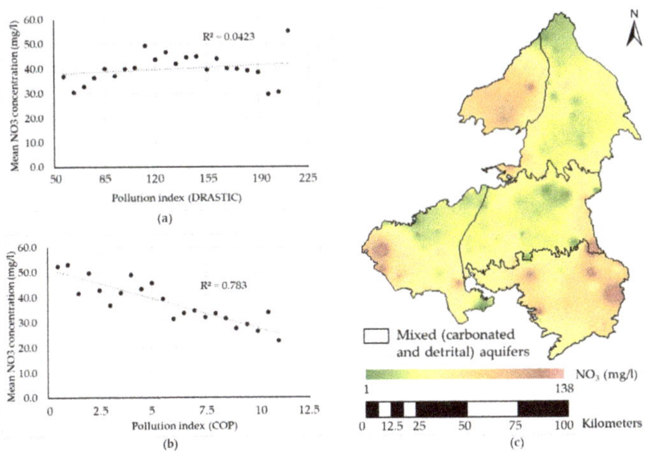

Figure 3. Validation of vulnerability maps in mixed aquifers: (**a**) DRASTIC + 5 × land use (LU); (**b**) COP + 5/LU; (**c**) mean nitrate concentration map (1975–2015).

2.3. Validation of Vulnerability Maps

Contaminant loads are linked to human activities and land use (LU). Therefore, in addition to aquifer vulnerability, these are a key factor in determining groundwater contamination [3,57]. In the study area, the intensive agriculture and the use of nitrogen fertilizers has provoked high levels of nitrates in many groundwater areas. Since nitrate is not naturally found in groundwater, it is considered to be a good indicator of contamination by human impact, especially in agricultural zones. The DRASTIC vulnerability map is validated by adding an LU factor to the DRASTIC index (Equation (1)), as proposed in other studies [27,58,59]. Making an analogy, the index defining protection against pollution could be based on the COP vulnerability (Equation (2)). In this case the inverse of the LU factor is considered since a higher COP index indicates lower vulnerability levels. Since the principal land use is agriculture, the pollution index and the protection-against-pollution index will maintain the factor of 5 used to weight the LU term, as was originally included in the DRASTIC pollution index by other authors [27,60]:

$$\text{Pollution index (DRASTIC)} = \text{DRASTIC index} + 5 \times \text{LU} \quad (1)$$

$$\text{Protection} - \text{against} - \text{Pollution index (COP)} = \text{COP index} + \frac{5}{\text{LU}} \quad (2)$$

The rates assigned to LU [3,27,33,34] are shown in Table 1.

Table 1. Rates of LU.

LU	Rate
Agro-forestry areas	7
Airports	2
Annual crops associated with permanent crops	8
Broad-leaved forest	2
Burnt areas	2
Complex cultivation patterns	8
Coniferous forest	2
Construction sites	2
Continuous urban fabric	10
Discontinuous urban fabric	10
Dump sites	9
Fruit trees and berry plantations	7
Industrial or commercial units	8
Inland marshes	1
Land principally occupied by agriculture, with significant areas of natural vegetation	5
Mineral extraction sites	3
Mixed forest	3
Natural grasslands	3
Non-irrigated arable land	5
Olive groves	6
Pastures	5
Permanently irrigated land	8
Road and rail networks and associated land	2
Sclerophyllous vegetation	3
Sparsely vegetated areas	3
Sport and leisure facilities	2
Transitional woodland-shrub	2
Vineyards	5
Water bodies	1

The correlations (R-squared coefficient) with the pollution index derived from DRASTIC and COP methods (Figure 3) were determined from the 214 observation points where nitrate concentration data was available (1974 to 2015 provided by the Spanish Geological Survey and the River Basin Authority). The DRASTIC pollution index (Figure 3a) is weakly correlated with the mean nitrate concentration (R^2 = 0.04) in the carbonate aquifers. However, the protection-against-pollution index linked to the COP map (Figure 3b) shows strong correlation (R^2 = 0.78). Accordingly, in this validation analysis, the COP vulnerability assessment for the study area provides a more reliable assessment than the DRASTIC index one.

As can be observed in Figures 2 and 3, both DRASTIC and COP indices suggest low vulnerability in some zones with high nitrate concentration (in the western part of the basin) that results from the intensive agricultural exploitation since early 1970's. Thus, groundwater contamination not only depends on intrinsic vulnerability, but also the land use, type of crops, type of irrigation, etc. For this reason, the most contaminated areas do not always correspond to the most vulnerable ones [31,32].

2.4. Methodology: Optimization of DRASTIC Method

In this paper we adapt the DRASTIC method to minimize differences with the COP vulnerability map for the five carbonate aquifers in the Upper Guadiana Basin, which has been proven to provide better results in those aquifers according to the validation analysis. The main objective is to obtain a harmonized method that allows vulnerability in carbonate and detrital aquifers to be assessed in a homogenous way, in a system comprising varying geological formations. The harmonization of criteria to assess groundwater vulnerability will make the results comparable at basin scale. To this end, the DRASTIC index is recalculated by varying the ranges and weights of its parameters (fulfilling certain constraints to maintain a rational definition), in order to minimize the differences with the COP vulnerability map for the five mixed (carbonate and detrital) aquifers in the Upper Guadiana Basin. Data mining techniques are then applied to identify the ranges and weights that provide an optimal solution.

Figure 4 shows the flowchart of the proposed optimization problem and the methodology applied. Two objective functions (O.F.) were tested: (A) to maximize the percentage of spatial coincidence (area, Si) between DRASTIC and COP vulnerability classes; and (B) to minimize the distance (di; see Equation (3)) between the vulnerability classes in DRASTIC and COP. The decision variables in this optimization problem are (1) the ranges (of the DRASTIC index and its parameters) and (2) the weights of the DRASTIC parameters.

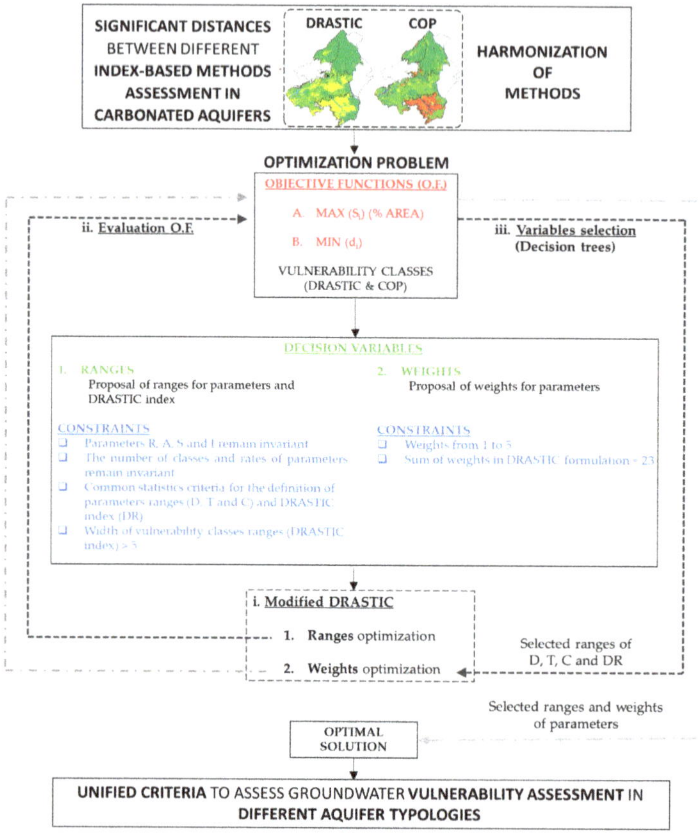

Figure 4. Flowchart of methodology.

The ranges of the DRASTIC parameters and index are proposed based on the available data in the study area and covering a wide range of hypothetical cases. The following constraints regarding the ranges proposal are imposed:

- The ranges of non-continuous parameters (A, S and I) were assumed to be invariant and we classify these parameters as proposed in [9]. We modify the DRASTIC index classification (DR) and the ranges of only three numerical parameters in DRASTIC: Depth to water (D), Topography (T) and Conductivity (C). The proposed ranges are shown in Table 2. Due to the narrow variability of the data in the study area the recharge ranges were not modified.
- The number of classes and rates adopted for all parameters are as proposed in [9]. We change only the distribution of numerical data within the ranges in order to adapt them to the characteristics of the case study.
- The modification of ranges is based on statistical criteria according to the distribution of data in the study area (quantile method, equal intervals, natural clusters in ArcGis and increasing the limit of the ranges by a constant value) [32,61].
- We established a minimum range amplitude of 5 for the DRASTIC vulnerability classes. We also established certain constraints in the weights of parameters:
- Only weights between 1 to 5 are considered;
- The sum of the weights had to be 23 for each combination (as the original proposal in [9]).

Table 2. Proposed classifications for the parameters and DRASTIC index.

Vulnerability	DR1	DR2	DR3	DR4	DR5	DR6	DR7	DR8	DR9	DR10	DR11	DR12	DR13	DR14	DR15	DR16
Very low	<70	<70	<70	<80	<80	<90	<90	<90	<100	<100	<100	<100	<110	<120	<120	<130
Low	80	90	100	90	100	100	100	110	110	115	120	130	120	130	130	140
Moderate	90	110	130	100	120	110	115	130	120	125	140	140	130	140	145	150
High	100	130	160	110	140	120	130	150	130	140	160	150	140	150	160	160
Very high	>100	>130	>160	>110	>140	>120	>130	>150	>130	>140	>160	>150	>140	>150	>160	>160
D (m) rate	D *	D1	D2	D3	D4	D5	D6	D7	D8	D9	D10	D11	D12	D13	D14	D15
10	1.5	5	5	10	10	13	15	15	20	20	25	25	30	30	30	35
9	5	10	10	15	20	21	20	30	25	40	30	50	35	55	60	70
7	10	15	20	20	30	30	25	45	30	60	35	75	40	80	90	105
5	15	20	40	25	40	37	30	60	35	80	40	100	45	100	120	140
3	23	25	80	30	50	54	35	75	40	100	45	125	50	127	150	175
2	30	30	160	35	60	88	40	90	45	120	50	150	55	160	180	210
1	>30	>30	>160	>35	>60	>88	>40	>90	>45	>120	>50	>150	>55	>160	>180	>210
T (%) rate	T *	T1	T2	T3	T4											
10	2	1	<3	4	9											
9	6	2	7	8	18											
5	12	4	14	12	27											
3	18	8	22	20	36											
1	>18	>8	>22	>20	>36											
C (m/day)	C *	C1	C2	C3	C4	C5	C6	C7	C8							
1	0.04–4	≤0.1	≤2	≤2	≤3	≤3	≤5	≤6	≤15							
2	4–12	3	5	6	6	12	12	15	30							
4	12–28	6	12	15	15	25	25	30	45							
6	28–40	15	25	30	30	40	40	75	60							
8	40–80	30	40	40	75	80	80	80	75							
10	>80	>30	>40	>40	>75	>80	>80	>80	>75							

* Original classification of DRASTIC parameters.

For each combination of parameter ranges and weights used to generate a DRASTIC map, the distance (d_i) or difference between vulnerability classes reported by DRASTIC and COP is calculated by Equation (3):

$$d_i = \sum \alpha_{jk} \times x_{jk} \quad (3)$$

where

- α_{jk} is the total area of "j" vulnerability class of COP overlapping with the "k" vulnerability class of DRASTIC;
- x_{jk} is a weight from 0 to 4 depending on the number of jumps from one vulnerability class to another.

In order to establish a dimensionless threshold to quantify the distance (d_i), it is expressed as a percentage of the maximum calculated distance (d_{max}) over all calculated DRASTIC indices:

$$d_i(\%) = \frac{d_i}{d_{max}} \times 100 \quad (4)$$

In order to reduce the number of calculations, we employ data mining techniques (decision trees) to select the values of the variables domain to be tested. The optimal solution is sought following two steps:

1. Ranges optimization:
    i. First, DRASTIC vulnerability maps are calculated modifying the ranges of parameters and the classification of the DRASTIC index. The weights of parameters are the same as proposed in [9].
    ii. All the DRASTIC indices are evaluated through the objective functions and the results are classified intro three categories (Table 3).
    iii. Decision trees are applied in order to find out the ranges for each parameter that gives the highest coincidence ($Max(S_i)$) and a lowest distance ($Min(d_i)$) between vulnerability classes assigned using DRASTIC and COP.
2. Weights optimization:
    i. In this second step, the weights of parameters are introduced as new variables to compute all the feasible combinations of weights and parameter ranges selected in the previous step. The DRASTIC index is calculated for all the combinations of weights and selected classifications in step 1.
    ii. The new set of DRASTIC maps are evaluated by means of the objective functions.
    iii. For each parameter, decision trees are applied again to determine the weight that yields greatest similarity between the DRASTIC and COP maps.

Table 3. Classification criteria of objective functions for the decision trees algorithm.

Objective Function	Value	Class
S_i (spatial coincidence)	<30%	1
	30–50%	2
	>50%	3
d_i (distance)	<30%	1
	30–50%	2
	>50%	3

The main objective of decision trees in this study is to identify the ranges and weights for each DRASTIC parameter involved in any combination that yields the maximum spatial coincidence (S_i)

and the minimum distance between vulnerability classes (d_i) (S_i = 3 or d_i = 1 according to Table 3). We establish these threshold values according to the distribution of results in the first set of combinations. Decision trees reduce the computational cost of the optimization problem and allow the most relevant variables to be identified in the vulnerability assessment in carbonate aquifers.

The CHAID (Chi-square Automatic Interaction Detection) algorithm [62,63] is applied in decision trees, considering the objective functions of the optimization problem (S_i and d_i) as dependent variables. The proposed ranges and weights of parameters and classifications of DRASTIC are the independent variables, and the chi-squared test of significance is used as the splitting criterion in the CHAID algorithm. Each dataset generated in steps 1 and 2 is partitioned into a training set (70%) and a testing set (the remaining 30%) in order to assess the performance of the models. The goodness-of-fit of the classification is evaluated using the precision index [51,64]:

$$Precision = \frac{\sum_{n=1}^{n=3} TP_n}{\sum_{n=1}^{n=3}(TP_n + FP_n)} \quad (5)$$

where:

- n = number of classes;
- TP_n = number of correctly recognized class examples in the class n;
- FP_n = number of examples incorrectly assigned to the class n;

Decision trees can represent the relationship between variables and output class using specific rules following the pathway from the root node to the terminal node [63,65,66]. A priori, the number of terminal nodes on the tree can determine the number of rules but it may generate a large number of irrelevant pieces of information. Only the most relevant variables (rules with the highest population for each decision tree with precision above 50%) whose terminal nodes in the tree are classified as S_i = 3 or d_i = 1 are selected to generate all the feasible combinations to compute the DRASTIC indices.

3. Results

3.1. Optimization of the DRASTIC Method

3.1.1. Ranges Optimization

In the first optimization step, we obtained a total of 11,520 different DRASTIC maps. The spatial coincidence (S_i) of the vulnerability classes between the DRASTIC and COP maps rose to 61.31% (from 55.75% in "original DRASTIC") and the minimum distance between vulnerability classes (d_i) fell to 20.85% ($d_{\text{original DRASTIC}}$ = 24.72%).

The selected ranges (from Table 2) for each parameter extracted from decision trees were as follows:

- DR11, DR12, DR14, DR15 and DR16 for DRASTIC classification;
- D*, D1, D2, D3 and D4 for Depth to water;
- T* and T1 for Topography;
- C7 and C8 for Conductivity.

The ranges proposed in [9] were selected for parameters D and T in the decision tress. For conductivity, the selected classifications (C7 and C8) assign lower rates to conductivity values.

3.1.2. Weights Optimization

The selected ranges for parameters and the DRASTIC index were combined, varying the weights of parameters between one and five. Due to a large number of generated combinations, we first considered weights one, three and five. This resulted in 35,700 new DRASTIC maps. The spatial coincidence (S_i) between vulnerability classes increased to 70.34% and the minimum distance between vulnerability classes (d_i) fell to 8.45%.

Decision trees were applied again to determine the optimum weight for each parameter. We aimed to determine if the value of the weight for each parameter provides relevant information in the optimization problem. We found that a weight equal to five did not appear in relevant rules in parameters D and T, whereas a weight equal to one did not appear for parameters R and S. All weights (one, three, and five) were found in rules for parameters A, I and C.

A new set of combinations of ranges and weights were computed to find an optimum between the gaps left due to constraints. We introduced the mid-weights for each parameter, discarding those not found in the relevant rules. We included the following weights for each parameter: $W_D = 2$; $W_R = 4$; $W_A = 2$ and 4; $W_S = 4$; $W_T = 2$; $W_I = 2$ and 4; $W_C = 2$ and 4. The total of combinations provide two optimal solutions:

- Optimum of O.F. $Min(d_i)$: $d_i = 8.45$; $S_i = 42.91$;
- Optimum of O.F. $Max(S_i)$: $d_i = 13.05$; $S_i = 70.34$;

The second solution was considered the best solution because the gain in S_i was higher than the loss in d_i. Finally, the best solution was refined by adjusting the classification of the DRASTIC index to increase the spatial coincidence with the COP map, obtaining the optimum DRASTIC. The classification of the optimum DRASTIC does not match with the ranges proposed originally in [9]. The objective functions take the following values for the optimum DRASTIC: $S_i = 76.75\%$; $d_i = 10.92\%$.

Figure 5 shows the dot-plot of the all the DRASTIC indices calculated. It reveals the efficacy of using decision trees in the methodology to reduce the number of combinations to be tested when seeking the optimal solution. Each set of combinations improves the objective functions. Black dots show those including the mid-weights for each parameter that produce improvement in the objective function "d_i", but not in "S_i". Red dots indicate the DRASTIC indices that provide $Max(S_i)$ and $Min(d_i)$.

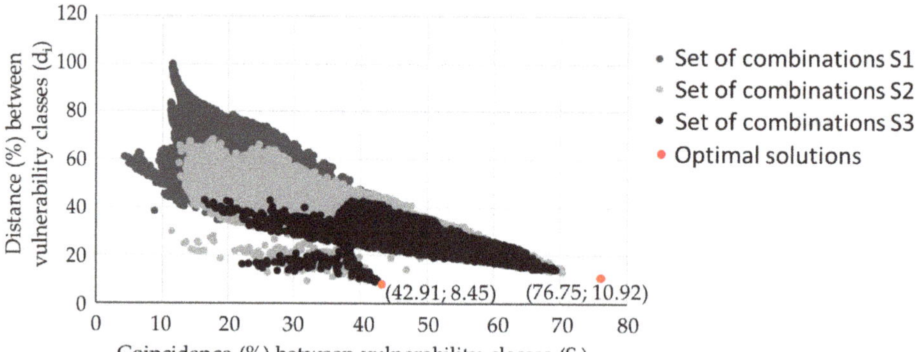

Figure 5. Results of objective functions for all DRASTIC combinations.

3.2. Analysis of Optimum DRASTIC (O-DRASTIC)

O-DRASTIC keeps the ranges proposed in [9] for parameters D and T. Only ranges of parameter C change in O-DRASTIC (C8 from Table 2). In general, the value of the conductivity rates is reduced in O-DRASTIC. The spatial distribution of conductivity ranges shows slight changes. The weights (W) of parameters of O-DRASTIC are shown in Table 4, compared with the original DRASTIC weights.

Table 4. Weights of parameters of original DRASTIC and optimum (O)-DRASTIC.

DRASTIC Parameters	D	R	A	S	T	I	C
W (original DRASTIC)	5	4	3	2	1	5	3
W (O-DRASTIC)	1	5	5	5	1	5	1

Results of O-DRASTIC reveal that Depth to water and Conductivity have no a significant impact on vulnerability for our case study area. The ranges of parameter C in O-DRASTIC also support this conclusion given that the new classification of this parameter reduces the rate assigned to conductivity values. This conclusion is confirmed by the single-parameter sensitivity analysis of O-DRASTIC, in which the parameter C takes a mean effective weight of 1.7% (compared to its empirical weight 4.3%). In general, in the study area, the carbonate aquifers have a scarcely developed karst. In fact, approximately half the study area shows lower rates in the conductivity map. Although conductivity takes high values in some zones, the ranges of this parameter change in O-DRASTIC with respect to the original DRASTIC. The new classification in O-DRASTIC assigns lower rates to the conductivity values. Therefore, this parameter losses influence, which is also reflected in the weight. Moreover, conductivity can be considered as implicit in the aquifer media parameter (A), which increased in weight in O-DRASTIC in our case study. The reduced weight of parameter D may be due to the large depth to water table over most of the study area. The mean effective weight obtained in the sensitivity analysis was 3.2% (empirical weight = 4.3%) Aquifer media and Soil media gained great significance in the vulnerability assessment, as well as Recharge, albeit by a reduced amount. Impact of vadose zone continues to be an important factor in O-DRASTIC.

The weights in O-DRASTIC are consistent with the concept of COP methodology, where the vulnerability assessment is based mainly on the degree of protection afforded by overlying layers and the way in which the water percolates (recharges) into the aquifer.

The main changes between the original and optimum DRASTIC vulnerability maps mostly occur in areas of "Very low" and "Moderate" vulnerability of the original DRASTIC, since these are the predominant classes in this vulnerability map. Remarkably, those changes do not always occur in one direction. "Moderate" vulnerability shifts towards "Low" or "Very high" vulnerability depending on the zone, whereas other "Moderate" vulnerability areas remain with the same class. In the same way, some "Very low" vulnerability zones jump up a vulnerability class to "Low", while other maintain the same class. We analyzed the differences in the distribution of the parameter rates in these areas.

Figure 6a shows that the main factor causing the vulnerability to change from "Moderate" to "Low" is Depth to water. Since the weight of this parameter changed from five to one in O-DRASTIC, zones with higher rates of D report a greater fall in the vulnerability value, leading the vulnerability class to drop by one level. This graph also shows how some areas with rates of three, five and seven change from "Moderate" to "High" vulnerability, which demonstrates that other parameters influence the "Very high" vulnerability class.

On the other hand, we can observe in Figure 6a that the rate of parameter S (soil media) is equal to eight over more than 80% of the area where vulnerability increased from "Moderate" to "Very high", corresponding with soils with a high organic content, and outcrops of limestone and dolomites. Furthermore, the rate of S is equal to four over nearly 90% of the area where vulnerability fell from "Moderate" to "Low". Zones where no change in vulnerability class was observed have soils with medium values.

A similar conclusion regarding Soil media is drawn in the areas where vulnerability rose from "Very low" to "Low" (Figure 6b). In these zones we can also observe higher rates of parameters A (aquifer media) and I (impact of vadose zone). These zones correspond to karstified areas and highly permeable layers.

The results highlight the influence of Aquifer media, Soil media and Impact of vadose zone in the vulnerability of carbonate aquifers.

These conclusions are supported by the single-parameter sensitivity analysis carried out for O-DRASTIC, which yields higher effective weights in parameters A, S and I and lower effective weights for parameters D and C.

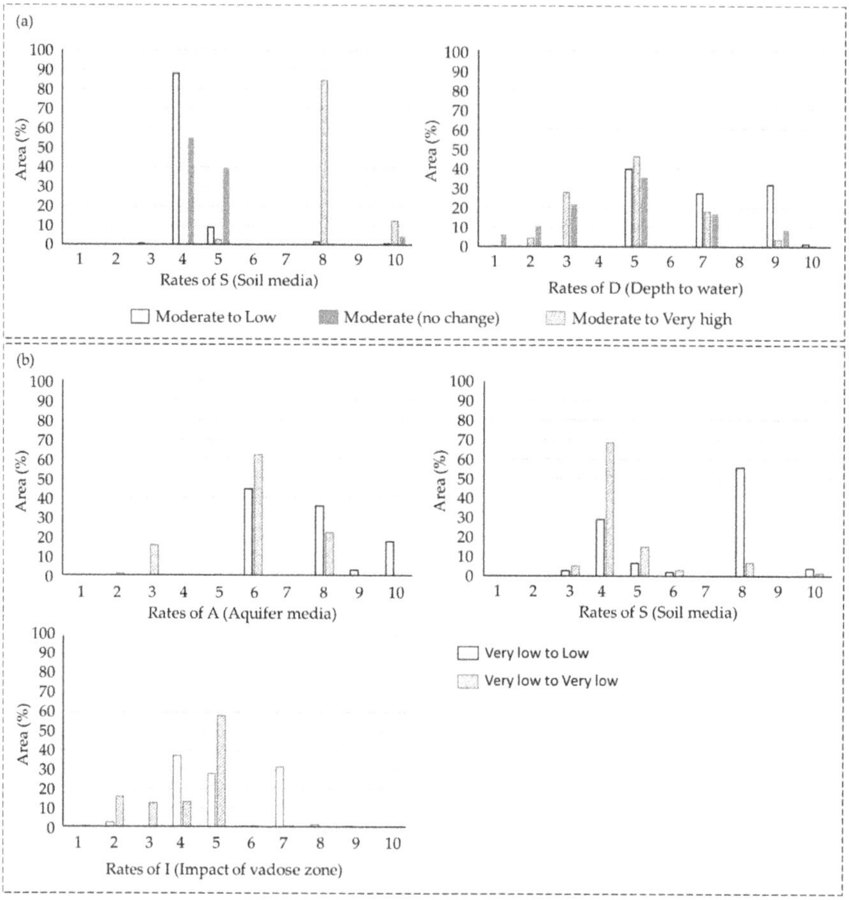

Figure 6. Distribution of rates of the different parameters within the area where significant changes are observed from (**a**) "Moderate" or from (**b**) "Very low" classes in the original DRASTIC to other vulnerability classes in O-DRASTIC.

In general, the new ranges of Conductivity in O-DRASTIC contribute to an overall drop in vulnerability class, but the weights of parameters cause a sharp jump in vulnerability class, especially in the southern part of the basin, where Aquifer media and Soil media have the greatest influence (higher rates in this area).

The values of the optimum DRASTIC (O-DRASTIC) vary between 52 and 178 and the optimal vulnerability classes are the following:

- "Very low": 52–107;
- "Low": 107–130;
- "Moderate": 130–138;
- "High": 138–146;
- "Very high": ≥146;

Figure 7a shows a better distribution of vulnerability values in O-DRASTIC within the COP vulnerability classes (compared to the original DRASTIC). Close similarities can also be appreciated between the O-DRASTIC and COP vulnerability maps (Figure 7b). The vulnerability classes in

O-DRASTIC overlap with COP over 76.75% of the basin, representing an improvement of 20% relative to the original DRASTIC. This spatial coincidence is particularly improved in the class of "Very high" vulnerability, with a 90% coincidence between COP and O-DRASTIC.

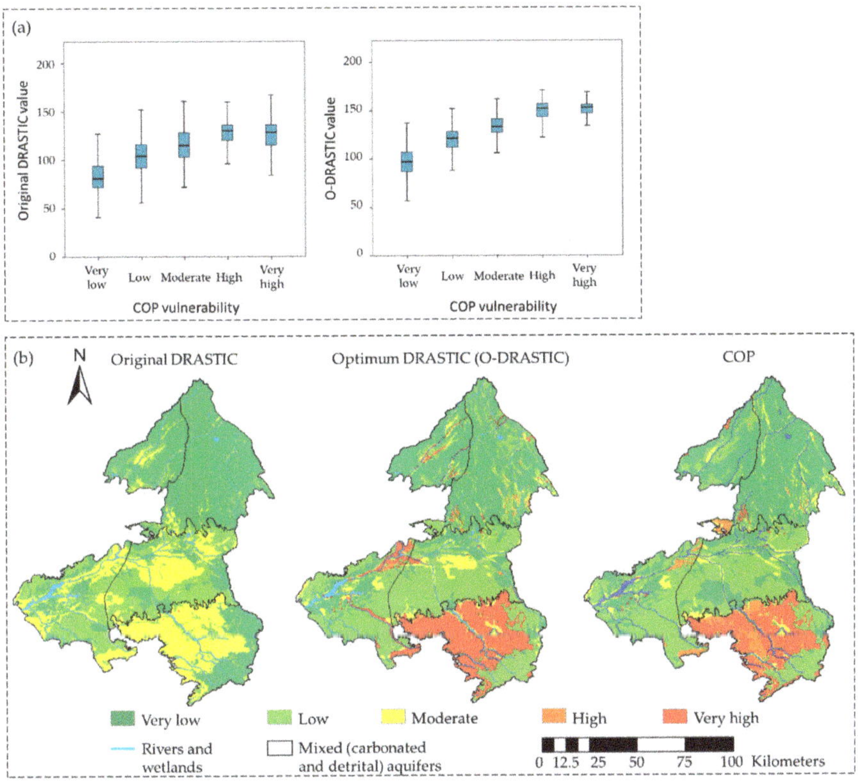

Figure 7. Distribution of vulnerability values of original DRASTIC and O-DRASTIC within vulnerability classes in COP (**a**); vulnerability maps for original DRASTIC, O-DRASTIC and COP (**b**).

O-DRASTIC was validated using the pollution index (Equation (1)) and the correlation with nitrate concentration ($R^2 = 0.653$) in carbonate aquifers improved with respect to the original DRASTIC).

Lastly, we assessed the vulnerability of the three detrital aquifers in the Upper Guadiana Basin by applying original DRASTIC and O-DRASTIC, and we validated the maps following Equation (1). Original DRASTIC showed a good correlation ($R^2 = 0.765$) in detrital aquifers but O-DRASTIC gave a significantly better correlation ($R^2 = 0.862$). Both methods (original DRASTIC and O-DRASTIC) perform a better vulnerability assessment in detrital aquifers than in carbonate aquifers.

Figure 8 shows the change that O-DRASTIC produces (compared to original DRASTIC) in terms of distance between vulnerability classes. Negative values mean the vulnerability class drops in O-DRASTIC compared to original DRASTIC, while positive values mean the vulnerability class in O-DRASTIC is higher than in original DRASTIC. Three zones are thus distinguished in the Upper Guadiana Basin: the southern area (Campo de Montiel) where vulnerability increases by one or two classes. This area is characterized by a higher recharge rate and a karstified aquifer. In the mid-basin (including Mancha Occidental I, Mancha Occidental II, Consuegra Villacañas and Rus-Valdelobos), vulnerability decreases generally by one class, although most of the area remains unchanged; in the northern zone (Lillo-Quintanar, Sierra de Altomira and La Obispalía) only small areas jump up a vulnerability class, and by only one level.

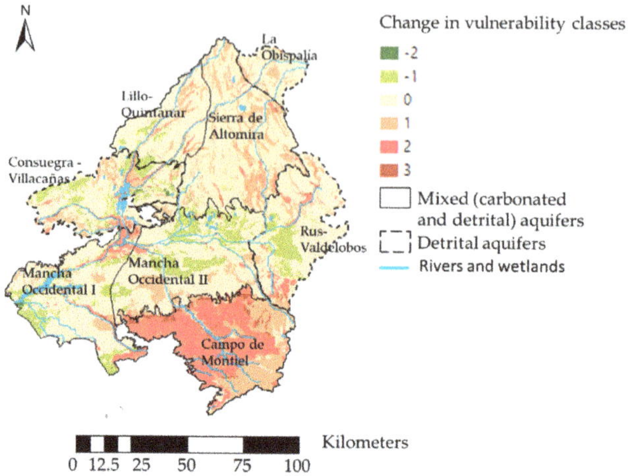

Figure 8. Spatial distribution of change in O-DRASTIC vulnerability classes in comparison to the original DRASTIC.

The largest difference in vulnerability class assigned is for carbonate aquifers. More than 17% of the carbonate aquifers area rises by two classes or more vulnerability classes. Only 7% of the detrital aquifers area jumps by this much, while 62% of the detrital aquifers show no change in the vulnerability class.

4. Discussion and Conclusions

This paper demonstrates that the DRASTIC method can be adapted to assess vulnerability in carbonate aquifers by undertaking some simple modifications of the weights and ranges of the parameters. Most recent studies to adapt the DRASTIC method to carbonate aquifers have aimed to transform DRASTIC by including and/or removing parameters that take the karst characteristics into account [12,28,29]. Other studies modified the classic assessment according to nitrate concentrations [27,30], though the most highly contaminated groundwater does not always imply higher vulnerability [31,32]. Instead of using either of these previous approaches, our study establishes a correspondence between DRASTIC and a vulnerability method that was specifically developed for karstic aquifers, the COP method, and therefore we avoid making conceptual changes in the original definition of DRASTIC method.

Our methodology is based on an optimization approach that identifies the ranges and weights of DRASTIC parameters that minimize the differences in the vulnerability assessment compared with the COP method, which was developed specifically for karstic aquifers. It allows us to identify the significance of the different DRASTIC parameters in the vulnerability assessment in karstic aquifers. Decision trees and spatial statistics analyses are combined to identify the ranges and weights of parameters that provide the optimal DRASTIC classes in terms of coincidence and distance with the COP vulnerability map. This optimization approach could be applied by minimizing the differences with respect to any other reference method for assessing vulnerability, whose results has been previously validated and considered as reasonable. Although many data mining techniques have been applied in groundwater vulnerability assessments [39,40,42], only a few studies have used decision trees in vulnerability studies [50,51]. They have been mainly used to assess other groundwater quality problems [65,67,68].

Our proposed method was applied in the Upper Guadiana Basin, an agricultural area that overlies carbonate and detrital aquifers. The socio-economic and hydrogeological particularities of the basin highlight the need to establish unified management measures at basin scale, not only regarding groundwater exploitation but also in terms of protecting the good quality of the groundwater resource.

Therefore, harmonization of criteria to assess groundwater vulnerability would allow comparison at basin scale and overcome the issue of dissonant results provided by contrasting vulnerability methods. Our approach assumes that the COP method provides a better approximation of the vulnerability in the case study. This assumption was tested by means of a validation analysis in which we show that the "protection-against-pollution" index (derived from COP and LU data) more closely correlates ($R^2 = 0.78$) with nitrate concentration than the DRASTIC "pollution index". The results of the validation show that the pollution index derived from the original DRASTIC for carbonate aquifers is not correlated with nitrates ($R^2 = 0.04$), whereas in detrital aquifers there is a close correlation (($R^2 = 0.76$). Other authors have previously pointed out that the original DRASTIC method does not significantly correlate with nitrate concentration in agricultural areas [45,59,69]. In contrast, the pollution index derived from O-DRASTIC shows significant correlation with nitrates for both, carbonate and detrital aquifers ($R^2 = 0.65$ and $R^2 = 0.86$, respectively).

The optimal solution (O-DRASTIC) obtained in this optimization problem shows that changing the range of the Conductivity parameter produces a small drop in the vulnerability class compared with the original DRASTIC but does not lead to a significant improvement in the objective functions (coincidence and distance between vulnerability classes). The reduced significance of Conductivity is also confirmed by the much-reduced weight assigned to this parameter in O-DRASTIC. Other DRASTIC adaptation for carbonate aquifers also pointed the reduced significance of Conductivity to assess the potential "protection-against-pollution" in karstic systems [12]. We also observed a reduced significance of the Depth to water table in our case study (WD = 1). Other research studies that modified the weights of DRASTIC parameters [3,69,70] also found the Depth to water parameter to be insignificant. Specifically in karstic aquifers Depth to water is not so relevant in protecting an aquifer from contamination because of the high transit velocity through the vadose zone [12]. Moreover, the low significance of in our case study may be due to fact that the water table lies below 30 m over most of the basin. The same argument was also stated in another case study [3]. The reduced significance of Depth to water and Conductivity is confirmed by a sensitivity analysis. Topography and Impact of vadose zone maintain their weights (W_T and W_I) as defined in the original DRASTIC, while the remaining parameters (Recharge, Aquifer media and Soil media) are given maximum weights ($W_R = W_A = W_S = 5$). The large increments in the weights given to Aquifer media and Soil media in O-DRASTIC show that they are the most significant factors controlling the vulnerability in karstic aquifers. Other authors concur that these parameters are the most significant [3,70,71]. The principal change in O-DRASTIC with respect to the original DRASTIC is in the weights of parameters, which highlights that these parameters embrace most of the uncertainty in the DRASTIC vulnerability assessment [72].

Our optimal solution provides an improvement of 20% in terms of coincidence between the vulnerability classes assigned by DRASTIC and COP. This improvement was achieved by applying spatial statistical analyses and decision trees, which allowed potential solutions to be obtained by exploring only a 0.1% of the total dimensionality of the defined optimization problem. The proposed method also helps to achieve a better understanding of the parameters and variables of the "equivalent detrital approach" that really influence vulnerability in this karstic system. This optimal solution was tested for the carbonate and detrital aquifers in our case study, but it should also be tested in other different aquifers with similar hydrogeological characteristics in order to prove its applicability in a broader context under different management framework.

In summary, results show that COP and O-DRASTIC report higher vulnerability classes than the original DRASTIC method over 36% of the total area overlying carbonate aquifers. The greatest differences between the original DRASTIC and O-DRASTIC are produced for the carbonate aquifers rather than the detrital aquifers. This confirms that the reliability of DRASTIC vulnerability assessment is significantly better for detrital aquifers than for karst aquifers. These underestimations of vulnerability in karstic aquifers when applying the classic DRASTIC is due to the physical particularities of these aquifers and their greater sensitivity to pollution [1,12,31].

Hypotheses, Limitations and Future Works

We have demonstrated the applicability of the method in the case of carbonate aquifers where the karst is not highly developed. Its applicability to well-developed karst aquifers also needs to be tested. The main adopted assumptions and limitations of the general methodology adopted are:

- The ranges of categorical non-continuous parameters (Aquifer media, Soil media and Impact of vadose zone) are not modified in this optimization procedure. We consider that the Delphi criteria proposed in [9] can be applied to establish the relative significance of each range with respect to potential pollution.
- Other algorithms and/or techniques (for example, a Random Forest algorithm) could be employed to achieve the goal in a more efficient way.
- We have not studied the whole domain of potential solutions, and a wider spectrum of parameter ranges could be tested to find other optimal solutions. Moreover, the optimization procedure provides local optimum solutions.
- Although decision trees help to reduce the dimensionality of the optimization problem, the methodology involves a large number of calculations, which might handicap extending the method to other case studies.
- The proposed methodology requires a previous validated vulnerability assessment in the study area in order to optimize the DRASTIC method.

Future work should be developed to verify the applicability of O-DRASTIC in other case studies, including aquifers with different physical (climatic and hydrogeological settings) and management particularities in order to withdrawal more generalized conclusions.

Author Contributions: Conceptualization, L.B.-R. and D.P.-V.; Data curation, L.B.-R.; Investigation, L.B.-R.; Methodology, L.B.-R. and D.P.-V.; Software, L.B.-R.; Supervision, D.P.-V.; Validation, L.B.-R.; Visualization, L.B.-R.; Writing—original draft, L.B.-R. and D.P.-V.; Writing—review and editing, L.B.-R. and D.P.-V. All authors have read and agreed to the published version of the manuscript.

Funding: This research did not receive any specific grant from funding agencies in the public, commercial, or not-for-profit sectors.

Acknowledgments: This paper was partially supported by the SIGLO-AN (RTI2018-101397-B-I00) project from the Spanish Ministry of Science, Innovation and Universities (Programa Estatal de I+D+I orientada a los Retos de la Sociedad), the GeoE.171.008-HOVER and the GeoE.171.008-TACTICproject from GeoERA organization funded by European Union's Horizon 2020 research and innovation program.

Conflicts of Interest: The authors declare that they have no known competing financial interests or personal relationships that could have appeared to influence the work reported in this paper.

Appendix A. DRASTIC and COP Methods

Appendix A.1. DRASTIC Method

DRASTIC method was developed by [9] to assess intrinsic groundwater vulnerability in any type of aquifer.

This method considers that there are seven parameters/variables influencing the vulnerability to contamination: Depth to water table (D), Net recharge (R), Aquifer media (A), Soil media (S), Topography (T), Impact of vadose zone (I) and Hydraulic conductivity (C). A rate of importance is assigned to the parameters according to the value or characteristics of each parameter (Table A1).

The values of the parameters are weighted to obtain the DRASTIC index, which is calculated following Equation (A1):

$$\text{DRASTIC index} = 5 \times D + 4 \times R + 3 \times A + 2 \times S + 1 \times T + 5 \times I + 3 \times C \tag{A1}$$

The DRASTIC index was originally classified into eight vulnerability levels according to some color codes [9] (Table A2) although it is usually grouped into five vulnerability classes that do not match with the original proposal in [9].

Table A1. Ranges and rates for DRASTIC parameters.

Groundwater Table Depth		
Original Ranges (ft)	**Transformed Ranges (m)**	**Original Ratings**
0–5	0–1.5	10
5–15	1.5–4.6	9
15–30	4.6–9.1	7
30–50	9.1–15.2	5
50–75	15.1–22.9	3
75–100	22.9–30.5	2
>100	>30.5	1
Net Recharge		
Original Ranges (inches)	**Transformed Ranges (mm)**	**Original Ratings**
0–2	0–50.8	1
2–4	50.8–101.6	3
4–7	101.6–177.8	6
7–10	177.8–254.0	8
>10	>254.0	9
Aquifer Media		
Original Ranges		**Original Ranges**
Massive shale		2
Metamorphic/Igneous		3
Weathered metamorphic/Igneous		4
Thin bedded sandstone, Limestone, Shale sequences		6
Massive sandstone		6
Massive limestone		6
Sand and gravel		8
Basalt		9
Karst limestone		10
Soil Media		
Original Ranges		**Original Ranges**
Thin or absent		10
Gravel		10
Sand		9
Peat		8
Shrinking and/or aggregated clay		7
Sandy loam		6
Loam		5
Silty loam		4
Clay loam		3
Muck		2
Nonshrinking and nonaggregated clay		1
Topography (% Slope)		
Original Ranges		**Original Ranges**
0–2		10
2–6		9
6–12		5
12–18		3
>18		1

Table A1. *Cont.*

Impact of Vadose Zone		
Original Ranges		**Original Ranges**
Silt/Clay		1
Shale		3
Limestone		6
Sandstone		6
Bedded limestone, sandstone, shale		6
Sand and gravel with significant silt and clay		6
Metamorphic/Igneous		4
Sand and gravel		8
Basalt		9
Karst limestone		10
Hydraulic Conductivity		
Original Ranges (GPD/FT^2)	**Transformed Ranges (m/s)**	**Original Ratings**
1–100	4.7×10^{-7}–4.7×10^{-5}	1
100–300	4.7×10^{-5}–1.4×10^{-4}	2
300–700	1.4×10^{-4}–3.3×10^{-4}	4
700–1000	3.3×10^{-4}–4.7×10^{-4}	6
1000–2000	4.7×10^{-4}–9.3×10^{-4}	8
>2000	$>9.3 \times 10^{-4}$	10

Table A2. Color codes for the DRASTIC index.

DRASTIC Index	Color Code
<79	Violet
80–99	Indigo
100–119	Blue
120–139	Dark Green
140–159	Light Green
160–179	Yellow
180–199	Orange
>200	Red

Appendix A.2. COP Method

The COP method was developed by [10] to assess intrinsic groundwater vulnerability in carbonate aquifers.

This method considers the properties of layers overlying the water table (O factor), the concentration of flow (C factor) and precipitation (P factor) as the main parameters influencing groundwater vulnerability in carbonate aquifers. The concept of this method is to assess the natural protection of groundwater determined by the overlying soils and the unsaturated zone, which may be modified by the infiltration process and climatic conditions.

Each factor is divided into subfactors, whose formulation is detailed explained in [10]. The factors are classified in ranges and the COP index calculated as the product of the three factors following the Equation (A2):

$$\text{COP index} = \text{C} \times \text{O} \times \text{P} \quad \text{(A2)}$$

The ranges of factors and the COP index are shown in Table A3.

Table A3. Values for COP parameters and vulnerability classes for the COP index.

C Factor		O Factor		p Factor		COP Index	
Ranges	Reduction of Protection	Ranges	Protection Value	Ranges	Reduction of Protection	Ranges	Vulnerability Classes
0–0.2	Very high	1	Very low	0.4–0.5	Very high	0–0.5	Very high
0.2–0.4	High	2	Low	0.6	High	0.5–1	High
0.4–0.6	Moderate	2–4	Moderate	0.7	Moderate	1–2	Moderate
0.6–0.8	Low	4–8	High	0.8	Low	2–4	Low
0.8–1.0	Very low	8–15	Very high	0.9–1	Very low	4–15	Very low

Appendix B. Data Source and Methodology to Calculate DRASTIC and COP

Table A4. Data source and methodology to calculate DRASTIC parameters.

Factor	Data Source	Methodology
D	Data from simulation flow model (River Basin Authority) (mean of the simulated data for each grid point from 1974 to 2015).	Spatial interpolation using IDW (Inverse Distance Weighted) of groundwater level data and reclassification into D index values.
R	Recharge time series calculated from SACRAMENTO model. Mean recharge value in the period 1974–2015.	Estimation of the mean net recharge taking into account the different hydrology cycle variables in the period 1974–2015.
A	Hydrogeological map of Spain 1:200,000.	Direct assignment of A index values for each hydrogeological unit delimitated.
S	Soil Map of Spain 1:1,000,000.	Direct assignment of S index values for each soil type.
T	Digital Terrain Model at 100 × 100 m cell size.	Calculation of the slope raster file and reclassification of values into T index values.
I	Lithostratigraphic map of Spain 1:200,000.	Direct assignment of I index values for each lithostratigraphic unit.
C	Flow model at 1000 × 1000 m cell size.	Spatial interpolation using IDW of conductivity data and reclassification into C index values.

Table A5. Data source and methodology to calculate COP parameters.

Factor	Subfactor	Data Source	Methodology
C	-	-	There are no catchment areas to swallow holes in these aquifers.
	Scenario 2	Vegetation from CORINE LAND COVER and slope from Digital Terrain Model (100 × 100 m cell size) Karstic features from previous research works & fieldwork and litthostratigraphic map 1:200,000.	Carbonate lithologies with low karstification are considered as fissured formations. Limestones and dolomites with high or very high permeability are considered as scarcely developed karst. For sv factor, the vegetation cover is considered high when more than 30% of the surface is covered. Assignment of the values for the karstic features, vegetation and slope according to COP methodology.
O	OS—Soil	Soil Map of Spain 1:1,000,000.	Assignment of the Os values after classify the different types of soil according to the COP methodology.
	OL—Lithology	Lithostratigraphic map of Spain 1:200,000.	Classification of each lithology according to COP methodology and determination of thickness of vadose zone from 3D flow model.
P	PQ—Precipitation quantity	Precipitation data from SPAIN02 [73] in the grid within the Upper Guadiana Basin (mean rainfall taking into account data above 0.5 mm/day).	Reclassification of the precipitation values into the PQ subfactor values, taking into account the average rainfall in the wet years. Precipitation series from SPAIN02 between 1974 and 2015 were used to extract the mean annual precipitation for wet years.
	PI—Temporal distribution	Precipitation data from SPAIN02 (number of rainy days in the grid within the Upper Guadiana Basin).	Counting of the number of rainy days above 0.5 mm for each cell in the SPAIN02 grid. For the estimate of the rainy days per year, meteorological historical series between 1974 and 2015 from SPAIN02 were analysed.

References

1. Kazakis, N.; Voudouris, K.S. Groundwater vulnerability and pollution risk assessment of porous aquifers to nitrate: Modifying the DRASTIC method using quantitative parameters. *J. Hydrol.* **2015**, *525*, 13–25. [CrossRef]
2. Kadkhodaie, F.; Moghaddam, A.A.; Barzegar, R.; Gharekhani, M.; Kadkhodaie, A. Optimizing the DRASTIC vulnerability approach to overcome the subjectivity: A case study from Shabestar plain, Iran. *Arab. J. Geosci.* **2019**, *12*, 527. [CrossRef]
3. Al-Hanbali, A.; Kondoh, A. Groundwater vulnerability assessment and evaluation of human activity impact (HAI) within the Dead Sea groundwater basin, Jordan. *Hydrogeol. J.* **2008**, *16*, 499–510. [CrossRef]
4. Javadi, S.; Kavehkar, N.; Mousavizadeh, M.H.; Mohammadi, K. Modification of DRASTIC model to map groundwater vulnerability to pollution using nitrate measurements in agricultural areas. *J. Agric. Sci. Technol.* **2011**, *13*, 239–249.
5. Neshat, A.; Pradhan, B.; Dadras, M. Groundwater vulnerability assessment using an improved DRASTIC method in GIS. *Resour. Conserv. Recycl.* **2014**, *86*, 74–86. [CrossRef]
6. Foster, S.S.D. Fundamental concepts in aquifer vulnerability, pollution risk and protection strategy. *Hydrol. Resour. Proc. Inf.* **1987**, *38*, 69–86.
7. Sonnenborg, T.O.; Scharling, P.B.; Hinsby, K.; Rasmussen, E.S.; Engesgaard, P. Aquifer Vulnerability Assessment Based on Sequence Stratigraphic and 39 Ar Transport Modeling. *Ground Water* **2015**, *54*, 214–230. [CrossRef]
8. Seifert, D.; Sonnenborg, T.O.; Scharling, P.; Hinsby, K. Use of alternative conceptual models to assess the impact of a buried valley on groundwater vulnerability. *Hydrogeol. J.* **2007**, *16*, 659–674. [CrossRef]
9. Aller, L.; Bennett, T.; Lehr, J.H.; Petty, R.J.; Hackett, G. *DRASTIC: A Standardized System for Evaluating Groundwater Potential Using Hydrogeologic Settings*; EPA/600/2-85/018; U.S. Environmental Protection Agency: Washington, DC, USA, 1987.
10. Vías, J.M.; Andreo, B.; Perles, M.J.; Carrasco, F.; Vadillo, I.; Jiménez, P. Proposed method for groundwater vulnerability mapping in carbonate (karstic) aquifers: The COP method. *Hydrogeol. J.* **2006**, *14*, 912–925. [CrossRef]
11. Moratalla, A.; Gómez-Alday, J.; Sanz, D.; Castaño, S.; Heras, J.D.L. Evaluation of a GIS-Based Integrated Vulnerability Risk Assessment for the Mancha Oriental System (SE Spain). *Water Resour. Manag.* **2011**, *25*, 3677–3697. [CrossRef]
12. Jiménez-Madrid, A.; Carrasco, F.; Martínez, C.; Gogu, R.C. DRISTPI, a new groundwater vulnerability mapping method for use in karstic and non-karstic aquifers. *Q. J. Eng. Geol. Hydrogeol.* **2013**, *46*, 245–255. [CrossRef]
13. Barbulescu, A. Assessing Groundwater Vulnerability: DRASTIC and DRASTIC-Like Methods: A Review. *Water* **2020**, *12*, 1356. [CrossRef]
14. Vallejos, A.; Andreu, J.M.; Sola, F.; Pulido-Bosch, A. The anthropogenic impact on Mediterranean karst aquifers: Cases of some Spanish aquifers. *Environ. Earth Sci.* **2015**, *74*, 185–198. [CrossRef]
15. Goldscheider, N.; Chen, Z.; Auler, A.S.; Bakalowicz, M.; Broda, S.; Drew, D.; Hartmann, J.; Jiang, G.; Moosdorf, N.; Stevanovic, Z.; et al. Global distribution of carbonate rocks and karst water resources. *Hydrogeol. J.* **2020**, *28*, 1661–1677. [CrossRef]
16. Ravbar, N.; Goldscheider, N. Comparative application of four methods of groundwater vulnerability mapping in a Slovene karst catchment. *Hydrogeol. J.* **2008**, *17*, 725–733. [CrossRef]
17. Plan, L.; Decker, K.; Faber, R.; Wagreich, M.; Grasemann, B. Karst morphology and groundwater vulnerability of high alpine karst plateaus. *Environ. Earth Sci.* **2008**, *58*, 285–297. [CrossRef]
18. Polemio, M.; Casarano, D.; Limoni, P.P. Karstic aquifer vulnerability assessment methods and results at a test site (Apulia, southern Italy). *Nat. Hazards Earth Syst. Sci.* **2009**, *9*, 1461–1470. [CrossRef]
19. Jiménez-Madrid, A.; Martínez-Navarrete, C.; Carrasco-Cantos, F. Groundwater Risk Intensity Assessment. Application to Carbonate Aquifers of the Western Mediterranean (Southern Spain). *Geodin. Acta* **2010**, *23*, 101–111. [CrossRef]
20. Marín, A.I.; Dörfliger, N.; Andreo, B. Comparative application of two methods (COP and PaPRIKa) for groundwater vulnerability mapping in Mediterranean karst aquifers (France and Spain). *Environ. Earth Sci.* **2011**, *65*, 2407–2421. [CrossRef]

21. Bagherzadeh, S.; Kalantari, N.; Nobandegani, A.F.; Derakhshan, Z.; Conti, G.O.; Ferrante, M.; Malekahmadi, R. Groundwater vulnerability assessment in karstic aquifers using COP method. *Environ. Sci. Pollut. Res.* **2018**, *25*, 18960–18979. [CrossRef]
22. Velázquez, D.P.; Sahuquillo, A.; Andreu, J. A two-step explicit solution of the Boussinesq equation for efficient simulation of unconfined aquifers in conjunctive-use models. *Water Resour. Res.* **2006**, *42*, 4205423. [CrossRef]
23. Pulido-Velazquez, D.; Sahuquillo, A.; Andreu, J.; Pulido-Velazquez, M. A general methodology to simulate groundwater flow of unconfined aquifers with a reduced computational cost. *J. Hydrol.* **2007**, *338*, 42–56. [CrossRef]
24. Llopis-Albert, C.; Pulido-Velazquez, D. Using MODFLOW code to approach transient hydraulic head with a sharp-interface solution. *Hydrol. Process.* **2014**, *29*, 2052–2064. [CrossRef]
25. Pulido-Velazquez, D.; Sahuquillo, A.; Andreu, J.; Pulido-Velazquez, M. An efficient conceptual model to simulate surface water body-aquifer interaction in conjunctive use management models. *Water Resour. Res.* **2007**, *43*, 07407. [CrossRef]
26. Velázquez, D.P.; Sahuquillo, A.; Andreu, J. Treatment on non-linear boundary conditions in groundwater modeling with Eigenvalue Methods. *J. Hydrol.* **2009**, *368*, 194–204. [CrossRef]
27. Panagopoulos, G.P.; Antonakos, A.K.; Lambrakis, N.J. Optimization of the DRASTIC method for groundwater vulnerability assessment via the use of simple statistical methods and GIS. *Hydrogeol. J.* **2006**, *14*, 894–911. [CrossRef]
28. Różkowski, J. Evaluation of intrinsic vulnerability of an Upper Jurassic karst-fissured aquifer in the Jura Krakowska (southern Poland) to anthropogenic pollution using the DRASTIC method. *Geol. Q.* **2007**, *51*, 17–26.
29. Mimi, Z.A.; Mahmoud, N.; Abu Madi, M. Modified DRASTIC assessment for intrinsic vulnerability mapping of karst aquifers: A case study. *Environ. Earth Sci.* **2011**, *66*, 447–456. [CrossRef]
30. Pacheco, F.; Pires, L.; Santos, R.; Fernandes, L.S. Factor weighting in DRASTIC modeling. *Sci. Total Environ.* **2015**, *505*, 474–486. [CrossRef]
31. Baalousha, H.M. Vulnerability assessment for the Gaza Strip, Palestine using DRASTIC. *Environ. Earth Sci.* **2006**, *50*, 405–414. [CrossRef]
32. Hamza, S.M.; Ahsan, A.; Imteaz, M.A.; Rahman, A.; Mohammad, T.A.; Ghazali, A.H. Accomplishment and subjectivity of GIS-based DRASTIC groundwater vulnerability assessment method: A review. *Environ. Earth Sci.* **2014**, *73*, 3063–3076. [CrossRef]
33. Saidi, S.; Bouri, S.; Ben Dhia, H. Groundwater vulnerability and risk mapping of the Hajeb-jelma aquifer (Central Tunisia) using a GIS-based DRASTIC model. *Environ. Earth Sci.* **2009**, *59*, 1579–1588. [CrossRef]
34. Antonakos, A.; Lambrakis, N. Development and testing of three hybrid methods for the assessment of aquifer vulnerability to nitrates, based on the drastic model, an example from NE Korinthia, Greece. *J. Hydrol.* **2007**, *333*, 288–304. [CrossRef]
35. Huan, H.; Wang, J.; Teng, Y. Assessment and validation of groundwater vulnerability to nitrate based on a modified DRASTIC model: A case study in Jilin City of northeast China. *Sci. Total Environ.* **2012**, *440*, 14–23. [CrossRef] [PubMed]
36. Thirumalaivasan, D.; Karmegam, M.; Venugopal, K. AHP-DRASTIC: Software for specific aquifer vulnerability assessment using DRASTIC model and GIS. *Environ. Model. Softw.* **2003**, *18*, 645–656. [CrossRef]
37. Hailin, Y.; Ligang, X.; Chang, Y.; Jiaxing, X. Evaluation of Groundwater Vulnerability with Improved DRASTIC Method. *Procedia Environ. Sci.* **2011**, *10*, 2690–2695. [CrossRef]
38. Sener, E.; Davraz, A. Assessment of groundwater vulnerability based on a modified DRASTIC model, GIS and an analytic hierarchy process (AHP) method: The case of Egirdir Lake basin (Isparta, Turkey). *Hydrogeol. J.* **2012**, *21*, 701–714. [CrossRef]
39. Fijani, E.; Nadiri, A.A.; Asghari-Moghaddam, A.; Tsai, F.T.-C.; Dixon, B. Optimization of DRASTIC method by supervised committee machine artificial intelligence to assess groundwater vulnerability for Maragheh–Bonab plain aquifer, Iran. *J. Hydrol.* **2013**, *503*, 89–100. [CrossRef]
40. Barzegar, R.; Moghaddam, A.A.; Baghban, H. A supervised committee machine artificial intelligent for improving DRASTIC method to assess groundwater contamination risk: A case study from Tabriz plain aquifer, Iran. *Stoch. Environ. Res. Risk Assess.* **2015**, *30*, 883–899. [CrossRef]

41. Jang, W.S.; Engel, B.A.; Harbor, J.; Theller, L. Aquifer Vulnerability Assessment for Sustainable Groundwater Management Using DRASTIC. *Water* **2017**, *9*, 792. [CrossRef]
42. Nadiri, A.A.; Gharekhani, M.; Khatibi, R. Mapping Aquifer Vulnerability Indices Using Artificial Intelligence-running Multiple Frameworks (AIMF) with Supervised and Unsupervised Learning. *Water Resour. Manag.* **2018**, *32*, 3023–3040. [CrossRef]
43. Dixon, B. Groundwater vulnerability mapping: A GIS and fuzzy rule based integrated tool. *Appl. Geogr.* **2005**, *25*, 327–347. [CrossRef]
44. Rodriguez-Galiano, V.F.; Luque-Espinar, J.; Chica-Olmo, M.; Mendes, M. Feature selection approaches for predictive modelling of groundwater nitrate pollution: An evaluation of filters, embedded and wrapper methods. *Sci. Total Environ.* **2018**, *624*, 661–672. [CrossRef] [PubMed]
45. Machiwal, D.; Jha, M.K.; Singh, V.P.; Mohan, C. Assessment and mapping of groundwater vulnerability to pollution: Current status and challenges. *Earth Sci. Rev.* **2018**, *185*, 901–927. [CrossRef]
46. Fusco, F.; Allocca, V.; Coda, S.; Cusano, D.; Tufano, R.; De Vita, P. Quantitative Assessment of Specific Vulnerability to Nitrate Pollution of Shallow Alluvial Aquifers by Process-Based and Empirical Approaches. *Water* **2020**, *12*, 269. [CrossRef]
47. Gogu, R.C.; Hallet, V.; Dassargues, A. Comparison of aquifer vulnerability assessment techniques. Application to the Nylon river basin (Belgium). *Environ. Earth Sci.* **2003**, *44*, 881–892. [CrossRef]
48. Worrall, F.; Besien, T.; Kolpin, D.W. Groundwater vulnerability: Interactions of chemical and site properties. *Sci. Total Environ.* **2002**, *299*, 131–143. [CrossRef]
49. Zwahlen, F. *COST Action 620 Vulnerability and Risk Mapping for the Protection of Carbonate (karst) Aquifers Final Report*; Office of the Official Publications of the European Communities: Brussels, Belgium, 2004; p. 297.
50. Rodriguez-Galiano, V.; Mendes, M.P.; Garcia-Soldado, M.J.; Chica-Olmo, M.; Ribeiro, L.F. Predictive modeling of groundwater nitrate pollution using Random Forest and multisource variables related to intrinsic and specific vulnerability: A case study in an agricultural setting (Southern Spain). *Sci. Total Environ.* **2014**, 189–206. [CrossRef]
51. Yoo, K.; Shukla, S.K.; Ahn, J.J.; Oh, K.; Park, J. Decision tree-based data mining and rule induction for identifying hydrogeological parameters that influence groundwater pollution sensitivity. *J. Clean. Prod.* **2016**, *122*, 277–286. [CrossRef]
52. Martínez-Santos, P.; Llamas, M.; Martinezalfaro, P. Vulnerability assessment of groundwater resources: A modelling-based approach to the Mancha Occidental aquifer, Spain. *Environ. Model. Softw.* **2008**, *23*, 1145–1162. [CrossRef]
53. Yustres, Á.; Botti, V.; Asensio, L.; Candel-Pérez, M.; García, B. Groundwater resources in the Upper Guadiana Basin (Spain): A regional modelling analysis. *Hydrogeol. J.* **2013**, *21*, 1129–1146. [CrossRef]
54. Conan, C.; De Marsily, G.; Bouraoui, F.; Bidoglio, G. A long-term hydrological modelling of the Upper Guadiana river basin (Spain). *Phys. Chem. Earth Parts A/B/C* **2003**, *28*, 193–200. [CrossRef]
55. FAO. *The FAO-Unesco Soil Map of the World*; Legend and 9 Volumes; UNESCO: Paris, France, 1981.
56. Confederación Hidrográfica del Guadiana. *Actualización y Calibración del Modelo de flujo de agua Subterránea de los Acuíferos del Alto Guadiana (FLUSAG)*; Ref: TEC0004594, Published Report; Dirección General del Agua: Madrid, Spain, 2018; 150p.
57. Ahmed, I.; Nazzal, Y.; Zaidi, F. Groundwater pollution risk mapping using modified DRASTIC model in parts of Hail region of Saudi Arabia. *Environ. Eng. Res.* **2017**, *23*, 84–91. [CrossRef]
58. Babiker, I.S.; Mohamed, M.A.; Hiyama, T.; Kato, K. A GIS-based DRASTIC model for assessing aquifer vulnerability in Kakamigahara Heights, Gifu Prefecture, central Japan. *Sci. Total Environ.* **2005**, *345*, 127–140. [CrossRef] [PubMed]
59. Stigter, T.Y.; Riberio, L.; Dill, A.M.M.C. Evaluation of an intrinsic and a specific vulnerability assessment method in comparison with groundwater salinization and nitrate contamination levels in two agricultural regions in the south of Portugal. *Hydrogeol. J.* **2006**, *14*, 79–99. [CrossRef]
60. McLay, C.; Dragten, R.; Sparling, G.; Selvarajah, N. Predicting groundwater nitrate concentrations in a region of mixed agricultural land use: A comparison of three approaches. *Environ. Pollut.* **2001**, *115*, 191–204. [CrossRef]
61. Mentzafou, A.; Panagopoulos, Y.; Dimitriou, E. Designing the National Network for Automatic Monitoring of Water Quality Parameters in Greece. *Water* **2019**, *11*, 1310. [CrossRef]

62. Kass, G.V. An Exploratory Technique for Investigating Large Quantities of Categorical Data. *J. R. Stat. Soc. Ser. C Appl. Stat.* **1980**, *29*, 119. [CrossRef]
63. Rokach, L.; Maimon, O. *Data Mining with Decision Trees Theory and Applications*; World Scientific: Toh Tuck Link, Singapore, 2007.
64. Sokolova, M.; Lapalme, G. A systematic analysis of performance measures for classification tasks. *Inf. Process. Manag.* **2009**, *45*, 427–437. [CrossRef]
65. Jeihouni, M.; Toomanian, A.; Mansourian, A. Decision Tree-Based Data Mining and Rule Induction for Identifying High Quality Groundwater Zones to Water Supply Management: A Novel Hybrid Use of Data Mining and GIS. *Water Resour. Manag.* **2019**, *34*, 139–154. [CrossRef]
66. Quinlan, J. Simplifying decision trees. *Int. J. Man-Mach. Stud.* **1987**, *27*, 221–234. [CrossRef]
67. Wheeler, D.C.; Nolan, B.T.; Flory, A.R.; Dellavalle, C.T.; Ward, M.H. Modeling groundwater nitrate concentrations in private wells in Iowa. *Sci. Total Environ.* **2015**, *536*, 481–488. [CrossRef] [PubMed]
68. Bui, D.T.; Khosravi, K.; Karimi, M.; Busico, G.; Khozani, Z.S.; Nguyen, H.; Mastrocicco, M.; Tedesco, D.; Cuoco, E.; Kazakis, N. Enhancing nitrate and strontium concentration prediction in groundwater by using new data mining algorithm. *Sci. Total Environ.* **2020**, *715*, 136836. [CrossRef] [PubMed]
69. Almasri, M.N. Assessment of intrinsic vulnerability to contamination for Gaza coastal aquifer, Palestine. *J. Environ. Manag.* **2008**, *88*, 577–593. [CrossRef] [PubMed]
70. Hasiniaina, F.; Zhou, J.; Guoyi, L. Regional assessment of groundwater vulnerability in Tamtsag basin, Mongolia using drastic model. *J. Am. Sci.* **2010**, *6*, 65–78.
71. Al Hallaq, A.H.; Abuelaish, B. Assessment of aquifer vulnerability to contamination in Khanyounis Governorate, Gaza Strip—Palestine, using the DRASTIC model within GIS environment. *Arab. J. Geosci.* **2011**, *5*, 833–847. [CrossRef]
72. Dizaji, A.R.; Hosseini, S.A.; Rezaverdinejad, V.; Sharafati, A. Groundwater contamination vulnerability assessment using DRASTIC method, GSA, and uncertainty analysis. *Arab. J. Geosci.* **2020**, *13*, 1–15. [CrossRef]
73. Herrera, S.; Fernández, J.; Gutiérrez, J.M. Update of the Spain02 gridded observational dataset for EURO-CORDEX evaluation: Assessing the effect of the interpolation methodology. *Int. J. Clim.* **2015**, *36*, 900–908. [CrossRef]

Publisher's Note: MDPI stays neutral with regard to jurisdictional claims in published maps and institutional affiliations.

water

MDPI

Article

Assessing the Groundwater Quality in the Liwa Area, the United Arab Emirates

Alina Barbulescu [1,*], Yousef Nazzal [2] and Fares Howari [2]

1 Department of Mathematics and Computer Science, Ovidius University of Constanta, 124 Mamaia Blvd., 900527 Constanta, Romania

2 College of Natural and Health Sciences, Zayed University, Abu Dhabi P.O. Box 144534, UAE; Yousef.nazzal@zu.ac.ae (Y.N.); fares.howari@zu.ac.ae (F.H.)

* Correspondence: alinadumitriu@yahoo.com

Received: 25 August 2020; Accepted: 29 September 2020; Published: 10 October 2020

Abstract: Last period groundwater quality raises big concerns all over the world since it is a limited source of drinkable water and for agricultural and industrial use. While the suitability of the groundwater of Liwa aquifer (Abu Dhabi Emirate) for agricultural use has been previously partially studied, not all the water parameters have been taken into account. Therefore, in this paper, we propose the study of 42 concentrations series of 19 groundwater parameters. We test the hypothesis that the water parameters series recorded at different locations are similar and group the samples in clusters. The main parameters that determine the differences between the clusters are determined by Principal Component Analysis (PCA). Finally, we use a quality index for assessing the water suitability for drinking. The conclusions emphasize the necessity of using more than one technique to evaluate water quality for different purposes and to cross-validate the results.

Keywords: groundwater; water parameters; k-means; principal component analysis; water quality index

1. Introduction

All over the world, billions of people suffer from water scarcity because less than 1% of the world's water is fresh and accessible [1]. The remediation of polluted aquifers is generally difficult and in many cases, is not possible. Therefore, the study of water quality is an essential study topic for water resources scientists [2–12] as a first step for taking informed measures for keeping it clean and as a warning for reducing its pollution [12].

MIKE-II, QUASAR, QUAL2E, and CE-QUALW2, SIMCAT, TOMCAT are water quality models providing comprehensive modeling of water quality conditions in river systems [13]. They are developed for particular purposes and none of them are best. For assessing the surface water quality, researchers [14–19] introduced water quality indices (WQI), the most known being CCMEWQI [16]. A review of the most important ones, containing their composition, structure, and comparison is provided in [14]. Other approaches use univariate statistical methods, as time series analysis, for describing the temporal [6–8,10] evolution of some water parameters. Bhat et al. [8] and Ioele et al. [20] employed multivariate statistical tools. ANOVA is utilized for evaluating the differences among the series of water quality variables recorded at the study sites. Cluster analysis (CA) allows selecting the groups of sites with similar characteristics (concentrations, pH) of water parameters. Principal Component Analysis (PCA) leads to the detection of the main water parameters that influence water quality [20]. Gad et al. [21] combined a drinking water quality index and four pollution indices, principal component analysis (PCA), partial least squares regression (PLSR), and stepwise multiple linear regression (SMLR) to evaluate the water quality for drinking purposes in the Nile Delta.

Assessing the groundwater quality became a study topic in 1968 when the "groundwater vulnerability" notion was introduced by Margat [22]. The definitions of this concept [22–24] aim at catching the interaction between a contaminant applied in the soil vicinity (or its surface) and the aquifer, during the pollutant's transportation by the rainwater. The physicochemical reactions and their effects on the groundwater depend on hydrogeological conditions and the pollutant characteristics, quantity, and exposure time [25–27].

The study of groundwater quality and the risk to pollutants' exposure is generally done by using the DRASTIC model, introduced by Aller et al. [28], DRASTIC-like methods (AVI [29], DIVERSITY [30], GOD [31], ISIS [32], PRAST [33], SINTACS [34–36], etc.), and their versions (DRAMIC [37], DRIST [38], DRAV [39], DRASTIC-LU [40], DRASIC-LU [41], SINTACS-LU [42,43], DRASTICA [44,45], etc.). For a review of these methods, the reader may refer to [46]. Specific methods for studying the karst aquifer vulnerability are also employed; among them are COP [47], EPIK [48], REKS [49], RISKE [50,51], PaPRIKa [52], and PI [53]. GIS is an important tool in these cases for drawing the vulnerability maps [54,55].

For assessing the groundwater suitability for drinking or agricultural use at different locations, scientists [56–59] use indices like Chloro-alkaline index (CAI), Saturation index (SI), Sodium Absorption Ratio (SAR), Residual Sodium Carbonate (RSC), Kelley's ratio, and magnesium hazard. A single WQI (single-factor pollution index (I), the nemerow pollution index (NI), heavy metal evaluation index (HEI), the degree of contamination (Cd)), or geostatistical methods can also be utilized to emphasize the groundwater pollution at a regional scale [60,61].

In this article, we propose a combined methodology for studying the groundwater characteristics at a regional scale. Firstly, we test the similarity of the series of water parameters (collected at different sites), then we cluster the sites with the same characteristics and perform the Principal Components Analysis (PCA) for extracting the significant components. Then, we assess the suitability of water for drinking using a water quality index. Finally, we compare the obtained results with those provided by the literature and conclude.

2. Study Area and Methods

2.1. Study Area

Abu Dhabi Emirate, the largest emirate of the United Arab Emirates, is situated along the Arabian Gulf, between 22.5° and 25° north latitudes and 51°–55° east longitudes. The study area, Liwa, belongs to Abu Dhabi Emirate. Water samples were collected in the northern part of the Liwa Crescent, between Madinat Zayed and Meziyrah. The distribution of the drilling wells is presented in Figure 1, whereas their coordinates are given in Table 1.

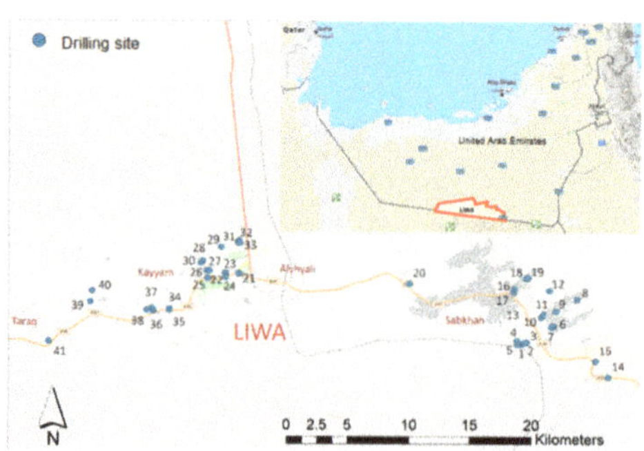

Figure 1. Locations of drilling sites.

Table 1. Coordinates of the drilling locations.

Sample No.	Coordinates		Sample No.	Coordinates		Sample No.	Coordinates	
	North	East		North	East		North	East
1	23.05.07	53.59.44	15	23.04.21	54.02.56	29	23.09.21	53.46.27
2	23.05.09	53.59.49	16	23.07.32	53.59.21	30	23.08.39	53.45.33
3	23.05.10	53.59.55	17	23.07.23	53.59.20	31	23.09.34	53.47.13
4	23.05.15	53.59.29	18	23.08.00	53.59.52	32	23.09.37	53.47.15
5	23.05.05	53.59.31	19	23.08.02	53.59.58	33	23.09.31	53.47.17
6	23.05.54	54.01.05	20	23.07.45	53.54.44	34	23.06.36	53.44.10
7	23.05.52	54.01.00	21	23.08.12	53.47.14	35	23.06.33	53.44.09
8	23.07.05	54.02.07	22	23.08.08	53.46.38	36	23.06.31	53.43.27
9	23.06.33	54.01.11	23	23.08.13	53.46.38	37	23.06.38	53.43.24
10	23.06.18	54.00.33	24	23.08.01	53.46.39	38	23.06.33	53.43.10
11	23.06.25	54.00.38	25	23.07.58	53.45.51	39	23.06.55	53.40.42
12	23.07.28	54.00.53	26	23.08.18	53.45.48	40	23.07.24	53.40.47
13	23.06.39	54.59.45	27	23.08.18	53.45.56	41	23.05.10	53.38.49
14	23.03.39	54.03.30	28	23.08.44	53.45.38			

The mean monthly temperature in the region is between 20 °C and 35 °C, with minima between 13 °C and 29 °C and maxima in the interval 31 °C–48 °C. The average humidity varies from 59% to 68%, with a maximum of about 79%. The maximum monthly average precipitation recorded from 2003 till 2017 was 16 mm (in December) and 10 mm (in January), without precipitation from May to October.

In the Liwa area, continental and shallow water marine sedimentary rocks were deposited from Cambrian to Quaternary. Liwa's aquifer lithology is composed of two essential stratigraphic units. The first one has a thickness between 100 m and 150 m and is formed by a Quaternary part, Holocene, and Pleistocene Aeolian fine to medium sands and interdunal deposits. The second one, with a thickness of over 350 m, is a Tertiary unit formed by mudstones, evaporites, and clastics of Miocene age [62]. The shallow aquifer formation consists of sand and sandstone, with a variable thickness underlain by siltstone, claystone, and evaporites. In the Liwa area, the altitude of the groundwater level is between 60 m and 107 m a.s.m.l. (above mean sea level). The shape of the groundwater table is concave down, elongated from East to West. Its top is situated approximately 25 km north of Mezairaa [63]. The gradient of the groundwater table is less than 0.5 m/km in the east-west direction, 0.5 m/km in the southern part, and more than 1 m/km in the northern region.

The principal aquifer of the Western Region, situated in the northern Liwa area, consists of the upper subunit of the Quaternary sediments. To the west, the aquifer extends to the Sabkha Matti area, while to the East, it borders the gravel plains located at the foot of the Oman Mountains. The average thickness of the principal aquifer varies between 30 m to 50 m. Overlaying sands dunes, forming a thick unsaturated zone, cover the aquifer. The lower subunit of the Quaternary sediments represents a fully saturated aquitard, situated above the aquiclude consisting of the Tertiary Lower Fars unit [62,63] area, while to the East, it borders the gravel plains located at the foot of the Oman Mountains. The average thickness of the principal aquifer varies between 30 m to 50 m. Overlaying sands dunes, forming a thick unsaturated zone, cover the aquifer. The lower subunit of the Quaternary sediments represents a fully saturated aquitard, situated above the aquiclude consisting of the Tertiary Lower Fars unit [62,63]. A schematic description of the Liwa aquifer and well cross-section are shown in Figure 2.

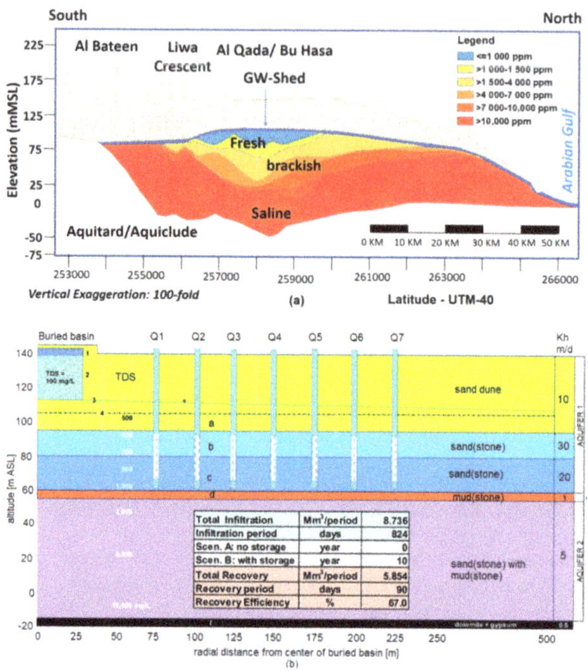

Figure 2. (**a**) A schematic description of the Liwa aquifer and (**b**) a well cross-section.

2.2. Experimental Study

For the present study, we collected groundwater samples from 41 wells situated in the Liwa zone in March 2018 and stored them in polyethylene bottles of 1 L capacity. We followed the methods of the American Public Health Association for the samples' preservation and analysis [64].

pH, electrical conductivity (EC), and total dissolved solids (TDS) were determined at the sampling sites using a pH-meter, a portable EC-meter, and a TDS-meter (Hanna Instruments, Ann Arbor, MI, USA). The sodium (Na^+), potassium (K^+), magnesium (Mg^{2+}), and calcium (Ca^{2+}) ions were determined by atomic absorption spectrophotometry (AAS), while the carbonate and bicarbonate were analyzed by volumetric methods. Sulfate (SO_4^{2-}) was estimated by the colorimetric and turbidimetric methods. The nitrate concentration was measured by ionic chromatography. Trace elements (Cd, Cr, Zn, Pb, Cu, Ni, Mn) were determined by Inductively Coupled Plasma spectrophotometer (ICP-OES, Agilent, CA, USA). One can find the results of the chemical analyses in [62].

2.3. Methodology

The statistical analysis performed on the series of water parameters consisted of the following.

➢ *Determine the variability of the water parameters at different locations*

For this aim, the basic statistics, the histograms, and the boxplots of the series were studied. Comparisons of the series values with the maximum admissible limits have also been performed.

➢ *Study of the similarity of the series collected at different sites*

For the rest of the study, except for the computation of the water quality index, data were standardized by dividing the concentration of each element by the maximum concentration of the element. Then, to test the hypothesis that there is no statistically significant difference between the water elements in the samples collected at different study places, the Kruskal-Wallis nonparametric test [65] was performed at a 5% significance level. If the null hypothesis was rejected, the Dunn post hoc test was applied to determine the pairs of dissimilar samples [66].

➢ *Perform data clustering*

The classification of the series into homogenous groups within which the patterns are the same was done by using the k-means clustering algorithm [67]. This algorithm determines groups of data series based on a criterion of error minimization, which computes the distance of instances to their representative values. The stages of the k-means algorithm are summarized below [68–70].

Let us consider that $Z = (z_1, z_2, ..., z_m), z_i \in \mathbf{R}^n, i = \overline{1, m}$ is the vector that contains the series concentrations measures at each location (column j is composed of data from site j).

(1) Firstly, the number of clusters, k, is selected.

This number is either introduced by the user or computed based on different algorithms. Among the methods used to determine the optimal number of clusters, the most popular are the elbow, the silhouette, and the gap statistic methods [71]. In this article, we employed the facilities of the R software, especially the NbClust, which provides 30 algorithms for the detection of the optimal k. The best number of clusters is selected according to the majority rule [72].

(2) The clusters' centroids $\vartheta_1, \ldots, \vartheta_k \in \mathbf{R}^n$ are initialized and the distances between the data points and the cluster centers are computed. Each point is assigned to the cluster that minimizes the distances from it to the clusters' centers.
(3) The new clusters' centers are determined, the procedure restarts from (2) and runs until no data point can be reassigned to another cluster. Then, stop.

➢ *Perform the Principal Component Analysis (PCA)* [73–75]

In hydrological studies, PCA is a tool for finding the water quality parameters that describe the processes that govern the groundwater chemistry and extract valuable information using only the significant variables. PCA is a statistical procedure of the multivariate analysis, designed for reducing the variables' number and replacing them with a few artificial ones (Principal Components—PCs). These PCs are independent factors that mainly explain the study phenomenon and sum up a significant amount of variance. There are many criteria used for extracting the principal components; among them, the most used are the Catell Scree plot, the Kaiser criterion (Kaiser), and the Explained Variance Criterion. The first PC accounts for the highest variability are emphasized on the Scree plot [74]. Kaiser criterion [76] takes into account the selection of those PCs that correspond to eigenvalues greater than 1 [77]. Here, we used both the Scree plot and the Kaiser criteria for detecting the PC.

The biplot shows the contributions of the variables on the PCs. The loadings emphasize the correlations between the input variables and the factors.

For a deeper insight into the PCA and its implementation in R software, the reader may see [74,75,78,79].

➢ *Assessing the suitability of water for drinking*

For this aim, we used a weighted arithmetic Water Quality Index (WQI) [80–82]. The WQI index is built as follows:

(1) choose the water parameters used in the computation;
(2) compute the quality rating (q_i) for each parameter by:

$$q_i = \frac{C_i - C_{ideal}}{S_i - C_{ideal}}, \quad (1)$$

where C_i is the actual concentration in sample i, $1/S_i$ is the admissible value, and $C_{ideal} = 0$ for all, but pH, for which $C_{ideal} = 7$;

(3) Compute the weight of each water parameter, i, by:

$$w_i = \frac{1/S_i}{\sum_{i=1}^{n}(1/S_i)}, \quad (2)$$

where n is the number of water parameters taken into consideration;

(4) Compute the WQI by:

$$WQI = \sum_{i=1}^{n}(w_i\, Q_i), \quad (3)$$

(5) Classify the water quality based on the interval in which the value is contained. The classes and the corresponding intervals are A (Excellent)—(0,25], B (Good)—(25,50], C (Poor)—(50,75], D (Very poor)—(76,100], and E (unsuitable) > 1000 [80].

(6) Compare the values of WQI for the samples contained in different clusters.

3. Results and Discussion

Table 2 contains the basic statistics of the study data series.

Table 2. Basic statistics of the water parameters.

	Minimum	Maximum	Average	Standard Deviation	Variation Coefficient (cv)
pH	6.1900	7.190	6.519	0.256	0.039
EC (μS/cm)	328.0000	3003.000	1478.488	648.574	0.439
TDS (mg/L)	136.0000	1565.000	863.049	354.590	0.411
Na+ (mg/L)	638.1750	5302.039	2923.174	1044.109	0.357
K+ (mg/L)	2.7043	17.203	8.964	3.121	0.348
Cl− (mg/L)	827.0140	9628.939	5670.833	2258.713	0.398
NO3− (mg/L)	0.4259	2.486	1.410	0.550	0.390
SO42− (mg/L)	4.1290	45.794	23.570	9.142	0.388
CO32− (mg/L)	14.4000	108.000	57.712	27.473	0.476
HCO3− (mg/L)	14.6400	236.680	87.546	48.842	0.558
Ca2+ (mg/L)	104.2060	1244.785	705.423	269.035	0.381
Mg2+ (mg/L)	21.5390	672.509	316.744	148.690	0.469
Cd (mg/L)	0.0001	0.002	0.001	0.000	0.762
Cr (mg/L)	0.0005	0.023	0.015	0.005	0.362
Cu (mg/L)	0.0009	0.004	0.002	0.001	0.433
Mn (mg/L)	0.0001	0.011	0.002	0.003	1.242
Ni (mg/L)	0.0005	0.004	0.002	0.001	0.563
Pb (mg/L)	0.0013	0.012	0.005	0.003	0.500
Zn (mg/L)	0.0004	0.052	0.005	0.010	2.135

High amplitudes of the registered values are found for almost all study elements, with very high standard deviations for EC, TDS, Na^+, Cl^-, Ca^{2+}, and Mg^{2+}. The higher the standard deviation is, the higher the variation of the values of a series about the mean is. The coefficients of variation show high variability of Zn, Mn, Cd, and Ni concentrations by comparison to the other series.

The boxplots of data series (Figure 3) show the existence of some outliers for pH, K^+, SO_4^{2-}, EC, HCO_3^-, Cd, Cu, Mn, Ni, Pb, and Zn series.

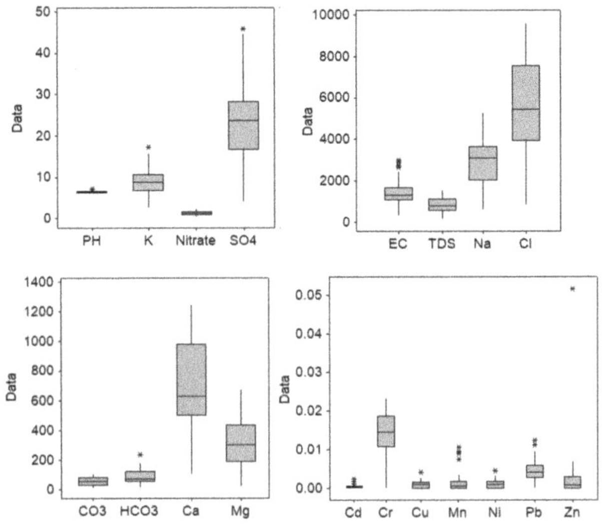

Figure 3. Boxplot of water parameters.

In the presence of outliers, the standard deviation becomes high. This is the situation, for example, for the EC series, which presents very high outliers (3003, 2995 µs/cm) compared to the other values, significantly augmenting the standard deviation.

Still, the pH remains in the admissible limits, its values being between 6.19 and 7.19. Also, the groundwater is not contaminated with Cd, Cr, Cu, Mn, Ni, and Zn, with the maximum concentrations of these elements being well below the admissible limits (0.003, 0.05, 2, 0.5, 0.07, and 3 mg/L, respectively). Only samples 9 and 11 contain Pb in concentrations (0.1117 and 0.01229 mg/L) greater than the prescribed limit (0.01 mg/L).

Comparing the determined values of the water parameters with the WHO's drinking water standards [57,83], it results that Na^+ and chlorides concentrations exceeded the admissible potability limits (200 mg/L and 250 mg/L, respectively) between 3.19 and 26.51 times (for Na^+) and between 4.13 and 48.14 times (for chlorides), respectively. In the aquifer system, sodium is mainly derived from the dissolution of salt minerals and silicate weathering [84]. EC has 80% of values above the permissible limit for drinkable water (1000 µs/cm). The variation of EC values could be explained by rock water interaction and anthropogenic influences, like agricultural run-off and wastewater discharges [57]. TDS has 80.5% of values above 600 mg/L, with 35.7% of water samples being in the category of highly undrinkable water (>1000 mg/L).

The concentrations of Ca^{2+} exceed the permissible limits (75 mg/L) between 1.39 and 16.60 times. The contribution of Ca^{2+} content in water is dependent on the solubility of $CaCO_3$ and $CaSO_4$ [85].

Figure 4 presents the histograms, showing asymmetric and non-uniform distributions of the water parameters. Zn, Cd, Mn, and Co series present the highest right—skewness, whereas Na^+ and K^+ have the lowest asymmetry.

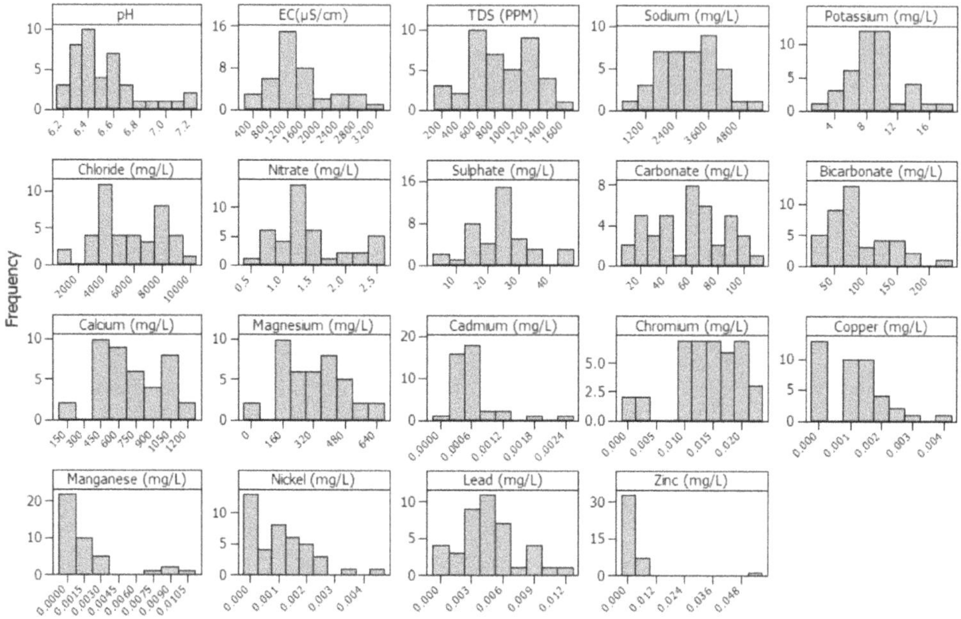

Figure 4. Histogram of water parameters.

The Kruskal-Wallis test rejected the null hypothesis that the data series obtained at different sites have the same distribution. The post hoc Dunn test emphasized the dissimilarities between the samples, listed in Table 3. The significance of the numbers' combinations is explained in the following. For example, in the first column and the fourth row, "4; 26, 27" means that Sample 4 and Sample 26 do not have the same distribution and Sample 4 and Sample 27 also do not have the same distribution. On another hand, in the first column and 12th row, "12; -" means that there is no dissimilarity between Sample 12 and all the samples with a label greater than 12. Being a diagonal table, the dissimilarities with the samples labelled with a number smaller than 12 are already displayed in the previous rows (and first column); in this example, these are 6, 7, 9, and 11.

Table 3. Diagonal table of dissimilarities that resulted from the Dunn test.

1; 21, 26, 27, 41	15; 21, 26, 27, 40, 41	29; -
2; 7, 9, 11	16; 21, 26, 27, 40, 41	30; 34
3; 7, 9, 11, 26	17; 21, 26, 27, 40, 41	31; -
4; 26, 27	18; 34, 35	32; 34
5; 18, 19, 21, 23, 25–27, 29, 30, 32, 33, 40, 41	19; 34	33; 34, 35
6; 12, 18, 19, 21–33, 40, 41	20; 21, 26, 27	34; 40, 41
7; 12, 18–33, 37, 38, 39, 40, 41	21; 34–37,39	35; 40, 41
8; 18, 21, 26, 27, 30, 32, 33, 40, 41	22; -	36; -
9; 12, 18, 19, 21–33, 40, 41	23; 34	37; -
10; 18, 21, 23, 25–27, 30, 32, 33, 40, 41	24; -	38; -
11; 12, 18, 19, 21–33, 40, 41	25; 34	39; -
12; -	26; 34–39	40; -
13; 21, 26, 27, 30, 32, 33, 40, 41	27; 34–37, 39	41; -
14; 18, 19, 21, 23, 25–27, 30, 32, 33, 40, 41	28; -	

One can also see that samples 40 and 41 are not similar to samples 5–11, 13–17, 34, 35; samples 18, 19, 21, 23, 25–27, 30, 32, 33 are not similar to 5 and 6; and samples 12, 18, 19, 21–33 are not similar to 7 and 9, and so on. These dissimilarities are originated in the concentrations of some groundwater elements.

To classify the water samples based on their chemical composition, clustering has been performed. The best number of clusters was found to be 3. The clusters are presented in Figure 5.

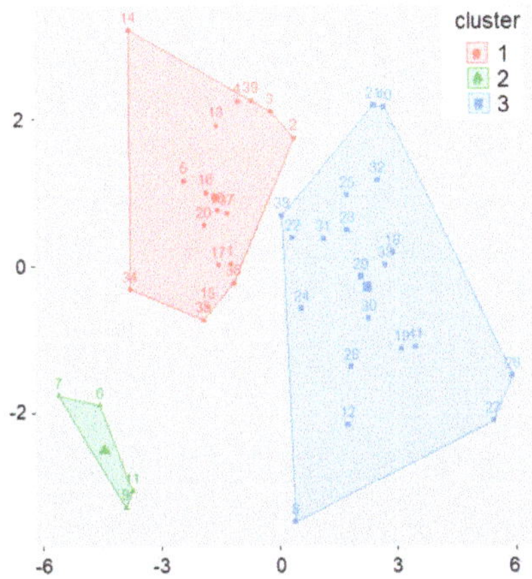

Figure 5. Clusters.

There is a clear separation between the clusters, as expected for a good clustering.

Comparing the results from Table 4 and Figure 5, one can remark their concordance. For example, 6, 7, 9, and 11 (situated in the second cluster) are not in the same cluster as 12, 18, 40, and 41 (situated in the third cluster); 15, 16, and 17 (situated in the first cluster) are not in the same cluster as 21, 26, 27, 40, and 41 (situated in the third cluster).

Table 4. Principal Component Analysis (PCA) Summary—the first nine principal components (PCs).

	PC1	PC2	PC3	PC4	PC5	PC6	PC7	PC8	PC9
Standard deviation	2.6936	1.6215	1.2924	1.1733	1.0402	0.9941	0.9708	0.8309	0.7715
Proportion of variance	0.3819	0.1384	0.0879	0.0725	0.0569	0.0520	0.0496	0.0363	0.0313
Cumulative proportion	0.3819	0.5203	0.6082	0.6806	0.7376	0.7896	0.8392	0.8755	0.9068

The sites in the first cluster are mainly situated along communication roads, so the effect of pollution is higher. The water in this cluster is highly undrinkable: TDS >1000 ppm, the sodium content is at least 12 times bigger than the admissible value (200 mg/L), the chloride content is more than 20 times higher than the upper limit (250 mg/L).

For the elements in the second cluster, pH is between 6.12–6.92, EC in the interval 2465–3003 (μS/cm), and TDS is between 1287 and 1585 ppm (higher than the values determined in the samples from the first cluster). The sodium and potassium concentrations in the samples from the second cluster are generally higher than those in the samples from the first one: the average values in the second (first) cluster are 3870 (3733) mg/L and 13.24 (10.74) mg/L, respectively. The same is true for the chlorides for which an average of 8630 (7277) mg/L is detected in the second (first) cluster.

Significant differences are found between the average values of carbonate, bicarbonate calcium, and magnesium, which are 87.6, 32.94, 1125, and 587 mg/L for the second cluster, whereas for the first one, they are 50.18, 106.59, 856, and 349 mg/L, respectively.

The mean concentrations of Cu and Zn (Pb) are 3 and 2.5 times higher (4 times smaller), respectively, in the second cluster by comparison to the first one.

The third cluster is characterized by the lowest TDS (603.35 pm), Na^+ (2045 mg/L), K^+ (6.7 mg/L), Cl^- (3714 mg/L), CO_3^{2-} (58.13 mg/L), and Ca^{2+} (494 mg/L) concentrations. The average concentrations of lead and zinc are about 1.7 times higher than those from the first cluster. Comparable concentrations of Mn are recorded in the second and third clusters, whereas concentrations of Cu are about 1.4 bigger in the second cluster than in the third one. The nitrates, sulfates, and Ni concentrations are the highest in the samples from the second cluster and the lowest in the last one.

The next step was to perform PCA for selecting the main components that determine the water quality. The Scree plot (Figure 6) was used for selecting the principal components (PCs). Figure 6 shows that the first two components have the highest contribution to explain the variance (38.2% and 13.8%).

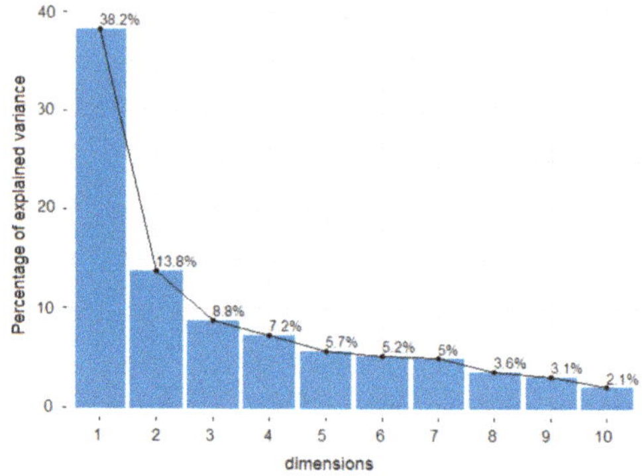

Figure 6. Scree plot.

Table 4 shows the summary of PCA (standard deviation, the proportion of variance, and cumulative proportion) for the first eight PCs. Generally, the selected principal components are those of which their corresponding eigenvalues are greater than 1. In this case, the first five components have eigenvalues greater than one and explain 73.76% of the variance (which is an acceptable percentage). Since the percentage of explained variation is high, we shall keep only PC1–PC5.

The contributions of the variables on the first two PCs are presented in the biplot (Figure 7).

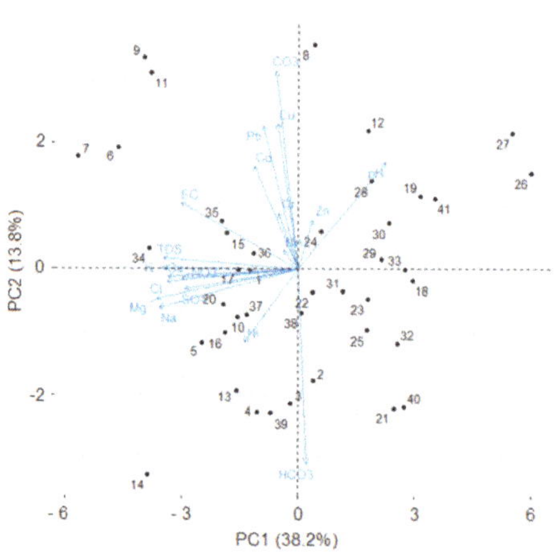

Figure 7. Biplot.

The distances from the variables to the origin represent the variables' quality on the factor map. The pairs (pH and Ni), (CO_3^{2-}, and HCO_3^-) are strongly negatively correlated (since the segments represented in Figure 7 are opposite). Sites 7, 6, 9, and 11 (Figure 7, top left-hand side) are opposite to 26 and 27 (Figure 7, top right-hand side), reconfirming the classification of the four sites (7, 6, 9, and 11) in a separate cluster. The same is true for sites 7, 6, 9, and 11 on one hand and 14 on the other one.

The variables whose distances to the origin are high are well represented on the factor map. In our case, the most significant contributions to PC1 are those of HCO_3^-, CO_3^{2-}, Cu, Pb, to PC2—Mg^{2+}, Na^+,

K^+, TDS (Figures 7 and 8, the first two bar charts) and on PC3—Cr, Mn, Cd, Zn, Pb, CO_3^{2-} (Figure 8, the last bar chart).

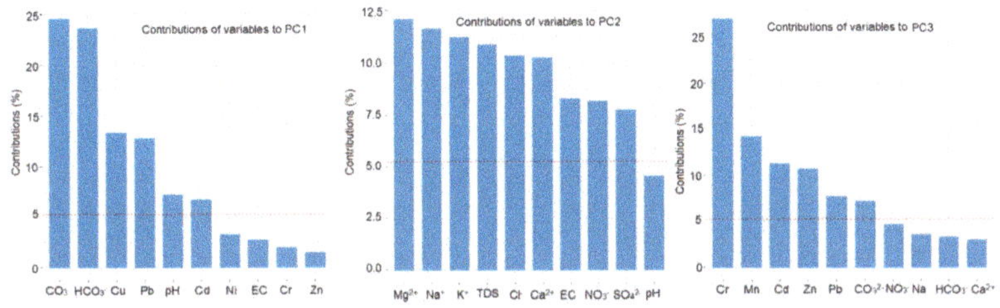

Figure 8. Contributions (%) of the variables on PC1, PC2, PC3.

The elements of the eigenvectors are called PC loadings [75]. The factor loadings associated with each of the variables in a given PC are the correlation between the original variable and the factor. The significant variables are those with loadings greater than 5%.

In Figure 8, the horizontal dashed line represents the 5% contribution to a PC. Nine elements have a contribution greater than 5% to PC1, six to PC2, and six to PC3.

Figure 9 contains the quality of representation of the variables on the factor map. It is another way to summarize the contributions of each element to the first five PCs. In our study, the significant ones are Mg^{2+} and Na^+ (on PC1), carbonates (on PC2), followed by the heavy metals: Cr (with the loading 0.52 on PC3), Mn (with the loading 0.509 on PC4), and Zn (with the loading 0.740 on PC5). Thus, we expect that they will differentiate the water samples. The high influence of HCO_3^- on PC1 could be explained by the recharge due to precipitation [86].

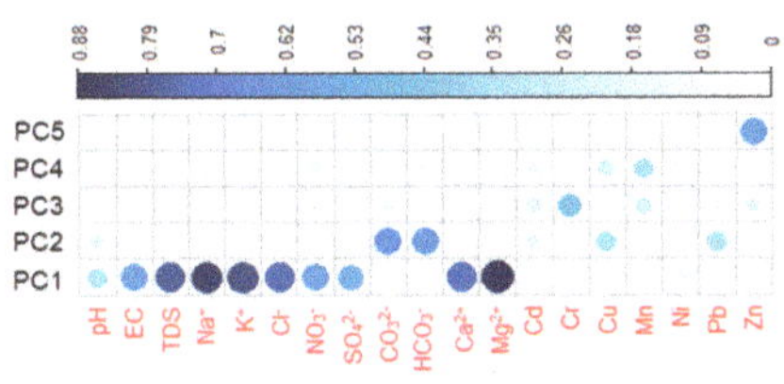

Figure 9. Quality representation of the variables on the factor map.

Finally, the values of the WQI index for all the sites are presented in Table 5.

The columns of this table contain:

- The numbers of the samples (columns 1 and 6);
- WQI 1—water quality index computed using EC, TDS, Na^+, K^+, Cl^-, NO_3^-, SO_4^{2-}, HCO_3^-, Ca^{2+}, Mg^{2+}, Cd, Cr, Cu, Mn, Ni, Pb, Zn data series;
- WQI 2—water quality index computed using all, but Cr, Mn, Zn;
- Cat 1 and Cat 2 represent the quality class of water corresponding to WQI 1 and WQI 2.

Table 5. WQI.

Sample	WQI 1	Cat.1	WQI 2	Cat.2	Sample	WQI 1	Cat.1	WQI 2	Cat.2
1	25.995	B	24.265	A	22	20.898	A	19.847	A
2	23.836	A	22.684	A	23	16.522	A	15.279	A
3	16.664	A	15.238	A	24	27.103	B	27.100	B
4	12.816	A	11.439	A	25	10.362	A	8.967	A
5	24.164	A	22.925	A	26	19.748	A	18.646	A
6	23.103	A	21.307	A	27	22.450	A	21.223	A
7	44.773	B	43.864	B	28	28.860	B	27.075	B
8	72.656	C	71.075	C	29	19.501	A	17.781	A
9	67.319	C	65.891	C	30	21.255	A	19.899	A
10	23.720	A	21.728	A	31	13.307	A	13.258	A
11	54.921	C	53.764	C	32	14.913	A	13.572	A
12	20.767	A	19.192	A	33	29.131	B	27.998	B
13	14.999	A	13.624	A	34	8.269	A	6.828	A
14	25.705	B	24.789	A	35	21.114	A	19.518	A
15	20.140	A	19.049	A	36	16.156	A	14.178	A
16	26.387	B	24.518	A	37	18.241	A	17.317	A
17	27.083	B	25.365	B	38	26.526	B	25.752	B
18	25.893	B	25.114	B	39	21.232	A	20.416	A
19	22.725	A	21.858	A	40	18.153	A	17.069	A
20	35.357	B	35.103	B	41	8.896	A	7.117	A
21	15.046	A	14.821	A					

Based on the WQI 1 (WQI 2), the quality of water for drinking can be classified as follows:

- for the samples from the first cluster, 70.6% (88.2%)—excellent, 29.41% (11.8%)—good;
- for the samples from the second cluster, 25% (25%)—excellent, 25% (25%)—good, 50% (50%)—poor;
- for the samples from the third cluster, 70% (70%)—excellent, 25% (25%)—good, 5% (5%)—poor.

After removing the concentrations of Cr, Mn, and Zn from the analysis, the modification of the percentage of samples in the first cluster classified in the category *Excellent* increased from 70.6% to 88.2%. The inclusion in the categories *Excellent*, *Good*, and *Poor* remain the same for all the samples in the second and third cluster when removing Cr, Mn, and Zn from the analysis. Therefore, we can conclude that these three elements have a significant influence on water quality.

4. Conclusions

In this article, we proposed an integrated approach for water quality evaluation and applied it to the data series containing the groundwater parameters measured in the Liwa area, the United Arab Emirates.

Firstly, the similarity of the water samples was determined using statistical tests. The results of the Dunn post hoc test determined pairs of samples that are not similar. The k-means algorithm was then used to determine the groups of samples with the same characteristics. We found three clusters, determined by the hydrogeological structure of the region and the anthropic activity. The clustering result is concordant with the dissimilarity test. This means that series that were found to be similar belong to the same cluster, while the dissimilar ones belong to different clusters.

The PCA shows that only five components out of 19 (analyzed) could be used to describe the water quality. The heavy metals have a significant influence on the first five PCs, so human activities impact the water quality. The samples included in the second cluster are the most polluted because they are extracted from places with heavy traffic and agricultural land use.

WQI was computed in two scenarios: taking into account all the elements and removing three of them. In both cases, the water quality was mainly excellent and good for the samples belonging to the first and third clusters. In the second scenario, the percentage of samples included in the category *Excellent* increased, showing the impact of removing the series of Cr, Mn, and Zn from the analysis. Thus, we can conclude that the existence of heavy metals impacts water quality.

When analyzing the TDS, of which its contribution is high on PC1, 37% of samples contain brackish water (TDS > 1000); the rest, is freshwater. On the other hand, when studying the suitability of the water for particular uses, like irrigation, specific indicators must be used, even if the WQI shows good water quality because the global index (utilized here) is computed as a weighted average of water parameters, with specific indices concentrating on particular water parameters (like Na, Mg, and Ca). Therefore, we recommend the use of different techniques and indicators that cross-validate each other for assessing the water quality for general and particular utilization.

Author Contributions: Conceptualization, A.B.; methodology, A.B., Y.N. and F.H.; software, A.B.; validation, A.B.; formal analysis, A.B.; investigation, A.B. and Y.N.; resources, Y.N. and F.H.; data curation, A.B.; writing—original draft preparation, A.B.; writing—review and editing, A.B.; visualization, A.B.; supervision, A.B.; project administration, Y.N.; funding acquisition, F.H. All authors have read and agreed to the published version of the manuscript.

Funding: This research was funded by the Research Office, Zayed University, United Arab Emirates, grant number R17081. The APC was funded by the Research Office, Zayed University, United Arab Emirates.

Conflicts of Interest: The authors declare no conflict of interest.

References

1. World Wildlife Fund. Available online: https://www.worldwildlife.org/initiatives/fresh-water (accessed on 19 August 2020).
2. Bărbulescu, A.; Barbeş, L. Assessment of surface water quality Techirghiol Lake using statistical analysis. *Rev. Chim. Bucharest* **2013**, *64*, 868–874.
3. Abbasnia, A.; Yousefi, N.; Mahvi, A.H.; Nabizadeh, R.; Radfard, M.; Yousefi, M.; Alimohammadi, M. Evaluation of groundwater quality using water quality index and its suitability for assessing water for drinking and irrigation purposes: Case study of Sistan and Baluchistan province (Iran). *Hum. Ecol. Risk Assess.* **2019**, *25*, 988–1005. [CrossRef]
4. Assaf, H.; Saadeh, M. Assessing water quality management options in the Upper Litani Basin, Lebanon, using an integrated GIS-based decision support system. *Environ. Modell. Softw.* **2008**, *23*, 1327–1337. [CrossRef]
5. Bărbulescu, A.; Barbeş, L. Assessing the Danube River water quality by statistical methods. *Environ. Earth Sci.* **2020**, *79*, 122. [CrossRef]
6. Bărbulescu, A.; Barbeş, L.; Dani, A. Statistical analysis of the quality indicators of the Danube River (in Romania). In *Frontiers in Water-Energy-Nexus—Nature-Based Solutions, Advanced Technologies and Best Practices for Environmental Sustainability, Advances in Science, Technology & Innovation (IEREK Interdisciplinary Series for Sustainable Development)*; Naddeo, V., Balakrishnan, M., Choo, K.H., Eds.; Springer: Cham, Switzerland, 2020; pp. 277–279. [CrossRef]
7. Bărbulescu, A.; Dani, A. Statistical analysis of the water quality of the major rivers in India. *Rom. Rep. Phys.* **2019**, *71*, 716.
8. Bhat, S.A.; Meraj, G.; Yaseen, S.; Pandit, A.K. Statistical Assessment of Water Quality Parameters for Pollution Source Identification in Sukhnag Stream: An Inflow Stream of Lake Wular (Ramsar Site), Kashmir Himalaya. *J. Ecosyst.* **2014**, *898054*. [CrossRef]
9. Chounlamany, V.; Tanchuling, M.A.; Inoue, T. Spatial and temporal variation of water quality of a segment of Marikina River using multivariate statistical methods. *Water Sci. Technol.* **2017**, *76*, 1510–1522. [CrossRef]
10. Dani, A.; Bărbulescu, A. Statistical Analysis of the Water Quality of the Major Rivers in India. In *Frontiers in Water-Energy-Nexus—Nature-Based Solutions, Advanced Technologies and Best Practices for Environmental Sustainability, Advances in Science, Technology & Innovation (IEREK Interdisciplinary Series for Sustainable Development)*; Naddeo, V., Balakrishnan, M., Choo, K.H., Eds.; Springer: Cham, Switzerland, 2020; pp. 281–283. [CrossRef]
11. Misaghi, F.; Delgosha, F.; Razzaghmanesh, M.; Myers, B. Introducing a water quality index for assessing water for irrigation purposes: A case study of the Ghezel Ozan River. *Sci. Total Environ.* **2017**, *689*, 109–116. [CrossRef]
12. Don-Pedro, K.N.; Oyewo, E.O.; Otitoloju, A.A. Trend of heavy metal concentration in Lagos Lagoon ecosystem, Nigeria. *West Afr. J. Appl. Ecol.* **2004**, *5*, 103–114. [CrossRef]

13. Cox, B.A. A review of currently available in-stream water-quality models and their applicability for simulating dissolved oxygen in lowland rivers. *Sci. Total Environ.* **2003**, *314–316*, 335–377. [CrossRef]
14. Bharti, N.; Katyal, D. Water Quality Indices Used for Surface Water Vulnerability Assessment. *Int. J. Environ. Sci.* **2011**, *2*, 154–173. Available online: https://ssrn.com/abstract=2160726 (accessed on 10 August 2020).
15. Bora, M.; Goswami, D.C. Water quality assessment in terms of water quality index (WQI): Case study of the Kolong River, Assam, India. *Appl. Water Sci.* **2017**, *7*, 3125–3135. [CrossRef]
16. Canadian Council of Ministers of the Environment 2001. CCME WATER QUALITY INDEX 1.0 User's Manual. Available online: Ceqg-rcqe.ccme.ca/download/en/138 (accessed on 15 August 2020).
17. Iticescu, C.; Georgescu, L.P.; Murariu, G.; Topa, C.; Timofti, M.; Pintilie, V.; Arseni, M. Lower Danube Water Quality Quantified through WQI and Multivariate Analysis. *Water* **2019**, *11*, 1305. [CrossRef]
18. Paun, I.; Chiriac, F.L.; Marin, N.M.; Cruceru, L.V.; Pascu, L.; Lehr, C.B.; Ene, C. Water quality index, a useful tool for evaluation of Danube River raw water. *Rev. Chim. Buchar.* **2017**, *68*, 1732–1739. [CrossRef]
19. Radu, V.M.; Ivanov, A.A.; Ionescu, P.; Gyorgy, D.; Diacu, E. Overall assessment of water quality on lower Danube River using multi-parametric quality index. *Rev. Chim. Buchar.* **2016**, *67*, 391–395.
20. Ioele, G.; De Luca, M.; Grande, F.; Durante, G.; Trozzo, R.; Crupi, C.; Ragno, G. Assessment of Surface Water Quality Using Multivariate Analysis: Case Study of the Crati River, Italy. *Water* **2020**, *12*, 2214. [CrossRef]
21. Gad, M.; Elsayed, S.; Moghanm, F.S.; Almarshadi, M.H.; Alshammari, A.S.; Khedher, K.; Eid, E.M.; Hussein, H. Combining Water Quality Indices and Multivariate Modeling to Assess Surface Water Quality in the Northern Nile Delta, Egypt. *Water* **2020**, *12*, 2142. [CrossRef]
22. Margat, J. *Ground Water Vulnerability to Contamination*; Bases de la Cartographie, Doc. 68 SGC198 HYD. BRGM: Orleans, France, 1968. (In French)
23. National Research Council. *Ground Water Vulnerability Assessment: Contamination Potential under Conditions of Uncertainty*; National Research Council National Academy Press: Washington, DC, USA, 1993; Available online: https://www.nap.edu/read/2050/chapter/3#17 (accessed on 10 August 2020).
24. Hirata, R.; Bertolo, R. Groundwater vulnerability in different climatic zones. In *Encyclopedia of Life Support Systems (EOLSS), Groundwater—Vol. II*; EOLSS Publications: Paris, France, 2009; Available online: https://www.eolss.net/Sample-Chapters/C07/E2-09-04-06.pdf (accessed on 20 May 2020).
25. Adams, B.; Foster, S.S.D. Land-surface zoning for groundwater protection. *Water Environ. J.* **1992**, *6*, 312–319. [CrossRef]
26. Gogu, R.C.; Pandele, A.; Ionita, A.; Ionescu, C. Groundwater vulnerability analysis using a low-cost Geographical Information System. In Proceedings of the MIS/UDMS Conference WELL-GIS WORKSHOP's Environmental Information Systems for Regional and Municipal Planning, Prague, Czech Republic, 20 November 1996; pp. 35–49.
27. Gogu, R.C.; Dassargues, A. Current trends and future challenges in groundwater vulnerability assessment using overlay and index methods. *Environ. Geol.* **2000**, *39*, 549–559. [CrossRef]
28. Aller, L.; Bennet, T.; Lehr, J.H.; Petty, R.J. *DRASTIC: Standardized System for Evaluating Ground Water Pollution Potential using Hydrogeologic Settings*; Office of Research Development, US EPA: Ada, OK, USA, 1985.
29. Van Stempvoort, D.L.; Evert, L.W. Aquifer vulnerability index: A GIS compatible method for groundwater vulnerability mapping. *Can. Water Resour. J.* **1993**, *18*, 25–37. [CrossRef]
30. Ray, I.A.; O'dell, P.W. DIVERSITY: A new method for evaluating sensitivity of groundwater to contamination. *Environ. Geol.* **1993**, *22*, 344–352. [CrossRef]
31. Foster, S.S.D. Fundamental concepts in aquifer vulnerability, pollution risk and protection strategy. *Hydrol. Resour. Proc. Inf.* **1987**, *38*, 69–86.
32. Civita, M.; De Regibus, C. Sperimentazione di alcune metodologie per la valutazione della vulnerabilità degli aquiferi. *Q. Geol. Appl. Pitagora Bologna* **1995**, *3*, 63–71.
33. Civita, M. *Le Carte Della Vulnerabilita Degli Acquiferi all Inquinamento: Teoria e Pratica*; Pitagora Editrice: Bologna, Italy, 1994; pp. 325–333.
34. Civita, M.; De Maio, M. *SINTACS. Un Sistema Parametrico per la Valutazione e la Cartografia Della Vulnerabilita' Degli Acquiferi All'inquinamento. Metodologia and Automatizzazione*; Pitagora: Bologna, Italy, 1997.
35. Civita, M.; de Maio, M. *SINTACS R5-Valutazione e Cartografia Automatica Della Vulnerabilità Degli Acquiferi All'inquinamento con il Sistema Parametrico*; Pitagora: Bologna, Italy, 2000; pp. 226–232.
36. Civita, M. The combined approach when assessing and mapping groundwater vulnerability to contamination. *J. Water Resour. Prot.* **2010**, *2*, 14–28. [CrossRef]

37. Wang, Y.; Merkel, B.J.; Li, Y.; Ye, H.; Fu, S.; Ihm, D. Vulnerability of groundwater in Quaternary aquifers to organic contaminants: A case study in Wuhan City, China. *Environ. Geol.* **2007**, *53*, 479–484. [CrossRef]
38. Chenini, I.; Zghibi, A.; Kouzana, L. Hydrogeological investigations and groundwater vulnerability assessment and mapping for groundwater resource protection and management: State of the art and a case study. *J. Afr. Earth Sci.* **2015**, *109*, 11–26. [CrossRef]
39. Zhou, J.; Li, G.; Liu, F.; Wang, Y.; Guo, X. DRAV model and its application in assessing groundwater vulnerability in arid area: A case study of pore phreatic water in Tarim Basin, Xinjiang, Northwest China. *Environ. Earth Sci.* **2010**, *60*, 1055–1063. [CrossRef]
40. Alam, F.; Umar, R.; Ahmad, S.; Dar, A.F. A new model (DRASTIC-LU) for evaluating groundwater vulnerability in parts of Central Ganga plain, India. *Arab. J. Geosci.* **2012**, *7*, 927–937. [CrossRef]
41. Khan, M.M.A.; Umar, R.; Lateh, H. Assessment of aquifer vulnerability in parts of Indo Gangetic plain, India. *Int. J. Phys. Sci.* **2010**, *5*, 1711–1720.
42. Secunda, S.; Collin, M.; Melloul, A.J. Groundwater Vulnerability Assessment Using a Composite Model Combining DRASTIC with Extensive Land Use in Israel's Sharon Region. *J. Environ. Manag.* **1998**, *54*, 39–57. [CrossRef]
43. Noori, R.; Ghahremanzadeh, H.; Kløve, B.; Adamowski, F.J.; Baghvand, A. Modified-DRASTIC, modified-SINTACS and SI methods for groundwater vulnerability assessment in the southern Tehran aquifer. *J. Environ. Sci. Health* **2019**, *54*, 89–91. [CrossRef] [PubMed]
44. Singh, A.; Srivastav, S.K.; Kumar, S.; Chakrapani, G.J. A modified-DRASTIC model (DRASTICA) for assessment of groundwater vulnerability to pollution in an urbanized environment in Lucknow, India. *Environ. Earth Sci.* **2015**, *74*, 5475–5490. [CrossRef]
45. Amadi, N.; Olasehinde, P.I.; Nwankwoala, H.O.; Dan-Hassan, M.A.; Okoye, N.O. Aquifer vulnerability studies using DRASTICA Model. *Int. J. Eng. Sci. Invent.* **2014**, *3*, 1–10.
46. Bărbulescu, A. Assessing the groundwater vulnerability: DRASTIC method and its versions: A review. *Water* **2020**, *12*, 1356. [CrossRef]
47. Andreo, B.; Ravbar, N.; Vias, J.M. Source vulnerability mapping, in carbonate (karst) aquifers by extension of the COP method: Application to pilot sites. *Hydrogeol. J.* **2009**, *17*, 749–758. [CrossRef]
48. Doerfliger, N.; Jeannin, P.Y.; Zwahlen, F. Water vulnerability assessment in karst environments: A new method of defining protection areas using a multi-attribute approach and GIS tools (EPIK method). *Environ. Geol.* **1999**, *39*, 165–176. [CrossRef]
49. Malik, P.; Svasta, J. REKS: An alternative method of Karst groundwater vulnerability estimation. In Proceedings of the XXIX Congress of the International Association of Hydrogeologists, Bratislava, Slovakia, 6–10 September 1999; pp. 79–85. [CrossRef]
50. Pételet-Giraud, E.; Dörfliger, N.; Crochet, P. RISKE: Méthode d'évaluation multicritère de la 1. vulnérabilité des aquifers karstiques. Application aux systèmes des Fontanilles et Cent-Fonts (Hérault, Sud de la France). *Hydrogéology* **2000**, *4*, 71–88.
51. Plagnes, V.; Théry, S.; Fontaine, L.; Bakalowicz, M.; Dörfliger, N. Karst vulnerability mapping: Improvement of the RISKE method. In Proceedings of the KARST 2005, Water Resources and Environmental Problems in Karst, Belgrade, Serbia, 14–19 September 2005.
52. Kavouri, K.; Plagnes, V.; Tremoulet, J.; Dörfliger, N.; Rejiba, F.; Marchet, P. PaPRIKa: A method for estimating karst resource and source vulnerability—Application to the Ouysse karst system (southwest France). *Hydrogeol. J.* **2011**, *19*, 339–353. [CrossRef]
53. Goldscheider, N. Karst groundwater vulnerability mapping: Application of a new method in the Swabian Alb, Germany. *Hydrogeol. J.* **2005**, *13*, 555–564. [CrossRef]
54. Al-Katheeri, E.S.; Howari, F.M.; Murad, A.A. Hydrogeochemistry and pollution assessmentof quaternary–tertiary aquifer in the Liwa area, United Arab Emirates. *Environ. Earth Sci.* **2009**, *59*, 581–592. [CrossRef]
55. Oroji, B.; Karimi, Z.F. Application of DRASTIC model and GIS for evaluation of aquifer vulnerability: A case study of Asadabad, Hamadan (western Iran). *Geosci. J.* **2018**, *22*, 843–855. [CrossRef]
56. Beyene, G.; Aberra, D.; Fufa, F. Evaluation of the suitability of groundwater for drinking and irrigation purposes in Jimma Zone of Oromia, Ethiopia. *Groundw. Sus. Dev.* **2019**, *9*, 100216. [CrossRef]

57. Nazzal, Y.; Ahmed, I.; Al-Arifi, N.S.N.; Ghrefat, H.; Zaidi, F.K.; El-Waheidi, M.M.; Batayneh, A.; Zumlot, T. A pragmatic approach to study the groundwater quality suitability for domestic and agricultural usage, Saq aquifer, northwest of Saudi Arabia. *Environ. Monit. Assess.* **2014**, *186*, 4655–4667. [CrossRef] [PubMed]
58. Nazzal, Y.; Howari, F.M.; Iqbal, J.; Ahmed, I.; Orm, N.B.; Yousef, A. Investigating aquifer vulnerability and pollution risk employing modified DRASTIC model and GIS techniques in Liwa area, United Arab Emirates. *Groundw. Sus. Dev.* **2019**, *3*, 567–578. [CrossRef]
59. Zaidi, F.K.; Jafri, M.K.; Naeem, M.; Ahmed, I. Reverse ion exchange as a major process controlling the groundwater chemistry in an arid environment: A case study from northwestern Saudi Arabia. *Environ. Monit. Assess.* **2015**, *187*, 607. [CrossRef]
60. Verma, P.; Singh, P.K.; Sinha, R.R.; Tiwari, A.K. Assessment of groundwater quality status by using water quality index (WQI) and geographic information system (GIS) approaches: A case study of the Bokaro district, India. *Appl. Water Sci.* **2020**, *10*, 27. [CrossRef]
61. Bodrud-Doza, M. Delineation of trace metals contamination in groundwater using geostatistical techniques: A study on Dhaka City of Bangladesh. *Groundw. Sustain. Dev.* **2019**, *9*, 100212. [CrossRef]
62. Iqbal, J.; Nazzal, Y.; Howari, F.; Xavier, C.; Yousef, A. Hydrochemical processes determining the groundwater quality for irrigation use in an arid environment: The case of Liwa Aquifer, Abu Dhabi, United Arab Emirates. *Groundw. Sustain. Dev.* **2018**, *7*, 212–219. [CrossRef]
63. GTZ/DCO/ADNOC. *Status Report Phases for Groundwater Assessment Project Abu Dhabi, United Arab Emirates, Vol. I-1: Exploration*, 21st ed.; ADNOC: Abu Dhabi, UAE, 2005.
64. Federation, Water Environmental, and American Public Health Association. *Standard Methods for the Examination of Water and Wastewater*, 21st ed.; American Public Health Association/American Water Works Association/Water Environment Federation: Washington, DC, USA, 2005.
65. Kruskal, W.H.; Wallis, A. Use of ranks in one-criterion variance analysis. *J. Am. Stat. Assoc.* **1952**, *47*, 583–621. [CrossRef]
66. Dunn, O.J. Multiple comparisons using rank sums. *Technometrics* **1964**, *6*, 241–252. [CrossRef]
67. Hartigan, J.A.; Wong, M.A. A K-Means Clustering Algorithm. *J. R. Stat. Soc. C* **1979**, *28*, 100–108.
68. Everitt, B.S.; Landau, S.; Leese, M.; Stahl, D. *Cluster Analysis*, 5th ed.; Wiley Series in Probability and Statistics: Chichester, UK, 2011.
69. Bărbulescu, A.; Nazzal, Y.; Howari, F. Statistical analysis and estimation of the regional trend of aerosol size over the Arabian Gulf Region during 2002–2016. *Sci. Rep.* **2018**, *8*, 9571. [CrossRef] [PubMed]
70. Hansen, P.; Nagai, E. Analysis of Global k-means, an Incremental Heuristic for Minimum Sum of Squares Clustering. *J. Classif.* **2005**, *22*, 287–310. [CrossRef]
71. Kassambara, A. Practical Guide to Cluster Analysis in R. Unsupervised Machine Learning. STHDA. Available online: http://www.sthda.com (accessed on 27 July 2020).
72. Charrad, M.; Ghazzali, N.; Boiteau, V.; Niknafs, A. Package 'NbClust'. NbClust: An R Package for Determining the Relevant Number of Clusters in a Data Set. *J. Stat. Softw.* **2014**, *61*, 1–36. Available online: http://www.jstatsoft.org/v61/i06/ (accessed on 27 July 2020). [CrossRef]
73. Abdi, H.; Williams, L.J. Principal Component Analysis. *WIRES Comput. Stat.* **2010**, *2*, 433–459. [CrossRef]
74. Jolliffe, I.T. *Principal Component Analysis*; Springer: New York, NY, USA, 2002.
75. Jolliffe, I.T.; Cadima, J. Principal component analysis: A review and recent developments. *Philos. Trans. R. Soc. A* **2016**, *374*, 20150202. [CrossRef] [PubMed]
76. Kaiser, H.F. The varimax criterion for analytic rotation in factor analysis. *Psychometrika* **1958**, *23*, 187–200. [CrossRef]
77. Kolsi, S.H.; Bouri, S.; Hachicha, W.; Dhia, H.B. Implementation and evaluation of multivariate analysis for groundwater hydrochemistry assessment in arid environments: A case study of Hajeb Elyoun–Jelma, Central Tunisia. *Environ. Earth Sci.* **2013**, *70*, 2215–2224. [CrossRef]
78. Hu, S.; Luo, T.; Jing, C. Principal component analysis of fluoride geochemistry of groundwater in Shanxi and Inner Mongolia, China. *J. Geochem. Explor.* **2012**, *135*, 124–129. [CrossRef]
79. Kassambara, A. Practical Guide to Principal Component Methods in R, STHDA. Available online: https://www.datanovia.com/en/product/practical-guide-to-principal-component-methods-in-r/ (accessed on 23 March 2020).
80. Tyagi, S.; Singh, P.; Sharma, B.; Singh, R. Assessment of Water Quality for Drinking Purpose in District Pauri of Uttarakhand, India. *Appl. Ecol. Environ. Sci.* **2014**, *2*, 94–99. Available online: http://pubs.sciepub.com/aees/2/4/2 (accessed on 18 June 2020). [CrossRef]

81. Tiwari, T.N.; Mishra, M. A preliminary assignment of water quality index to major Indian rivers. *Indian J. Environ. Prot.* **1985**, *5*, 276–279.
82. Behmanesh, A.; Feizabadi, Y. Water quality index of Babolrood River in Mazandaran, Iran. *Int. J. Agr. Crop Sci.* **2013**, *5*, 2285–2292.
83. WHO. Guidelines for Drinking-Water Quality, 4th Edition, Incorporating the 1st Addendum. 2017. Available online: https://www.who.int/water_sanitation_health/publications/drinking-water-quality-guidelines-4-including-1st-addendum/en/ (accessed on 25 March 2020).
84. Saber, M.; Abdelshafy, M.; El-Ameen, M.; Faragallah, A.; Abd-Alla, M.H. Hydrochemical and bacteriological analyses of groundwater and its suitability for drinking and agricultural uses at Manfalut district, Assuit, Egypt. *Arab. J. Geosci.* **2014**, *7*, 4593–4613. [CrossRef]
85. Ziani, D.; Abderrahmane, B.; Boumazbeur, A.; Benaabidate, L. Water quality assessment for drinking and irrigation using major ions chemistry in the semiarid region: Case of Djacer spring, Algeria. *Asian J. Earth Sci.* **2017**, *10*, 9–21. [CrossRef]
86. Fianko, J.R.; Osae, S.; Adomako, D.; Achel, D.G. Relationship between land use and groundwater quality in six districts in the eastern region of Ghana. *Environ. Monit. Assess.* **2009**, *153*, 139–146. [CrossRef]

water

MDPI

Article

Sources, Influencing Factors, and Pollution Process of Inorganic Nitrogen in Shallow Groundwater of a Typical Agricultural Area in Northeast China

Xinqiang Du [1,2], Jing Feng [1,2], Min Fang [1,2,3] and Xueyan Ye [1,2,*]

1 Laboratory of Groundwater Resources and Environment, Ministry of Education, Jilin University, Changchun 130021, China; duxq@jlu.edu.cn (X.D.); fengjing20@mails.jlu.edu.cn (J.F.); fangmin312@foxmail.com (M.F.)
2 College of New Energy and Environment, Jilin University, Changchun 130021, China
3 The Institute of Smart Water, Wuhan University, Wuhan 430072, China
* Correspondence: yexy@jlu.edu.cn; Tel.: +86-130-2913-2162

Received: 21 October 2020; Accepted: 14 November 2020; Published: 23 November 2020

Abstract: As one of the largest agricultural areas, the Sanjiang Plain of Northeast China has faced serious inorganic nitrogen pollution of groundwater, but the sources and the formation mechanism of pollution in the regional shallow groundwater remain unclear, which constrains the progress of pollution control and agricultural development planning. An investigation on potential nitrogen sources, groundwater inorganic nitrogen compounds (NH_4^+, NO_3^-, NO_2^-), and topsoil total nitrogen concentration (TN) was conducted in a typical paddy irrigation area of Sanjiang Plain. Multivariate statistical analysis combined with geospatial-based assessment was applied to identify the sources, determine the governing influencing factors, and analyze the formation process of inorganic nitrogen compounds in shallow groundwater. The results show that the land use type, oxidation-reduction potential (Eh), groundwater depth, NO_2^- concentration, and electrical conductivity (EC) are highly correlated with the NO_3^- pollution in groundwater, while DO and Eh affected the distribution of NH_4^+ most; the high concentrations of NO_3^- in sampling wells are most likely to be found in the residential land and are distributed mainly in densely populated areas, whereas the NH_4^+ compounds are most likely to accumulate in the paddy field or the lands surrounded by paddy field and reach the highest level in the northwest of the area, where the fields were cultivated intensively with higher fertilization rates and highest values of topsoil TN. From the results, it can be concluded that that the NO_3^- compounds in groundwater originated from manure and domestic waste and accumulated in the oxidizing environment, while the NH_4^+ compounds were derived from N fertilization and remained steady in the reducing environment. NO_2^- compounds in groundwater were the immediate products of nitrification as a result of microorganism activities.

Keywords: inorganic nitrogen; shallow groundwater; multivariate statistical analysis

1. Introduction

Nitrogen pollution in groundwater, especially in areas of intensive agriculture, has aroused widespread concern throughout the world [1]. It has been reported that the excess of the main inorganic nitrogen compounds in drinking water (NO_3^-, NH_4^+, NO_2^-) are detrimental to human health [2–4].

The mechanism of the formation process of inorganic nitrogen compounds in groundwater is complex due to the variety of nitrogen sources and the intricacy of influencing factors in the environment. Numerous studies have reported that the inorganic nitrogen compounds in groundwater could originate from nitrogen-based fertilizer, manure, as well as domestic and industrial pollution. Besides, the discharge from septic tanks, leaking sewers, and eutrophic surface water can all contribute [5]. In some cases, the atmospheric nitrogen deposition [6] and soil organic nitrogen mineralization [7] can also play important roles in nitrogen pollution in groundwater. In addition, some environmental factors such as pH, oxidation-reduction potential (Eh), soil organic matter, and bacterial activity [8–11] can greatly affect the species and amount of nitrogen in groundwater.

In previous research, several methods have been developed to identify nitrogen sources, among which the chemical analysis and stable isotope methods are the most traditional ones. Moreover, some new tracing methods and comprehensive methods have also been developed. As for chemical analysis, it has been widely used as an auxiliary identifying method because some characteristic ions (e.g., Ca^{2+}, Mg^{2+}, and SO_4^{2-}) in groundwater can carry information about their origins [12], and some halides (Cl^-, Br^- and I^-) usually have something to do with anthropogenic activities and remain relatively conserved in the subsurface environment [13–15]. The ratios of NO_3^-/Cl^-, Cl^-/Br^-, and I^-/Na^+ in groundwater are also key identifiers of the origin, by which Katz [16], Panno [17], and Pastén-Zapata [13] successfully identified the sources of nitrate coming from wastewater. Nevertheless, the types and concentrations of characteristic components vary widely in different pollution sources and change greatly during the physical and chemical reactions occurring in the subsurface environment, limiting the ability of chemical analyses to achieve accurate pollution source results.

In a similar manner, the stable isotope ratios of nitrogen (δ15N, δ18O) and boron (δ11B) are effective indicators of the pollution sources since different sources of nitrogen often share distinct isotopic compositions [18]. However, the use of single isotope tracers often cannot discriminate the sources correctly. This is because isotope ratios between sources have overlap values and the nitrification, denitrification, and other reactions that nitrogen may experience in the subsurface environment could cause the isotope values to deviate from theoretical ones, impacting the accuracy of the results. Hence, the nitrate-nitrogen and nitrate-oxygen dual-isotope methods have become a powerful tool in nitrate source identification since Kendall [19] reviewed the distribution of δ15N and δ18O values of various sources. The dual-isotope method can not only improve the accuracy of source differentiation but also make it possible to quantify the contribution of different sources to the pollution [20–22].

In recent years, some new types of tracers have been applied to identify sources of nitrogen pollution in groundwater. Nakagawa et al. [23] used coprostanol, which is produced by bacterial reduction of cholesterol in the gut of higher animals, as an indicator to investigate nitrate sources of pollution for an aquifer in Shimabara, Nagasaki, Japan, and verified that coprostanol had the potential for nitrate source identification by comparing the results with those obtained by the dual-isotope method. This indicates that it is essential to develop and adopt some new types of tracers as additional tools to support the dual-isotope method, which will make the identification process more efficient and accurate.

The single identification methods inevitably have some limitations in application due to the complexity of nitrogen formation processes, and some comprehensive methods have shown advantages in recent years, namely the adoption of multitracers and the integration of source appointment with the analysis of relevant factors (e.g., land use types). Moreover, some researchers have used geospatial-based assessment [24,25], groundwater age interpretation [26], and microbial community analysis [27] as auxiliaries to enhance the investigation of nitrogen sources. Multivariate statistics [28,29] is a powerful tool to integrate all the identification methods and relevant studies together, and thus the factor analysis [30], principal component analysis [31,32], clustering analysis [13,30], and factorial correspondence analysis [14] have been widely applied in hydrogeological research. Nevertheless, such studies have either focused on the origin and fate of the pollutants or the reactions experienced by nitrogen before or after leaching into groundwater, but both lines of study have seldom been combined in an analysis of the whole formation process of pollution systematically. As such deep insight into the essential nature of the pollution process is lacking.

There is no doubt that agricultural nonpoint source nitrogen pollution in groundwater is a serious worldwide problem since the pollution behaviors are intricate and disordered. Thus, to make clear the sources and pollution process of this kind of nitrogen pollution, it is necessary to investigate the potential pollution sources and relevant information such as the population density and amounts of fertilizer application at the early stage of source identification. Meanwhile, statistical analysis methods like multivariate statistics and geospatial statistics should be utilized to explore the intrinsic connection between nitrogen concentration and other factors that may contribute to pollution. Moreover, the tracing methods such as chemical analysis and stable isotope methods can provide more direct and powerful support to the analysis process.

The Sanjiang Plain in Northeast China (Figure 1), which is one of the most important national food production bases, has been intensively developed for agriculture since the 1950s. It has experienced four instances of large-scale reclamation, during which the large areas of wetland were adapted into paddy fields, and large amounts of fertilizers were applied to the soil every year. At the same time, the deterioration of groundwater quality in this area had become the key factor that limited the sustainable development of local water supply and agricultural planting. Recent studies reported that the Sanjiang Plain has faced the risk of serious inorganic nitrogen pollution of groundwater in some regions [33,34]. However, the information about nitrogen sources and behavior in this area is limited, and the formation process of nitrogen pollution in regional groundwater remains unclear, which will undoubtedly constrain the progress of its control and impact the large-scale plan of agricultural development.

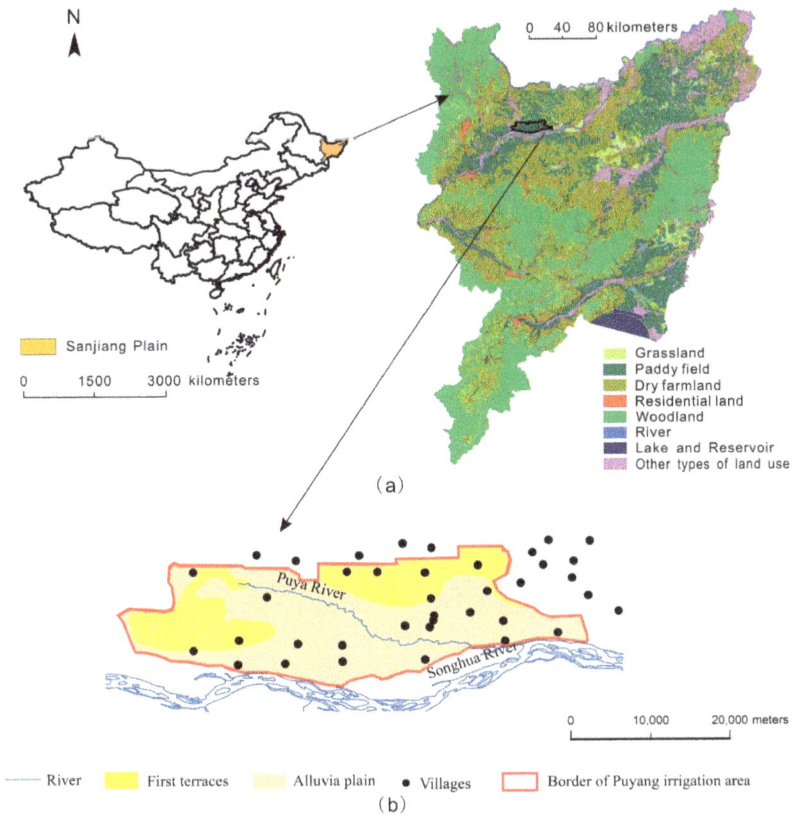

Figure 1. *Cont.*

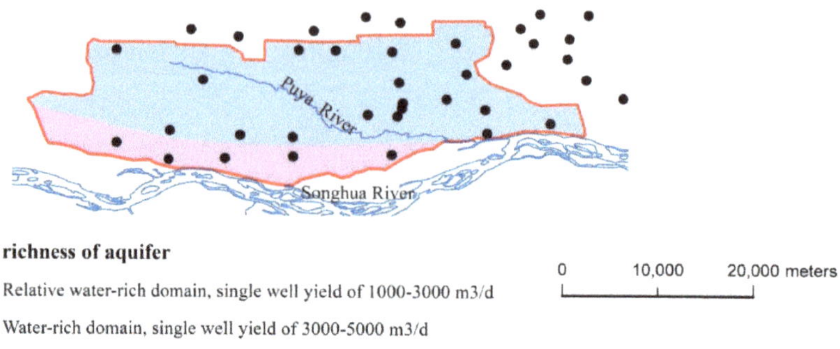

Figure 1. Location (**a**), water richness (**b**), and landform pattern (**c**) of Puyang irrigation area.

In this study, the survey was conducted in a typical paddy irrigation area of Songhua River watershed on the distribution of the potential nitrogen sources, groundwater inorganic nitrogen compounds (nitrate, ammonia, and nitrite), and topsoil total nitrogen concentration. Then, multivariate statistics and the geospatial-based assessment were combined to identify the nitrogen sources and the governing factors affecting pollution. After describing the methods and detailing the results, this paper discusses the formation process of inorganic nitrogen in groundwater. The results of this study provide a scientific basis for pollution control in the irrigation area and promote the future development of the agricultural security of the Sanjiang Plain.

2. Study Area

2.1. Geographic Location and Climate

Puyang irrigation area is located between Luobei and Suibin counties (47°12′54′′–47°22′04′′ N, 131°00′32′′–131°31′51′′ E) of the Sanjiang Plain and falls on the north bank of the lower reach of Songhua River (Figure 1). It has a cold temperate continental monsoon climate with a mean annual rainfall of 535.5 mm and a mean annual evaporation of 694.4 mm (E601). Most of the rainfall occurs from June to September, which accounts for about 70% of annual precipitation. The annual air temperature ranges from −19.3 °C (January) to 21.7 °C (July).

2.2. Geology and Hydrology Characteristics

Lying in the Songhua River alluvial plain, the characteristic landform pattern consists of first terraces and flood plains (Figure 1). The thickness of the quaternary strata ranges from 270 to 280 m, and the thickness of the aquifer is between 50 and 200 m. From top to bottom, the aquifer lithology can be separated as medium sand, medium-coarse sand, and a sand and gravel layer, which developed from the upper, middle, and lower Pleistocene respectively. There are no stable aquitard layers between aquifers, and the hydraulic conductivity ranges from 9.6 to 16.1 m/day. According to the embedment features and dynamic characteristic of groundwater, the regional groundwater can be divided into feeble confined water and phreatic water; the former is mainly distributed along the sides of the valley plain, which belong to the first terraces, and has a buried depth of 3–5 m, whereas the latter is mainly distributed in the flood plain area and shares a buried depth between 1 and 3 m (Figure 1).

The groundwater of the irrigation area is mainly recharged by precipitation and lateral groundwater runoff and is also recharged by surface water seepage, floodwaters, and irrigation water infiltration at some times and in some regions. Lateral groundwater runoff is also the main form of groundwater discharge. The irrigation area has flat terrain, and the groundwater flows slowly with the hydraulic gradient of 1/5000 to 1/10,000. In accordance with the terrain slope aspect, the groundwater flow direction is generally from southeast to northwest.

The groundwater table of a year meets the lowest level during the periods of March and April and then rises shortly after along the river bank due to the spring flood. It rises greatly during July and August as it is recharged by precipitation infiltration and reaches the peak level during August and September. From late October or early November to the end of April or early May, the ground surface is frozen and the groundwater table is in a state of decline.

3. Materials and Methods

3.1. Pollution Source Investigation

Due to the major activity of the area being rice farming and industrial pollution being absent, a field survey that focused on the domestic and agricultural situations was conducted to investigate the potential nitrogen sources in the Puyang irrigation area. The amount and distribution of towns, villages, population, livestock, fertilization, sewage systems, and landfills were all included in the survey.

3.2. Sampling and Analysis

A total of 78 groundwater samples and 19 soil samples were collected in August 2017. Most of the groundwater sampling points were evenly distributed within the domain of Puyang irrigation area, whereas some densely populated villages and previously monitored pollution areas were sampled much more intensively. Some outside wells adjacent to the irrigation area were also sampled to properly map the distribution of nitrogen pollution intensity. The types of sampling wells mainly consisted of domestic wells and irrigation wells. Because the depths of the two types of wells were significantly different (the depths of the domestic wells were generally less than 20 m, while those of the irrigation wells were about 30 m), the adjacent domestic and irrigation wells were both sampled in order to better understand how the well depth affected the nitrogen pollution of the groundwater. The pH, electrical conductivity (EC), dissolved oxygen (DO), and oxidation-reduction potential (Eh) of the groundwater were measured in situ using a portable multiparameter meter (HQ40d, Hach, Loveland, CO, USA), which was previously calibrated. All water samples for chemical analysis were filtered with a 0.45-μm filter before laboratory analysis. The inorganic nitrogen, including NH_4^+, NO_3^-, NO_2^-, was analyzed by ion chromatography. In addition, the use type of the land where the sampling wells were located, the well depth, and the potential pollution source conditions were recorded, and the water depth of the sampling wells was measured in situ.

Topsoil samples were collected to analyze the concentrations of the total organic and inorganic nitrogen (TN). The position of sampling points was close to the water sampling well, covering various land use types and geomorphic units. TN was analyzed by Jilin University Testing Center.

3.3. Statistical Methods and Graphical Representation

The average value, median value, and standard deviation of the pH, EC, DO, and Eh were evaluated to depict the results of the chemical analysis of groundwater. Nonparametric testing was used because inorganic nitrogen data in the study were not normally distributed. The concentrations of NH_4^+ and NO_3^- of water samples in this study were grouped by well depth and land use types. Mann–Whitney U test was used to determine whether there was a significant difference ($\alpha = 0.05$) between every two groups ($n > 10$) of data [35], which made the basis of the regrouping of the data.

To understand the spatial distribution characteristics of nitrogen pollution in groundwater, the pollution intensity of NH_4^+ and NO_3^- of the study area were mapped according to the ordinary kriging interpolation in ArcGIS (ESRI, Redlands, CA, USA) software.

3.4. Statistical Methods and Graphical Representation

Some indicators of nitrogen pollution sources in groundwater such as land use type and Eh, which also reflect the formation process that the nitrogen load experiences in the subsurface environment, are highly correlated among themselves. Multivariate statistical analysis could provide insight into the relationship between variables. In this study, factor analysis (FA) was conducted to highlight the main factors that determined the nitrogen concentration in groundwater. Then, the correspondence analysis (CA) was combined to further analyze and the conclusions drawn from FA. Both the FA and CA were performed in SPSS 20.0 software.

FA is widely applied for data reduction in hydrochemical and hydrogeological studies [31] and has also been used in nitrogen source appointment in recent years [14]. The main process of FA includes establishing an orthogonal factor model, selecting common factors, and performing factor rotation. In this study, eight quantitative variables (nitrogen concentration, Eh, EC, well depth, DO, water depth) and three qualitative variables (land use type, water richness of aquifer, landform pattern) were selected. Prior to FA, the qualitative variables were transformed into ordinal ones based on the practical situation and the regrouped results from the significance test, and all the variables were standardized to eliminate the effects of dimension.

CA is a multivariate analysis method that can reduce the original variables to a small number of orthogonal factors that by definition are independent [36]. It can not only study variables and samples simultaneously but also study both the qualitative and quantitative variables by dividing them into classes [36]. The correlation of variables can be depicted by a correspondence analysis plot, of which the vicinity of points could reflect the close level between variables. Due to the weakness of the analysis of the qualitative variables by FA, which could only distinguish between the effects of the two land use type groups on the nitrogen pollution in this study, CA was proposed to determine the qualitative variables, especially the land use type's effect on nitrogen pollution in the study area, and further analyze the correlation between the main influencing factors and identify the sources of nitrogen pollution. The variables of CA in this study include the concentrations of NH_4^+ and NO_3^-, land use type, well depth, and Eh. Among them, the concentrations of NH_4^+ and NO_3^- were divided into three classes, while Eh and the well depth were divided into two groups.

4. Results

4.1. Distribution of Potential Nitrogen Sources

According to the pollution source investigation, there was one densely populated town and 19 villages in the irrigation area. The main residential population in the villages had moved to the town in 2010, leaving all of the villages but four uninhabited most of the year except during the transplanting and harvesting periods. The northern suburb of the town had 70–80 households inhabited by farmers. Only the town and one village in the area had been equipped with operational sewage systems and centralized garbage disposal facilities. One hoggery and one cattle farm were operating in the area, and several heads of livestock were being raised in the inhabited villages.

Obviously, most of the irrigation area land surface was covered by paddy fields, which had an area of 273.5 km^2 according to the statistical data. The rate of fertilization of the farmland is shown in Figure 2.

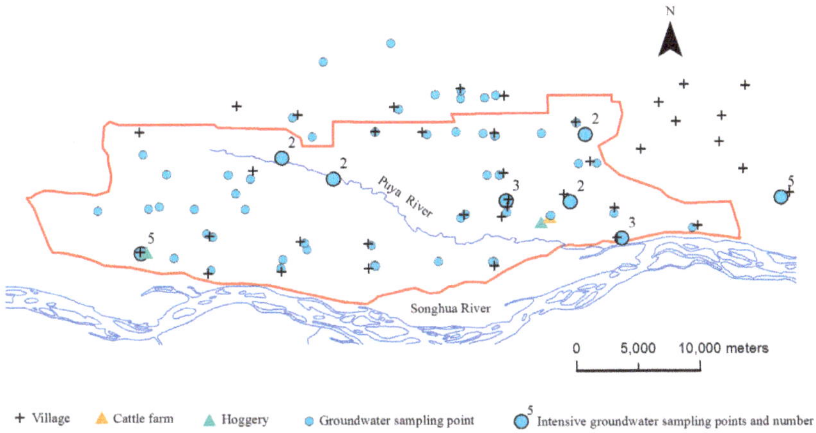

Figure 2. Location of groundwater and topsoil samples and distribution of potential nitrogen sources.

4.2. Distribution of NH_4^+ and NO_3^- in Different Well Depths and Land Use Types

The chemical analysis results of groundwater samples are shown in Table 1, and it can clearly be observed that the nitrogen concentrations under different well depths and land use types are significantly different (as shown in Figures 3 and 4). Considering that the nitrogen concentrations at different depths show a very obvious difference, the depth of sampling wells could be divided into two groups: greater than 20 m (WO_{20}), which is the depth of irrigation wells, and less than 20 m (WU_{20}), which is the depth of drinking water supply wells. The use type of the land where the sampling wells were located could be divided into six categories: concentrated residential land; paddy field; warehouse; and decentralized residential land surrounded by paddy fields, livestock farm, and vegetable field.

Table 1. Descriptive statistics of the chemical analysis of groundwater.

	Average Value	Median Value	Minimum Value	Maximum Value	Standard Deviation
NH_4^+-N (mg/L)	0.31	0.20	0.00	2.50	0.44
NO_3^--N (mg/L)	1.8	0.3	0.2	11.6	2.8
NO_3^--N (mg/L)	0.015	0.006	0.001	0.215	0.033
EC (μs/cm)	325.4	244.5	83.6	1296.0	223.3
DO (mg/L)	3.94	3.52	1.52	9.85	1.69
Eh (mv)	30.1	24.6	−108.6	162.6	76.1
pH	6.97	6.92	6.03	7.61	0.30

According to Mann-Whitney U test, the concentration of NH_4^+ in WO_{20} was significantly higher than that in WU_{20}; the opposite was found for NO_3^-. The significant differences in NH_4^+ concentration under different land use types can be summarized as follows: paddy field > concentrated residential land, decentralized residential land > concentrated residential land, and paddy field > warehouse. For NO_3^-, these differences were as follows: concentrated residential land > paddy field, concentrated residential land > decentralized residential land, warehouse > paddy field, and warehouse > decentralized residential land (Table 2). The water samples from livestock farm and vegetable field were not included in the test due to the small sample size. As seen in Figure 3, the distribution of NH_4^+ and NO_3^- in livestock farm is similar to that in decentralized residential land; the NH_4^+ concentration in vegetation field is lower than that in other land use types, and the distribution of NO_3^- in vegetation field is similar as that in concentrated residential land and warehouse.

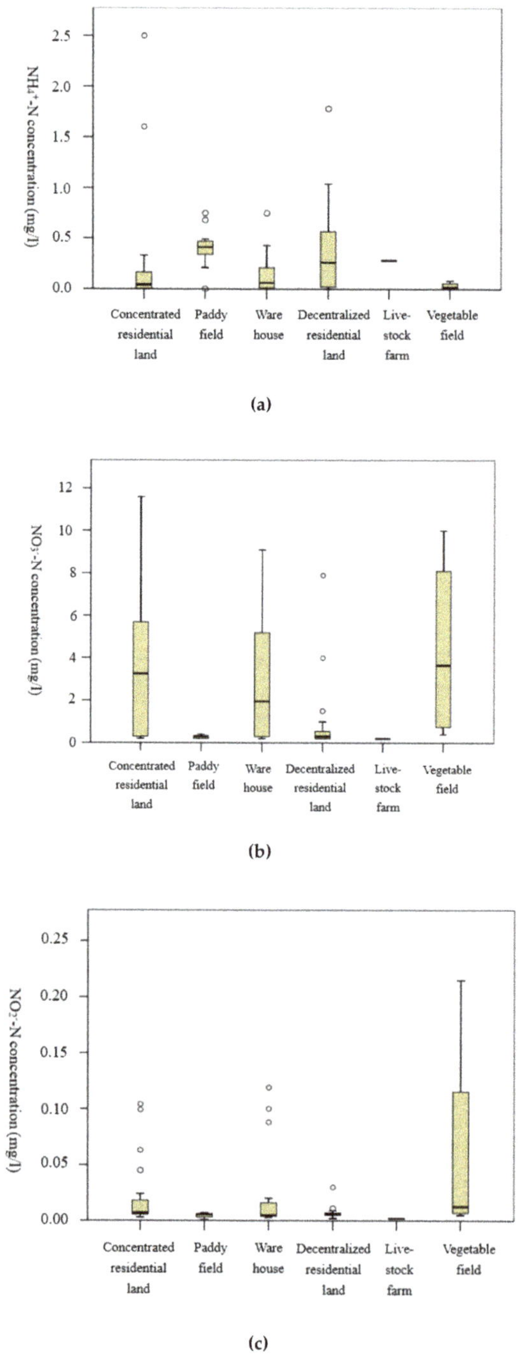

Figure 3. Box plot of NH_4^+ (**a**), NO_3^- (**b**) and NO_2^- (**c**) concentration in groundwater from different land use types.

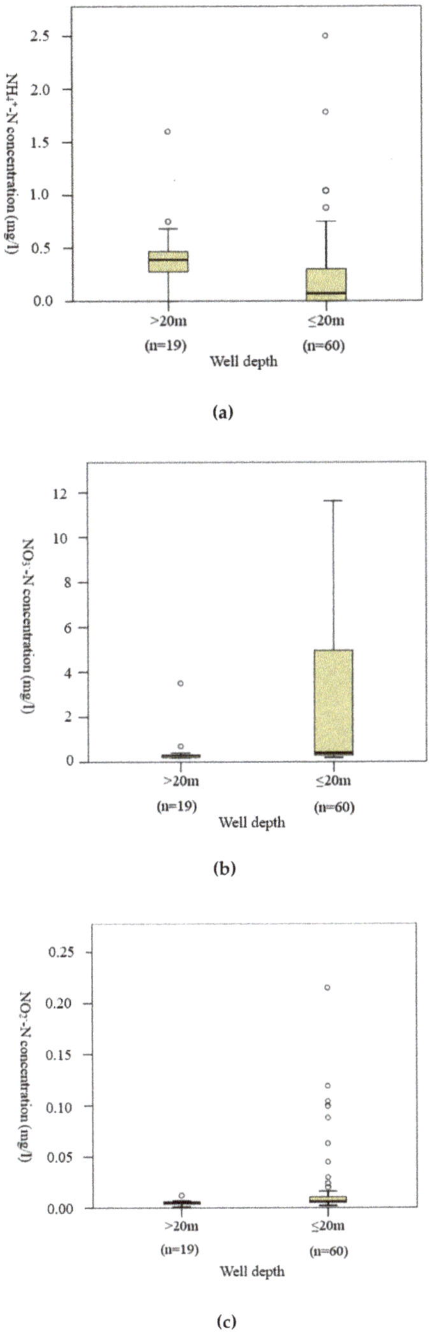

Figure 4. Box plot of NH_4^+ (**a**), NO_3^- (**b**) and NO_2^- (**c**) concentration in groundwater from different well depths.

Table 2. The differences in NH_4^+ and NO_3^- concentrations under different categories of land use type and well depth.

	Categories		Significance Level (*p*)	Differences between Categories (Confidence Interval of 95%)
NH_4^+	Land use type	CR [a] vs. P [b]	0.001	P > CR
		CR vs. W [c]	0.421	not significantly different
		CR vs. DR [d]	0.044	DR > R
		P vs. W	0.002	P > W
		P vs. DR	0.302	not significantly different
		W vs. DR	0.108	not significantly different
	Well depth	WU_{20} vs. WO_{20}	0.004	$WO_{20} > WU_{20}$
NO_3^-	Land use type	CR vs. P	0.000	CR > P
		CR vs. W	0.508	not significantly different
		CR vs. DR	0.003	CR > DR
		P vs. W	0.004	W > P
		P vs. DR	0.134	not significantly different
		W vs. DR	0.037	W > DR
	Well depth	WU_{20} vs. WO_{20}	0.005	$WU_{20} > WO_{20}$

[a] Concentrated residential land; [b] paddy field; [c] warehouse; [d] decentralized residential land.

Based on the test results of the distribution of NH_4^+ and NO_3^- in different land use types, the land use type could be reclassified into two groups, with one including concentrated residential land, warehouse, and vegetation field and the other including paddy field, decentralized residential land, and livestock farm.

4.3. *Multivariate Statistical Analysis*

4.3.1. Factor Analysis

The values of the variables in FA are shown in Table 3. A correlation test of the standardized variables was conducted, and the correlation matrix and the correlation heat map are shown in Table 4 and Figure 5, respectively. The correlation coefficients between variables were not large due to the interactions between multiple variables, thus indicating that it is essential to highlight the common information in variables using FA. The KMO test statistic of variables was 0.655 (greater than 0.5), and the Sig value of Bartlett sphericity test statistic was less than 0.01, which indicated that the variables were significantly correlated and suitable for FA.

Table 3. Assigned values of the variables for factor analysis.

Quantitative Variables	Value	Qualitative Variables		Value
NH_4^+	The actual value of analysis	Land use type	Concentrated residential land, warehouse, vegetable field	2
NO_3^-			Paddy field, decentralized residential land, livestock farm	1
NO_2^-		Water richness of aquifer	Water-rich	2
Eh			Relatively water-rich	1
DO		Landform pattern	Alluvial plain	2
EC			First terrace	1

Table 4. Correlation matrix of variables in factor analysis.

	Well Depth	Water Depth	Eh	DO	EC	NH_4^+	NO_3^-	NO_2^-	Land Use Type	Water Richness	Landform Pattern
Well depth	1.000	0.296	−0.335	0.184	−0.354	0.143	−0.368	−0.212	−0.379	−0.037	−0.032
Water depth		1.000	−0.227	0.063	−0.267	0.191	−0.291	−0.157	−0.196	−0.572	−0.254
Eh			1.000	−0.030	0.384	−0.597	0.605	0.313	0.335	0.200	0.236
DO				1.000	0.025	−0.052	0.089	−0.063	0.010	−0.158	0.039
EC					1.000	−0.157	0.592	0.264	0.305	0.246	0.207
NH_4^+						1.000	−0.381	−0.193	−0.239	−0.111	−0.237
NO_3^-							1.000	0.572	0.479	0.169	0.001
NO_2^-								1.000	0.311	0.181	0.045
Land use type									1.000	0.299	−0.006
Water richness										1.000	0.353
Landform pattern											1.000

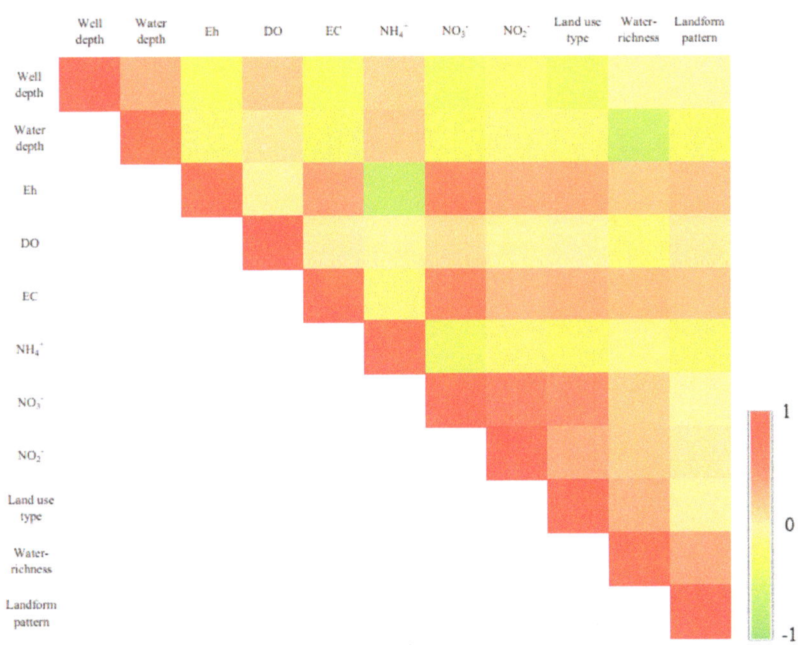

Figure 5. Correlation heat map of variables in factor analysis.

The first three factors (PC1, PC2, and PC3) contribute to 27.67%, 17.20%, and 12.38% of the variance, respectively, accounting for 57.24% of the total. From the load matrix and loading diagram of the rotation factor (Table 5 and Figure 6), the first factor (PC1) has a strong positive correlation with the concentration of NO_3^-, and this factor could be regarded as the NO_3^- pollution influencing index. The second factor (PC2) exhibits strong to moderate positive correlation with landform pattern and water richness of aquifer and negative correlation with water depth, and this factor could be considered as the groundwater cycling index. The last factor (PC3) shows the strongest negative correlation with the concentration of NH_4^+ and could be useful as a NH_4^+ pollution index.

Table 5. Load matrix of the rotation factor of the first three principal factors.

Variables	Principal Factors		
	PC1	PC2	PC3
NO_3^-	0.855	0.038	0.261
Land use type	0.671	0.104	−0.034
Well depth	−0.660	−0.073	0.254
NO_2^-	0.636	0.028	0.037
Eh	0.622	0.230	0.475
EC	0.609	0.248	0.103
Landform pattern	0.160	0.817	−0.137
Water richness	−0.112	0.712	0.370
Water depth	−0.288	−0.705	0.113
NH_4^+	−0.358	−0.214	−0.644
DO	−0.122	−0.200	0.638
Eigenvalue	3.043	1.892	1.362
Cumulative % of variance	27.667%	44.864%	57.244%

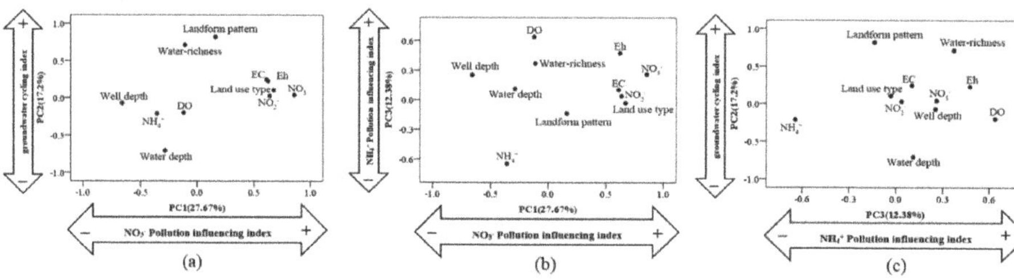

Figure 6. Loading diagram of the variables for principle factors. (**a**) Loading of the variables for PC1 and PC2. (**b**) Loading of the variables for PC1 and PC3. (**c**) Loading of the variables for PC3 and PC2.

The loading of variables for PC1 and PC3 depicted in Figure 6b reflects the variables' correlation with NO_3^- and NH_4^+ pollution and the distances between variables directly. From the x-axis (representing NO_3^- pollution index), it can be seen that PC1 has a strong to moderate correlation with land use type, well depth, the concentration of NO_2^-, Eh, and EC. The positive correlation with land use type means that greater NO_3^- pollution risks exist for groundwater in concentrated residential land, warehouse, and vegetation field than the others. EC has a moderate positive correlation with PC1 (0.609), indicating that the NO_3^- and EC may arise together in groundwater. In the meantime, the Eh, which represents the redox potential of groundwater, is positively correlated with PC1 (0.622), suggesting that an oxidizing environment is essential for NO_3^- in groundwater. The positive correlation with the concentration of NO_2^- reflects the transformation between NO_3^- and NO_2^-. On the other hand, PC1 shows a moderate negative correlation with well depth (−0.660), which means that the groundwater in shallow, oxidizing wells has more potential to accumulate NO_3^-. As for the y-axis (representing NH_4^+ pollution index), the concentration of NH_4^+ lies on the opposite end of the axis from DO and Eh, indicating that NH_4^+ exists in the anoxic and reducing groundwater environment. The weak positive correlation of PC3 with the water richness of aquifer demonstrates that the distribution of NH_4^+ in the regional aquifer is also affected by the dilution effect.

4.3.2. Correspondence Analysis (CA)

The values and categories of the variables are shown in Table 6, and the classes of nitrogen concentration (NO_3^-, NH_4^+), Eh, well depth, and land use type in the CA are projected on the primary plane in Figure 7. The first two factorial axes which explain 90.60% of the total data variance were retained in the primary plane. The first factor (x-axis) explained 58.74% of the total variance, where the reducing and oxidizing environments are oppositely located and could be divided into two parts, the left part representing the reducing environment and the right one representing the oxidizing environment. The WU_{20}, high to moderate concentration of NO_3^- (NN3, NN2), and the low concentration of NH_4^+ (AN1) are located on the right side, whereas the WO20, low concentration of NO_3^- (NN1), and high to moderate concentration of NH_4^+ (AN3, AN2) are on the left side, indicating that the deep aquifer is rich in reducing substances such as ferrous and organic matter, and the oxidizing groundwater has more potential to induce NO_3^-, while NH_4^+ is more likely to accumulate in the reducing environment, which is consistent with the FA explanation. Various land use types are distributed on the different sides of x-axis and reflect the well construction preference of local people. The wells in paddy field were constructed for irrigation, so the well depth is relatively deep to meet the need for water supply, and the groundwater is generally in a reducing condition. The wells situated in concentrated residential land mainly provide domestic water and were constructed with higher quality standards; the wells in such land are relatively shallow (the local deep aquifer is rich in iron), and the groundwater is in an oxidizing environment.

Table 6. Assigned values of the variables for correspondence analysis.

Variables	Categories	Value
NH_4^+-N	<0.20 mg/L	AN1
	≥0.20 and <0.50 mg/L	AN2
	≥0.5 mg/L	AN3
NO_3^--N	<1.0 mg/L	NN1
	≥1.0 and <5.0 mg/L	NN2
	≥5.0 mg/L	NN3
Eh	<0 mv	Reducing environment
	≥0 mv	Oxidizing environment
Well depth	≤20 m	WU_{20}
	>20 m	WO_{20}
Land use type	Concentrated residential land, paddy field, warehouse, decentralized residential land, vegetable field, livestock farm	

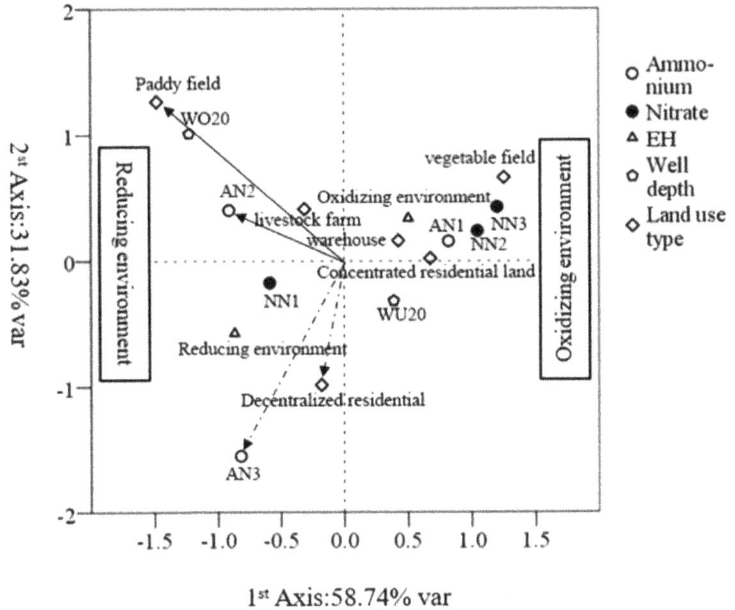

Figure 7. Primary factorial plane of correspondence analysis (CA) based on the variables of NH_4^+, NO_3^-, Eh, well depth, and land use type.

The distances between the variable classes in the plane can be used to interpret the essential relevance between each of them. The high to moderate concentration of NO_3^- is in the vicinity of vegetation field and concentrated residential land, which suggests that these two land use types are most likely to result in NO_3^- pollution in groundwater. Warehouses are closest to the oxidizing environment and are in the vicinity of NN3 and NN2, representing another place vulnerable to NO_3^- pollution.

Measuring the angle of the vector is another method to facilitate interpretation. Concerning the NH_4^+ classes, AN3 has the smallest angle with decentralized residential land which was surrounded by paddy field, and AN2 has the smallest angle with paddy field, suggesting that paddy field has the largest potential to impose NH_4^+ pollution.

4.4. Spatial Distribution of Nitrogen Concentration in Groundwater

The concentrations of NH_4^+ and NO_3^- in the wells of different depths were significantly different ($p < 0.01$); the spatial distribution of NH_4^+ and NO_3^- in the wells with the depths of ≤20 and >20 m was mapped using the method of ordinary kriging interpolation. Figure 8 shows that the concentrations of NH_4^+ in WU_{20} were generally smaller than those in WO_{20}, which was in accordance with the above-mentioned test ($p = 0.004$). The spatial area where NH_4^+-N concentrations exceeded the WHO criteria (0.5 mg/L) in the WO_{20} was larger than that of WU_{20}. However, the positions of the NH_4^+ excessive area were distributed similarly for both well depths. Generally, the highest concentration of NH_4^+ occurred in the northwest of the irrigation area, and the central part of the area also showed the risk of NH_4^+ pollution.

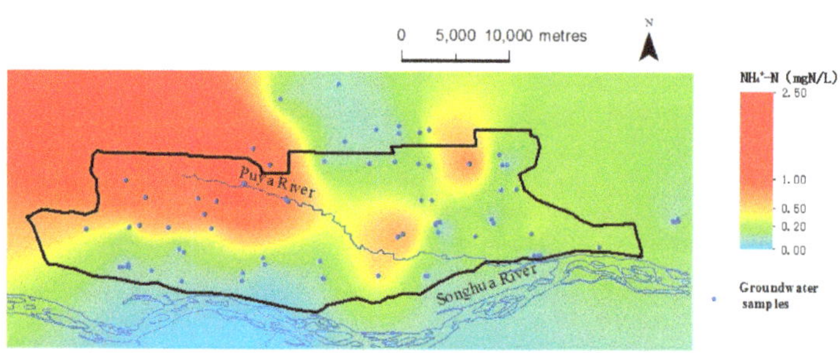

(**a**)

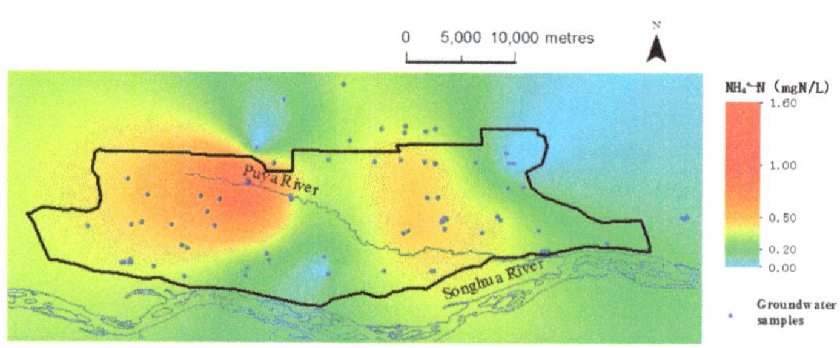

(**b**)

Figure 8. Spatial distribution of NH_4^+-N concentration in groundwater using ordinary kriging: (**a**) WO_{20}, (**b**) WU_{20}.

Figure 9 shows that the NO_3^- concentrations in WU_{20} were generally higher than those in WO_{20}, as verified by the above-mentioned test ($p = 0.005$). Large areas of the NO_3^- concentration in WU_{20} of the irrigation area exceeded 3 mg/L and were believed to be a result of anthropogenic sources as shown in previous studies, whereas NO_3^--N concentrations in WO_{20} were generally lower than 3 mg/L. The areas of excessive NO_3^- in WU_{20} were mostly distributed around the densely populated town and villages in a patchy shape or located in vegetation fields of places that were not densely populated.

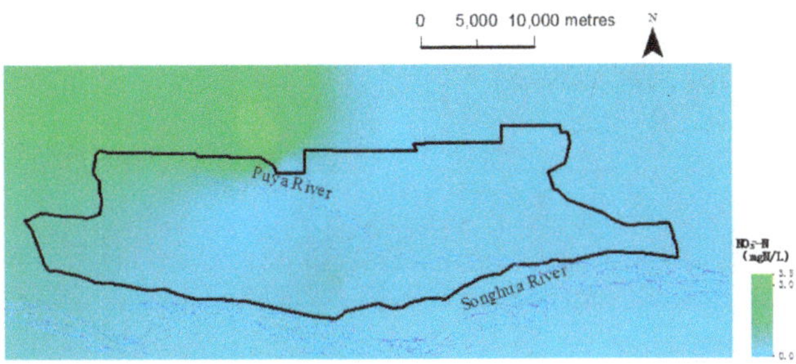

(a)

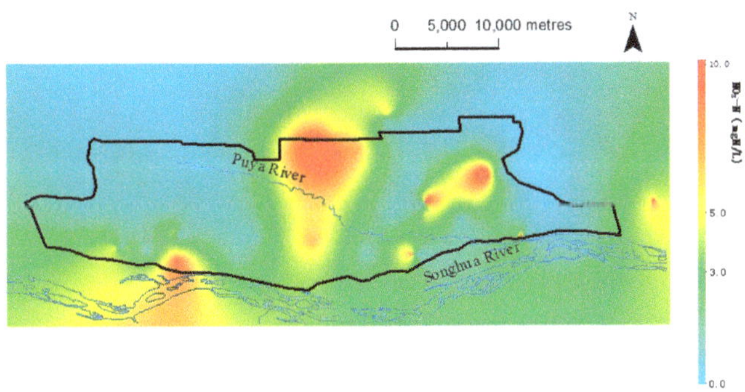

(b)

Figure 9. Spatial distribution of NO_3^--N concentration in groundwater using ordinary kriging: (**a**) WO_{20}, (**b**) WU_{20}.

4.5. Distribution of TN Concentration in Topsoil

The distribution of TN concentration in topsoil is depicted in Figure 10, which shows that the average TN concentration of fields (dry land, paddy field, vegetation field) was higher than that of natural land. As for natural land, the TN concentration in forest land was higher than that in riverbank and wetland, due to the N deposition of plant litter.

Figure 11 shows that the TN concentration in topsoil was not significantly relevant to the NH_4^+-N concentration in the vicinity of groundwater sampling wells, while the location of the largest value of topsoil TN was in the vicinity of the highest NH_4^+-N concentration well; both were located in the northwest of the area. The TN values of topsoil located near the groundwater wells with low NO_3^--N concentrations were in random distribution, whereas the moderate to high concentrations of NO_3^--N in groundwater (>1.0 mg/L) appeared to be relevant to the TN value of nearby topsoil (Figure 11).

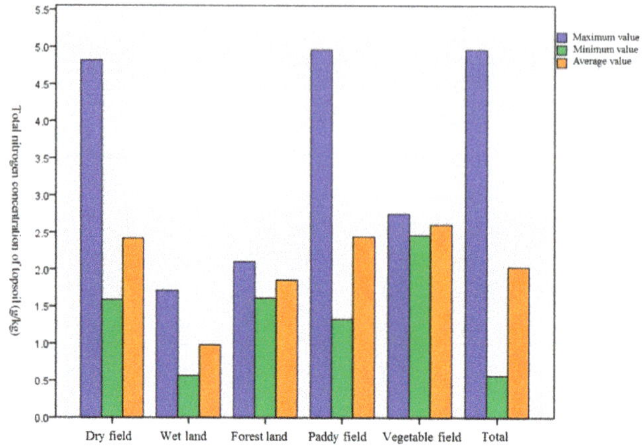

Figure 10. Descriptive statistics of total nitrogen (TN) concentration in topsoil of various land use types.

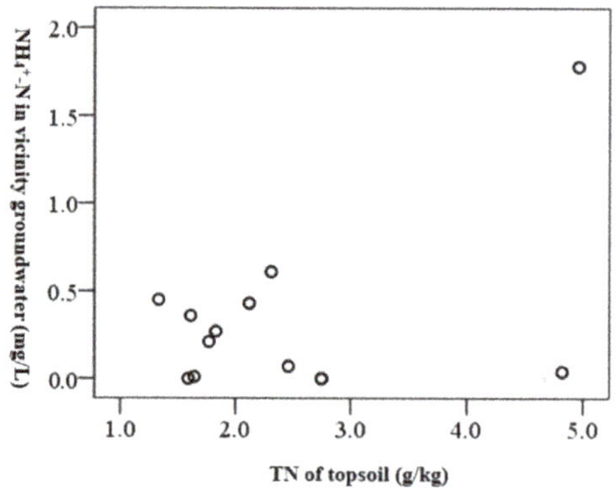

(a)

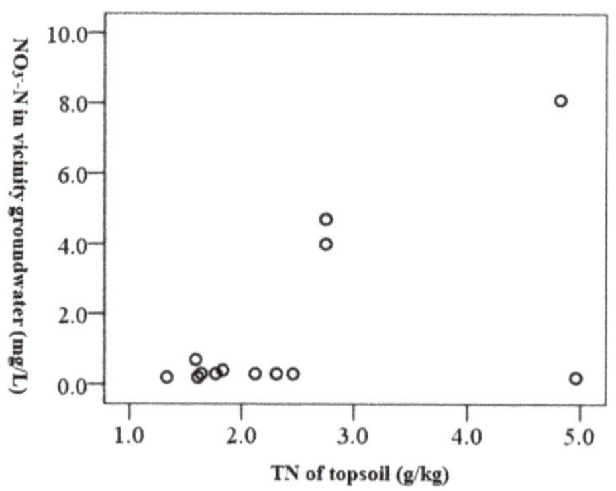

(b)

Figure 11. Scatter plot of topsoil TN and the NH_4^+-N (**a**) and NO_3^--N (**b**) concentrations of groundwater in nearby sampling wells.

5. Discussion

5.1. Nitrogen Source Appointments

According to the results of the potential nitrogen source investigation in the study area, the dominating N sources include the excess N of fertilization (mainly composed of urea and ammonia), domestic sewage, and manure. Regarding the different land use types of paddy field, concentrated residential land, decentralized residential land, warehouse, and livestock farm, the N fertilization mainly contributed to the pollution in paddy fields and the land surrounded by paddy fields, such as decentralized residential land and warehouse, and the domestic sewage and manure pollution mainly occurred in residential land, vegetable field, and livestock farm, also occurring in decentralized residential land.

Figure 10 shows that the moderate to high concentrations of NO_3^--N in groundwater (>1.0 mg/L) appeared to be relevant to the TN value of nearby topsoil, proving that the NO_3^- pollution in groundwater was a result of the surface nitrogen infiltration. The FA results indicate that greater NO_3^- pollution risks exist for groundwater in concentrated residential land, warehouse, and vegetation field than groundwater in other land use types. It is expressed in detail by CA that the vegetation field and the concentrated residential land are most likely to result in NO_3^- pollution in groundwater. This indicates that the NO_3^- in groundwater of the irrigation area originated from domestic sewage and manure. This conclusion can be supported by FA results suggesting that the EC, which is regarded as an indicative index of wastewater, rises together with NO_3^-. Considering the spatial distribution of NO_3^- concentration in groundwater, the areas of excessive NO_3^- are mostly distributed around the densely populated town and villages or located in vegetation fields, which demonstrates again that it is the domestic sewage and manure that generate NO_3^- pollution in groundwater.

The CA results also indicate that paddy field has the greatest potential to impose NH_4^+ pollution, from which it can be concluded that the NH_4^+ in groundwater mainly came from the fertilizer N excess. Concerning the spatial distribution of NH_4^+ in groundwater, the highest concentration of NH_4^+ occurred in the northwest of the irrigation area. According to the fertilizer application investigation, the fertilizer rate in the north part of the irrigation area was larger than that in the south, and the northwest was the intensive agricultural district of the area; these findings are in accordance with the spatial distribution characteristics of NH_4^+ in groundwater and support the conclusion that NH_4^+ in groundwater originated from fertilizer. Meanwhile, the highest value of TN in topsoil also occurred in the northwest of the area, which was in the vicinity of the well with the highest concentration of NH_4^+. It is further suggesting that fertilizer was the main contributor to both the soil N and the NH_4^+ in groundwater.

5.2. Governing Factors Determining the Nitrogen Distribution in Groundwater

The nitrogen components in groundwater are a result of nitrogen emission and a series of physical, chemical, and biochemical reactions in the surface and subsurface environment. Besides the nitrogen sources, the specific characteristics of aquifers and vadose zones such as their permeability and thickness, the soil medium, and the environmental factors (e.g., dissolved oxygen and reducing matter, temperature, and soil water content) can all influence the distribution of nitrogen in groundwater. In this study, some comprehensive and accessible indexes (land use type, water richness of the aquifer, landform pattern, Eh, EC, DO, well depth, and water depth) were selected to facilitate the analysis of the influential elements and determine the governing factors. Among the selected indexes, the land use type and EC were indicative of the pollution sources; water richness of the aquifer, water depth, and landform pattern can represent the specific characteristics of aquifers and vadose zones; and the Eh, DO, and well depth reflect the oxidizing and reducing matter in the environment.

The FA results have highlighted the main factors that determine the nitrogen concentration in groundwater. The results show that the NO_3^- pollution influencing index (PC1) has a strong to moderate positive correlation with land use type, NO_2^- concentration, Eh, and EC and a negative correlation with the well depth, indicating that both the nitrogen sources and the redox environment are important for the development of NO_3^- pollution in groundwater of the study area. NO_3^- is stable in the oxidizing environment, but denitrification (the reduction of NO_3^- to N_2 and NO_2^-) happens as a result of microbial action when groundwater conditions become reducing; as such, Eh has a positive correlation with NO_3^- distribution. Moreover, in the results of CA, the high to moderate concentrations and low concentrations of NO_3^- belong to oxidizing and reducing environments, respectively, again proving that the redox environment is one of the main factors affecting NO_3^- distribution. The groundwater in shallow wells has a higher potential to accumulate NO_3^-. This is due to the mixing of the groundwater from different depths, which dilutes the polluting shallow water, and it is also due to the reducing matter in deeper aquifers making the denitrification possible and attenuating NO_3^-, which has been verified in the results of FA.

The results of FA also suggest that the DO and Eh are the most important factors that determine the NH_4^+ concentration. In an aerobic and oxidizing environment, NH_4^+ is easily oxidized, thus making it difficult to keep the NH_4^+ loading stable in the groundwater. Besides, the dilution effect is another factor affecting NH_4^+ distribution by diminishing the concentration of it. For the irrigation area, the NH_4^+ sources (mainly from fertilizer) are not highly variable in spatial distribution, and thus the variety of pollutants on the surface contribute little to the NH_4^+ difference in groundwater. The anoxic and reducing environment is the dominant factor that determines whether NH_4^+ can exist in a stable state and the concentration at which it exists in groundwater.

5.3. Formation Process of Inorganic Nitrogen in Groundwater

The inorganic nitrogen in groundwater is a result of surface nitrogen emission and the physical, chemical, and biochemical reactions that the nitrogen load experiences in the subsurface environment. According to the above-mentioned analysis, the nitrogen speciation and concentration are greatly affected by the redox environment of the aquifer, which is represented by the combination of Eh, DO, and well depth. When the groundwater was in the oxidizing condition, the NH_4^+ concentration was low, and the NO_3^- concentration was determined by nitrogen loading. When the groundwater was in a reducing environment, the NO_3^- concentration was fairly low, and the NH_4^+ concentration was determined by the amount of fertilizer application. The high levels of NO_2^- were accompanied by high concentrations of NO_3^-, as an immediate product of nitrification. The formation process of inorganic nitrogen pollution in groundwater can be summarized as follows:

(1) NH_4^+-N pollution: The paddy field, of which the soil was generally in a reducing environment due to the standing water, was mainly treated with ammonium fertilizer and urea, which easily transforms into ammonium; thus, the nitrogen loading was mostly in the form of NH_4^+-N. Previous studies have mentioned that NH_4^+ is apt to be assimilated by vegetation [37] and volatilization [38], and the excess NH_4^+ would be absorbed by soil materials to a great extent [39]. This greatly attenuates NH_4^+-N content before leaching into the groundwater. The NH_4^+-N that leaches into groundwater has two different fates: one is to remain stable as NH_4^+-N if the groundwater is in a reducing environment, while the other is to be transformed into NO_3^- or NO_2^- if the groundwater is in an oxidizing aquifer. The threshold concentration of NO_3^--N in groundwater is much greater than that of NH_4^+-N, so there will not be enough oxidized NO_3^- to lead to pollution, but the NH_4^+-N in a reducing environment has a large potential risk.

(2) NO_3^--N and NO_2^--N pollution: The soil of residential land is commonly in an oxidizing environment, and thus the nitrogen emission from manure and sewage water is mainly in the form of organic N and NO_3^--N. These kinds of nitrogen are not lessened as much as NH_4^+-N in the vadose zone, and most of them will leach into groundwater. Afterward, if the groundwater is in an oxidizing environment, NO_3^--N will remain stable, and organic nitrogen will be transformed into NO_3^--N or NO_2^--N by mineralization and nitrification with microorganisms, causing the NO_3^--N and NO_2^--N pollution of groundwater. If the groundwater is in a reducing environment, NO_3^--N will be transformed into N_2O and N_2 and attenuated to a large extent.

6. Conclusions

Groundwater inorganic nitrogen and topsoil total nitrogen were analyzed in the Puyang irrigation area of Sanjiang Plain, and a pollution source investigation was conducted to identify the sources, influencing factors, and formation process of inorganic nitrogen pollution in regional shallow groundwater. In the study area, the potential nitrogen sources are fertilizer, manure, rural domestic waste, and septic system leakage, while atmospheric nitrogen deposition was not considered in this study. For all of the land use types evaluated, the land use types could be reclassified into two groups, with one including concentrated residential land, warehouse, and vegetation field and the other including paddy field, decentralized residential land, and livestock farm. These groups were determined by the distribution characteristic of inorganic nitrogen, where the former might have higher NO_3^- and lower NH_4^+ concentration than the latter. As for the well depth, the concentration of NH_4^+ in WO_{20} was found to significantly higher than that in WU_{20} by Mann-Whitney U test. The opposite relationship was found for NO_3^-.

The results of multivariate statistical analysis showed that the land use type, well depth, NO_2^- concentration, Eh, and EC were highly related to the NO_3^- pollution, and the high concentration of NO_3^- was likely to be found in vegetation field and concentrated residential land and was associated with an oxidizing environment; the NH_4^+ pollution had the strongest correlation with DO and Eh, and the reducing environment, decentralized residential land, and paddy field had more potential to impose NH_4^+ pollution. These results highlight that the nitrogen sources and the redox environment determine the distribution of NO_3^- and the redox environment governs the distribution of NH_4^+ in the shallow groundwater of the irrigation area.

The NH_4^+ pollution area was mainly distributed in the northwest of the area, where the fertilizer application rate was much higher and the highest value of topsoil TN was found, supporting the conclusion drawn from multivariate statistical analysis that the NH_4^+ in groundwater originated from fertilizer. As for the high concentration of NO_3^- in groundwater, which was mainly situated around the densely populated villages and towns and was relevant to the TN value of nearby topsoil, this was thought to come from manure and domestic waste.

The formation process of inorganic nitrogen pollution in shallow groundwater of the area can be summarized as follows: (1) the NH_4^+ from fertilizer was greatly attenuated by volatilization, plant uptake, and soil matter absorption and then accumulated in a reducing aquifer or was transformed into NO_3^- and NO_2^- by nitrification in an oxidizing aquifer with microorganisms; (2) the organic nitrogen and NO_3^- in manure and domestic waste were leached, losing little on the surface, to the vadose zone, where they remained steady as NO_3^--N in the oxidizing groundwater or were attenuated by microorganism-caused denitrification in the reducing groundwater.

Author Contributions: The research was conducted mainly by X.D. and M.F.; the draft manuscript was finished mainly by J.F. and M.F., and it was revised mainly by X.Y. All authors have read and agreed to the published version of the manuscript.

Funding: This research was funded by National Key Research and Development Program (2017YFC0406002) and National Natural Science Foundation of China (41672231).

Conflicts of Interest: The authors declare no conflict of interest.

References

1. Denk, T.R.A.; Mohn, J.; Decock, C.; Lewicka-Szczebak, D.; Harris, E.; Butterbach-Bahl, K.; Kiese, R.; Wolf, B. The nitrogen cycle: A review of isotope effects and isotope modeling approaches. *Soil Biol. Biochem.* **2017**, *105*, 121–137. [CrossRef]
2. Galal-Gorchev, H. WHO guidelines for drinking-water quality. *Water Supply* **1993**, *11*, 1–16.
3. Johnson, P.T.J.; Townsend, A.R.; Cleveland, C.C.; Glibert, P.M.; Howarth, R.W.; Mckenzie, V.J.; Rejmankova, E.; Ward, M.H. Linking environmental nutrient enrichment and disease emergence in humans and wildlife. *Ecol. Appl.* **2010**, *20*, 16–29. [CrossRef] [PubMed]

4. Townsend, A.R.; Howarth, R.W.; Bazzaz, F.A.; Booth, M.S.; Cleveland, C.C.; Collinge, S.K.; Dobson, A.P.; Epstein, P.R.; Holland, E.A.; Keeney, D.R.; et al. Human Health Effects Changing Nitrogen Cycle. *Front. Ecol. Environ.* **2003**, *1*, 240–246. [CrossRef]
5. Wakida, F.T.; Lerner, D.N. Non-agricultural sources of groundwater nitrate: A review and case study. *Water Res.* **2005**, *39*, 3–16. [CrossRef]
6. Wang, W.; Song, X.; Ma, Y. Identification of nitrate source using isotopic and geochemical data in the lower reaches of the Yellow River irrigation district (China). *Environ. Earth Sci.* **2016**, *75*, 1–13. [CrossRef]
7. Du, Y.; Ma, T.; Deng, Y.; Shen, S.; Lu, Z. Sources and fate of high levels of ammonium in surface water and shallow groundwater of the Jianghan Plain, Central China. *Environ. Sci. Process. Impacts* **2017**, *19*, 161–172. [CrossRef]
8. Liu, C.; Yu, D.; Wang, Y.; Chen, G.; Tang, P.; Huang, S. A novel control strategy for the partial nitrification and anammox process (PN/A) of immobilized particles: Using salinity as a factor. *Bioresour. Technol.* **2020**, *302*, 122864. [CrossRef]
9. Pandey, C.B.; Kumar, U.; Kaviraj, M.; Minick, K.J.; Mishra, A.K.; Singh, J.S. DNRA: A short-circuit in biological N-cycling to conserve nitrogen in terrestrial ecosystems. *Sci. Total Environ.* **2020**, *738*, 139710. [CrossRef]
10. Peng, T.; Feng, C.; Hu, W.; Chen, N.; He, Q.; Dong, S.; Xu, Y.; Gao, Y.; Li, M. Treatment of nitrate-contaminated groundwater by heterotrophic denitrification coupled with electro-autotrophic denitrifying packed bed reactor. *Biochem. Eng. J.* **2018**, *134*, 12–21. [CrossRef]
11. Rivett, M.O.; Buss, S.R.; Morgan, P.; Smith, J.W.N.; Bemment, C.D. Nitrate attenuation in groundwater: A review of biogeochemical controlling processes. *Water Res.* **2008**, *42*, 4215–4232. [CrossRef] [PubMed]
12. Babiker, I.S.; Mohamed, M.A.A.; Terao, H.; Kato, K.; Ohta, K. Assessment of groundwater contamination by nitrate leaching from intensive vegetable cultivation using geographical information system. *Environ. Int.* **2004**, *29*, 1009–1017. [CrossRef]
13. Pastén-Zapata, E.; Ledesma-Ruiz, R.; Harter, T.; Ramírez, A.I.; Mahlknecht, J. Assessment of sources and fate of nitrate in shallow groundwater of an agricultural area by using a multi-tracer approach. *Sci. Total Environ.* **2014**, *470–471*, 855–864. [CrossRef]
14. Wang, Y.; Li, Y.; Li, Y.; Liu, F.; Liu, X.; Gong, D.; Ma, Q.; Li, W.; Wu, J. Intensive rice agriculture deteriorates the quality of shallow groundwater in a typical agricultural catchment in subtropical central China. *Environ. Sci. Pollut. Res.* **2015**, *22*, 13278–13290. [CrossRef]
15. Murgulet, D.; Tick, G.R. Understanding the sources and fate of nitrate in a highly developed aquifer system. *J. Contam. Hydrol.* **2013**, *155*, 69–81. [CrossRef] [PubMed]
16. Katz, B.G.; Eberts, S.M.; Kauffman, L.J. Using Cl/Br ratios and other indicators to assess potential impacts on groundwater quality from septic systems: A review and examples from principal aquifers in the United States. *J. Hydrol.* **2011**, *397*, 151–166. [CrossRef]
17. Panno, S.V.; Hackley, K.C.; Hwang, H.H.; Greenberg, S.E.; Krapac, I.G.; Landsberger, S.; O'Kelly, D.J. Characterization and identification of Na-Cl sources in ground water. *Ground Water* **2006**, *44*, 176–187. [CrossRef]
18. Xue, D.; Botte, J.; De Baets, B.; Accoe, F.; Nestler, A.; Taylor, P.; Van Cleemput, O.; Berglund, M.; Boeckx, P. Present limitations and future prospects of stable isotope methods for nitrate source identification in surface- and groundwater. *Water Res.* **2009**, *43*, 1159–1170. [CrossRef]
19. Kendall, C. *Tracing Nitrogen Sources and Cycling in Catchments*; Elsevier B.V.: Amsterdam, The Netherlands, 1998; Volume 23, pp. 441–442. [CrossRef]
20. Xue, D.; De Baets, B.; Van Cleemput, O.; Hennessy, C.; Berglund, M.; Boeckx, P. Use of a Bayesian isotope mixing model to estimate proportional contributions of multiple nitrate sources in surface water. *Environ. Pollut.* **2012**, *161*, 43–49. [CrossRef]
21. Moore, J.W.; Semmens, B.X. Incorporating uncertainty and prior information into stable isotope mixing models. *Ecol. Lett.* **2008**, *11*, 470–480. [CrossRef]
22. Parnell, A.C.; Inger, R.; Bearhop, S.; Jackson, A.L. Source partitioning using stable isotopes: Coping with too much variation. *PLoS ONE* **2010**, *5*, e9672. [CrossRef] [PubMed]
23. Nakagawa, K.; Amano, H.; Takao, Y.; Hosono, T.; Berndtsson, R. On the use of coprostanol to identify source of nitrate pollution in groundwater. *J. Hydrol.* **2017**, *550*, 663–668. [CrossRef]

24. Huang, J.; Xu, J.; Liu, X.; Liu, J.; Ramsankaran, R.; Wang, L.; Su, W. Geospatial based assessment of spatial variation of groundwater nitrate nitrogen in Shandong intensive farming regions of China. *Sens. Lett.* **2012**, *10*, 491–500. [CrossRef]
25. Delgado, J.A.; Ascough, J.C.; Lighthart, N.; Neer, D. Potential use of a new nitrogen trading tool to assess nitrogen management practices to protect groundwater quality. *Comput. Electron. Agric.* **2020**, *169*, 105195. [CrossRef]
26. Anornu, G.; Gibrilla, A.; Adomako, D. Tracking nitrate sources in groundwater and associated health risk for rural communities in the White Volta River basin of Ghana using isotopic approach ($\delta^{15}N$, $\delta^{18}O\text{-}NO_3$ and ^3H). *Sci. Total Environ.* **2017**, *603–604*, 687–698. [CrossRef]
27. Kim, H.; Kaown, D.; Mayer, B.; Lee, J.Y.; Hyun, Y.; Lee, K.K. Identifying the sources of nitrate contamination of groundwater in an agricultural area (Haean basin, Korea) using isotope and microbial community analyses. *Sci. Total Environ.* **2015**, *533*, 566–575. [CrossRef]
28. Liu, F.; Zhao, Z.; Yang, L.; Ma, Y.; Xu, Y.; Gong, L.; Liu, H. Geochemical characterization of shallow groundwater using multivariate statistical analysis and geochemical modeling in an irrigated region along the upper Yellow River, Northwestern China. *J. Geochemical Explor.* **2020**, *215*, 106565. [CrossRef]
29. Liu, F.; Qian, H.; Shi, Z.; Wang, H. Long-term monitoring of hydrochemical characteristics and nitrogen pollution in the groundwater of Yinchuan area, Yinchuan basin of northwest China. *Environ. Earth Sci.* **2019**, *78*, 1–15. [CrossRef]
30. Lambrakis, N.; Antonakos, A.; Panagopoulos, G. The use of multicomponent statistical analysis in hydrogeological environmental research. *Water Res.* **2004**, *38*, 1862–1872. [CrossRef]
31. Matiatos, I. Nitrate source identification in groundwater of multiple land-use areas by combining isotopes and multivariate statistical analysis: A case study of Asopos basin (Central Greece). *Sci. Total Environ.* **2016**, *541*, 802–814. [CrossRef]
32. Arauzo, M.; Martínez-Bastida, J.J. Environmental factors affecting diffuse nitrate pollution in the major aquifers of central Spain: Groundwater vulnerability vs. groundwater pollution. *Environ. Earth Sci.* **2015**, *73*, 8271–8286. [CrossRef]
33. Ma, H.W.; Li, X.; Cui, J.; Yang, Z. A preliminary study on the Organic and Inorganic pollution of groundwater in Jiansanjiang Farm in Sanjiang Plain. *Geol. Resour.* **2014**, *23*, 38–41. (In Chinese)
34. Lu, L.; Cheng, H.; Pu, X.; Liu, X.; Cheng, Q. Nitrate behaviors and source apportionment in an aquatic system from a watershed with intensive agricultural activities. *Environ. Sci. Process. Impacts* **2015**, *17*, 131–144. [CrossRef] [PubMed]
35. Lockhart, K.M.; King, A.M.; Harter, T. Identifying sources of groundwater nitrate contamination in a large alluvial groundwater basin with highly diversified intensive agricultural production. *J. Contam. Hydrol.* **2013**, *151*, 140–154. [CrossRef] [PubMed]
36. Andrade, A.I.A.S.S.; Stigter, T.Y. Multi-method assessment of nitrate and pesticide contamination in shallow alluvial groundwater as a function of hydrogeological setting and land use. *Agric. Water Manag.* **2009**, *96*, 1751–1765. [CrossRef]
37. Nacry, P.; Bouguyon, E.; Gojon, A. Nitrogen acquisition by roots: Physiological and developmental mechanisms ensuring plant adaptation to a fluctuating resource. *Plant Soil* **2013**, *370*, 1–29. [CrossRef]
38. Shan, L.; He, Y.; Chen, J.; Huang, Q.; Wang, H. Ammonia volatilization from a Chinese cabbage field under different nitrogen treatments in the Taihu Lake Basin, China. *J. Environ. Sci.* **2015**, *38*, 14–23. (In Chinese) [CrossRef]
39. Brauns, B.; Bjerg, P.L.; Song, X.; Jakobsen, R. Field scale interaction and nutrient exchange between surface water and shallow groundwater in the Baiyang Lake region, North China Plain. *J. Environ. Sci.* **2015**, *45*, 60–75. (In Chinese) [CrossRef]

water

MDPI

Article

The Impact of Human Interventions and Changes in Climate on the Hydro-Chemical Composition of Techirghiol Lake (Romania)

Carmen Maftei [1], Constantin Buta [2] and Ionela Carazeanu Popovici [3,*]

1 Civil Engineering Faculty, Transilvania University of Brasov, 29 Eroilor Blvd, 500036 Brașov, Romania; carmen.maftei@unitbv.ro or cemaftei@gmail.com

2 Civil Engineering Faculty, Ovidius University of Constanta, 124 Mamaia Blvd, 8700 Constanta, Romania; buta.constantin@univ-ovidius.ro or costi_buta@yahoo.com

3 Applied Science and Engineering Faculty, Ovidius University of Constanta, 124 Mamaia Blvd, 8700 Constanta, Romania

* Correspondence: icarazeanu@univ-ovidius.ro; Tel.: +40-74-888-5377

Received: 18 May 2020; Accepted: 21 July 2020; Published: 12 August 2020

Abstract: The aim of this study is to establish the potential effect of changes in climate and anthropic interventions made over time on the hydro-chemical properties of the Techirghiol Lake. Located in the littoral region of the Black Sea, Techirghiol Lake is the most hypersaline lake of Romania—well-known for the therapeutic properties of the saline water and sapropelic mud. Long-term time series of salinity and water level were investigated in relation to the lake water inputs (precipitation, overland flow and groundwater), to chemical parameters (pH, DO and BOD5) and also to the hydraulic works designed and built in the region. The obtained results reveal a degradation of this ecosystem in the period of 1970–1998, when the extensive irrigation practice in the proximity of the lake had a negative effect on the water budget of Techirghiol Lake (an increased freshwater input through runoff and seepage), followed by a major decrease of the lake's salinity.

Keywords: human intervention; changes in climate; salty lake

---

1. Introduction

Climate change has a considerable impact on ecosystems, affecting air temperature, the amount of precipitation, the frequency and intensity of extreme events, the sea level, etc. In the past decades, many studies have been conducted on saline lakes which show not only the importance of saline lakes in the economies, but also the impact of climate change on the water level and chemical content [1–3]. The study conducted by Valero-Garces et al. [4] on the saline lakes from Spain highlights the influences of agricultural practices, particularly of irrigation, on the lake's hydrological behavior. Webster et al. [5] have examined the influence of the increasing trend of drought on semiconservative cations, Ca^+ and Mg^+, in seven lakes from Northern Wisconsin, and have concluded that the high evaporation rates, combined with the decreased amount of precipitation, caused an increase of cation concentrations in all lakes. Recent studies conducted in Poland concerning the influence of many factors (climatic, hydrologic, morphometric) on lake temperatures have shown that the lake response to factor modifications depends on the local conditions and lake characteristics [6,7]. All over the world, the studies conducted had the same conclusion: the saline lakes are threatened by climate change and by the various anthropogenic activities, which lead to dramatic changes in lakes chemistry and dynamics [1,8–10]. The analysis of water chemistry in relation to environmental factors allows a better understanding of the process variability and is very useful for researchers and deciders in the field of water management and monitoring [7,11].

This study presents an analysis of the potential effect of changes in climate and anthropic interventions on the hydro-chemical properties of Techirghiol Lake, located in the littoral region of the Black Sea. We have combined historical knowledge of human activities and management of the lake and the surrounding areas with a compilation of data detailing precipitation, river discharge and more. The first part of this study is focused of the study area and its main characteristic elements (climatic, geologic, hydrogeologic and hydrologic). In the second part, the chemical composition and water quality of Techirghiol Lake were investigated in the context of climatic and anthropogenic impact using Romanian methodologies and regulation.

2. Materials and Methods

This section is divided in two parts: the first part is dedicated to the presentation of the study area and its main characteristic elements (climatic, geologic, hydrogeologic, hydrologic and lake water chemistry) and the second one features the methods used.

2.1. Study Area and Its Characteristics

Techirghiol Lake is a result of the latest paleogeographic evolution of the Black Sea, dictated by the evolution of the sea level over time [12–14], which contributed to the development of the present shoreline. Accordingly, the coastal development and the sand-belt formation have completely isolated Techirghiol Lake. Located on the Black Sea coast, 16 km south from Constanta City (Figure 1), Techirghiol Lake is mainly known for the curative properties of its sapropelic mud and hypersaline water. Here, a veritable tourism economy has developed since 1899, around balneological treatment and medical rehabilitation.

The catchment area of Techirghiol Lake is situated in the South Dobrogea Plateau, having a surface of approximately 160 km^2. The lake is 8 km long, with the maximum width of 4.4 km and a water depth varying between 1.5 m and 9.5 m. The maximum water depth was recorded at 9.75 m and the average water depth is 3.6 m [15].

The studied area has a relief consisting in a not very tall plateau (+70–80 m), with a slope which descends to the sea that ends abruptly with a 30-m-high cliff (Figure 1).

The Techirghiol Lake area is situated in a temperate–continental climatic zone, which is influenced by the Black Sea. The region is characterized by an average annual temperature of approximately 11 °C and an annual rainfall amount of about 400 mm [16]. The data recorded at Constanta meteorological station were chosen in order to analyze the influence of climatic parameters on the lake's behavior.

From a geological point of view, Techirghiol Lake area is situated in the South Dobrogea Plateau. The South Dobrogea Plateau basement layer consists of granitic gneiss and crystalline shale. Above this basement layer, this sector integrates three main sedimentary geological systems: Sarmatian limestone, red clay mixed with gypsum and loess deposits. The presence of faults and the sedimentary structure of Techirghiol Lake area have determined the development of several deep complex aquifers, among which a free surface aquifer situated in Sarmatian limestone and a pressure aquifer located in limestone and dolomitic deposits [17].

From the hydrological point of view, Techirghiol Lake is situated at the confluence of several important valleys (Figure 2), most of them with an intermittent flow. In 1910, Pascu [18] identified four important valleys that drained the Techirghiol catchment: Carlichioi (Biruinta) Valley, Techirghiol Valley, Muzurat (today Urlichioi) valley and Tuzla Valley. In 1976, Breier identified three important valleys: Techirghiol, Tuzla and Carlichioi (Biruinta) valleys (Figure 2). Today, the hydrological regime of the main hydrographic networks is very different, and the important valleys are barred by different hydraulic works (dams and penstock) in order to prevent the entrance of freshwater in the lake (Figure 2). The hydrological features of the lake are related to the evolution of the lake's level, which is strongly influenced by the aquifer input and by the discharge of the valleys.

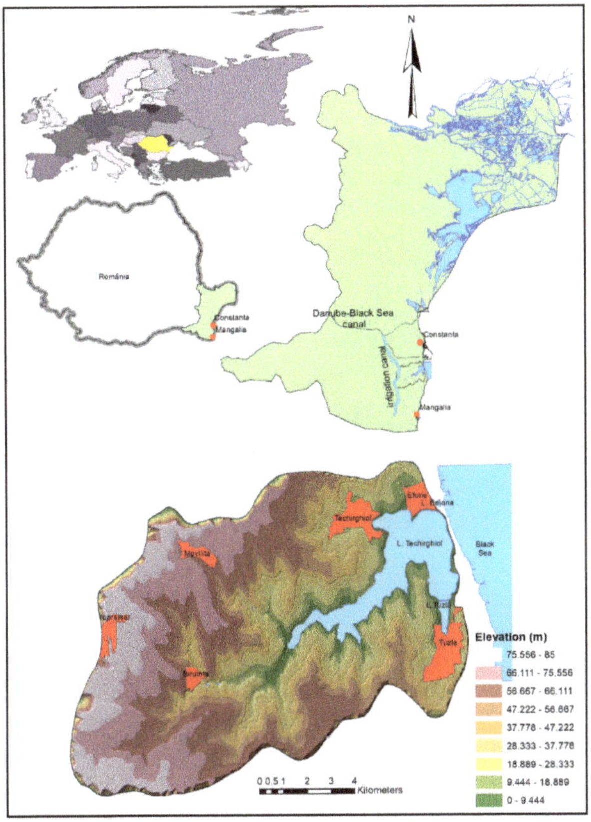

Figure 1. Location of Techirghiol Lake.

Techirghiol Town is well known as a balneotherapy center due to the importance of sapropelic mud and saline water used in therapeutic treatment. Two important centers have been developed here: the Techirghiol Balneotherapy Center (in 1899) and the Eforie Balneotherapy Center (in 1923). To sustain the economic development of this area, water supplies for the localities were established around the lake in the period of 1953–1956. During this period, wastewater was discharged into the lake. Since 1956, wastewater has been treated by a wastewater treatment plant built in South Eforie Town. Treated water is discharged first into the Tuzla pond, and then into Techirghiol Lake. In 1969, important hydraulic works were made in the Lake Techirghiol catchment: (i) 12 km west of the lake is situated the principal irrigation channel "Basarabi—Negru Voda", which loses 60% of the water through infiltration; (ii) 8 km north of the lake is located the "Danube—Black Sea"-navigable channel. In 1971, an irrigation system built in the area was put into operation. Since 1976, water from the treatment plant has been introduced into the irrigation system. In order to eliminate the effects of irrigation on the lake's parameters—and due to the fact that the stoppage of irrigation was incompatible with the state policies of that period (Dobrogea Region being an arid area where crops cannot grow in optimal conditions without irrigation)—the National Water Administration took at that time a series of measures to limit the effects of irrigation. First, in 1972–1973 and then in 1983, water from the lake was pumped directly into the sea. The protection works were carried out in three stages, which were completed in 2005. In the first stage during 1977–1979, all groundwater observation drillings were equipped with pumps and the wastewater discharge into the lake was forbidden. During the second stage (1980–1983), another 11 groundwater drillings were equipped with pumps (intercepted water was used for water supply) and on the rivers Biruinta, Izvoarele and Gospodarie were built dams (behind the dams were placed two pumping stations). The third stage began in 1988 with the construction of the Techirghiol dam—which ended in 1991—and the drainage of the freshwater from behind the dam through a pipe (diameter of 1400 mm and a length of 9.1 km) into Belona Lake (near Eforie Town). A number of small dams were also built in all of the small valleys to intercept the freshwater and evacuate it. As a result of all these hydraulic works, the water surface of Techirghiol Lake decreased. Now, the studied area is divided into three zones: the freshwater area—Biruinta, Izvoarele, Gospodariei lakes, the brackish water area—Zarguzon Lake and the saline water area—Techirghiol Lake (Figure 2).

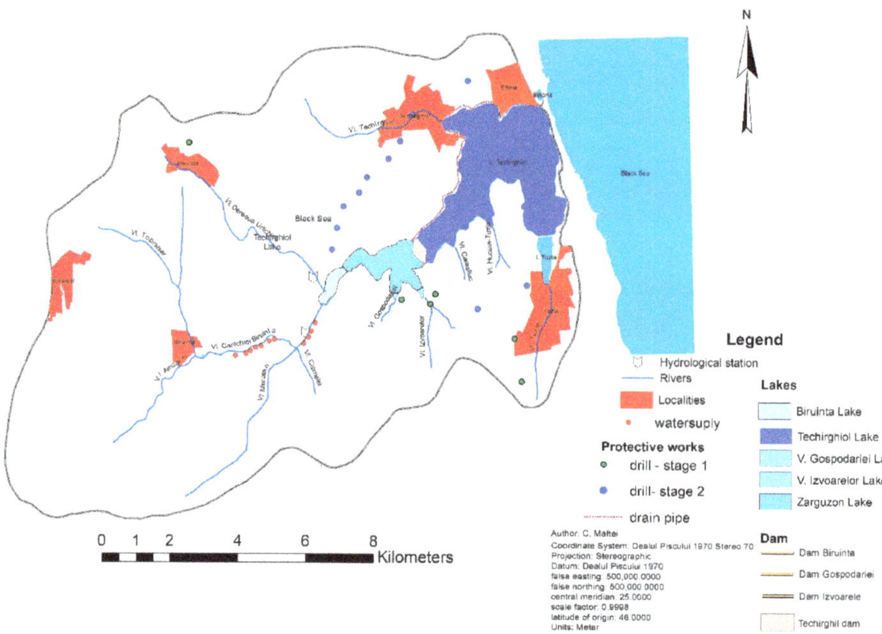

Figure 2. Digital Elevation Model of Techirghiol Lake basin.

The main characteristics investigated in this study which influence the Techirghiol water budget are annual precipitation, overland flow and groundwater. These data are obtained from government reports spanning the period of 1953–2015.

The lake water chemistry parameters investigation that covers salinity, pH, dissolved oxygen (DO) and biochemical oxygen demand (BOD) is based on the data provided by the Romanian Water Administration—Dobrogea Littoral Branch. The data were obtained from various sources, such as government reports, old published papers or unpublished reports [19–25]. The systematic measurement started in 1993, but several government works [21,22] provide some values for these chemical parameters before this period. The investigated period is 1993–2015.

2.2. Methodologies

The methodology used for the analysis of hydrological data is described by Kundzewicz and Robsson [26] and is based on the following steps: (1) obtainment and preparation of a suitable dataset; (2) exploratory analysis of the data and (3) application of statistical tests. Concerning the first step the datasets were performed by INHGA (Romanian National Institute of Hydrology and Water Management), so they are expected to be reliable and free of gross errors, given that the gauging process was supervised by professional personnel. A set of statistical tests which detect step-change in the mean or median of a series was used in previous studies [16,27–31], namely Pettitt, Buishand test, Lee and Heghinian test. In order to detect multiple changes in time series data, the segmentation procedure of Hubert and changing point analysis (based on CUSUM procedure) were used and presented in previous studies we have already mentioned. Some results will be provided in the following paragraph in correlation with other investigated parameters.

The chemical composition and water quality of Techirghiol Lake were investigated in the context of climatic and anthropogenic impact using Romanian methodologies and regulation [19,20]. According to the methodology, Techirghiol Lake is a heavily modified water body. In this respect, for each chemical element mentioned above, the methodology establishes the limits and the ecological status/potential. Three ecological potential classes are identified for heavily modified water bodies: (i) maximum ecological potential (PEM), (ii) good ecological potential (PEB), (iii) moderate ecological potential (PEMo). The range of variation of each class was developed by a series of research institutes and experts.

3. Results and Discussion

To determine the effects of changes in climate in the Dobrogea region and thus on the behavior of Lake Techirghiol, the results obtained in the studies previously mentioned are capitalized [16,27–30]. To conclude: (T—temperature) a break point is identified in 1997–1998 and the mean annual temperature increased by 0.8 °C in the 1997–2015 period—compared to the period of 1953–1997, which is in concordance with the estimation made for Europe by different reports [32–35]. (P–precipitation) Figure 3 shows the variability of rainfall amount from 1953 to 2015.

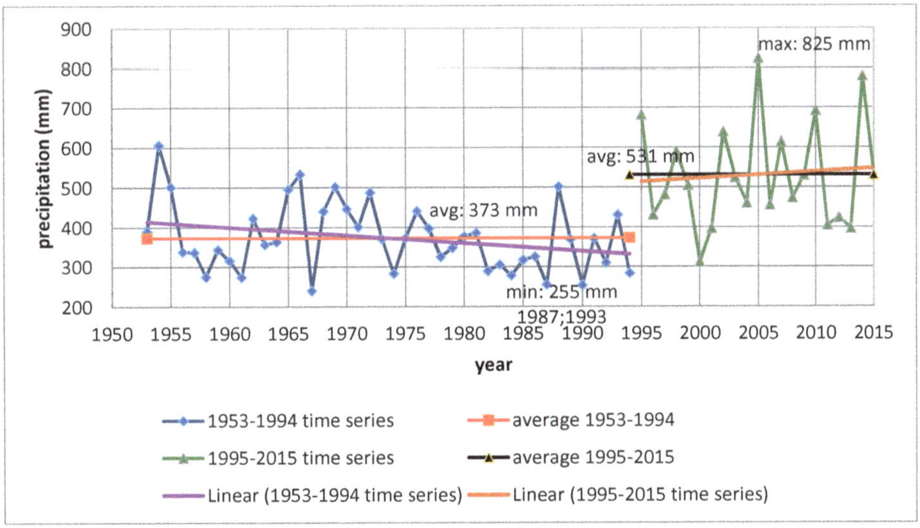

Figure 3. Precipitation variation (1953–2015 period).

The annual precipitation value varies between a minimum value of 255 and a maximum value of 825 mm. The multiannual rainfall value for the entire study period is 425 mm. According to [27,35], precipitations have a break point in 1994. For the period of 1953–1994 we observe a decreasing trend and after it an increasing trend (Figure 3), and the mean annual precipitation increased from 373 to 531 mm. Starting with 1995, the annual precipitation has been above the multiannual precipitation except the following years: 2000 and 2001, 2011 and 2013.

The overland flow (OvF) is presented in Figure 4. This flow is provided by two major river valleys: Biruinta and Urlichioi.

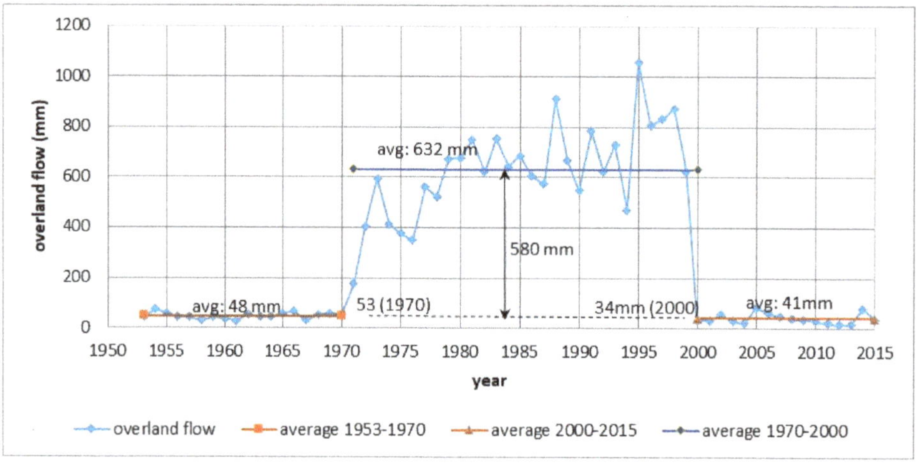

Figure 4. Overland flow variation [35].

According to [35], the overland flow time data series presents three break points in 1971, 1978 and 2000. The first was observed in 1970 when the irrigation system became operational (in the period of 1953–1970 the average overland flow rate was 48 mm). The third one is observed in 2000. After 2000 the overland flow value returned to the average value of the 1953–1970 period (41 mm). We consider that this breakpoint is in relation to anthropic intervention: 1997–1998 was the last time freshwater was introduced into the main irrigation channel that crosses the lake's catchment: in 1991, the Techirghiol dam entered into operation. In the period of 1970–2000, the average overland flow increased to 632 mm (the increase was about 14 times relative to the previous period—580 mm). However, the maximum value of overland flow was recorded in 1995 (1058 mm).

The groundwater supply (GW) is presented in Figure 5. The values varied between a minimum value of 98.2 mm (2015) and a maximum value of 1206.9 mm recorded in 1985. It is noted that the groundwater input time series is divided in three subseries [35]. During the period of 1953–1969, the groundwater input value did not exceed 534 mm. Since 1970, this value has increased on average about 1.5 times. The average value for the period of 1970–2000 was 754.7 mm. Since 2000, as a result of finalizing the works proposed in the third stage, the groundwater input value decreased, reaching the minimum value (98 mm) in 2015.

In Figure 6 is represented the variation of the main inputs (P + OvF + GW) in Techirghiol Lake as average per period. The periods marked by the human interventions and the breakpoint in precipitation data series is highlighted. Analyzing the results obtained we could conclude that the hydraulic works built until 1970, especially the irrigation system, changed the water budget of Techirghiol Lake after 1971. In the 1971–1978 periods, the overland flow increased from 46 to 411 mm on average. In this period, the overland flow represented 31% of the total budget. The first protective works performed in 1977–1979 did not influence the overland flow and groundwater regime in the sense that it diminished. On the contrary, in the period of 1979–1983 (only five years), the average values of these parameters increased and represented 80% of the total water budget. We could conclude that the hydraulic works did not have the expected effect, given the average values of groundwater increase from 692 to 1040 mm approximately, in the period of 1984–1986.

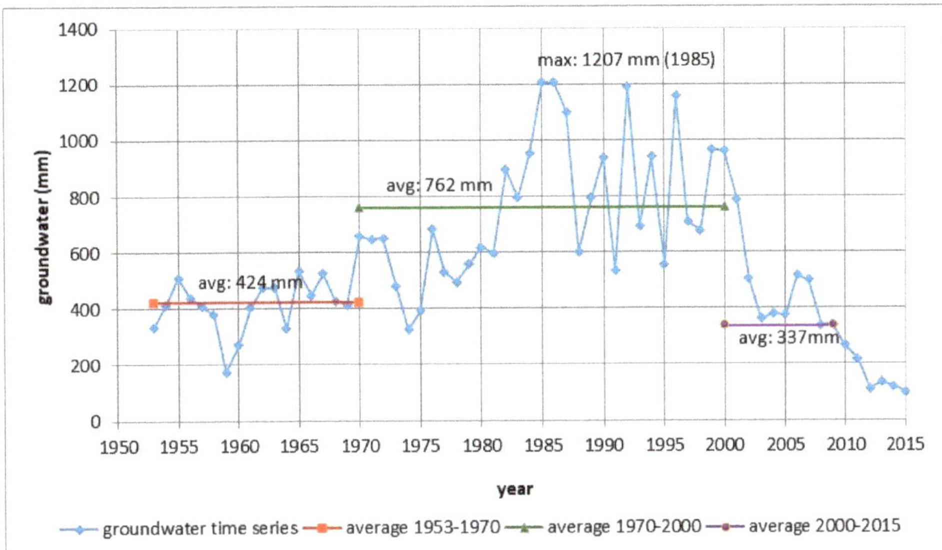

Figure 5. Groundwater input variation.

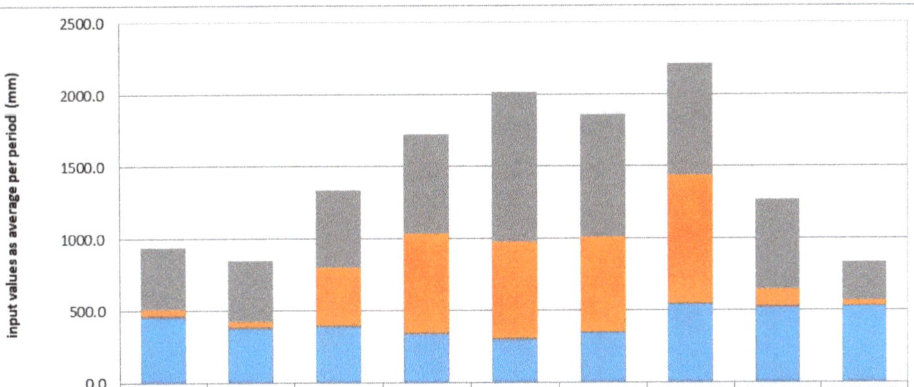

	1953-1956	1957-1970	1971-1978	1979-1983	1984-1986	1987-1994	1995-1998	1998-2005	2005-2015
groundwater	424.8	423.3	529.6	692.9	1040.7	848.9	773.8	620.3	264.7
overland flow	55.9	46.1	410.8	695.4	671.8	663.7	893.6	126.4	39.3
rainfall	458.7	381.7	393.2	341.1	306.6	347.2	546.2	523.4	530.4

period (year)

Figure 6. Variation of the min input values (average/period) in the Techirghiol water budget.

The hydraulic works performed in the period of 1983–1986 failed to bring new improvements to the Techirghiol water budget. The average values of overland flow are maintained at the level of the period of 1987–1994, while the average values of groundwater flow have decreased by 200 mm. As previously mentioned, starting with 1995 the value of precipitation increased. In the following period, the average values are maintained at 43% of the total water budget. The protective works started in 1988 and were finalized in 1991 and 2005, the stoppage of the irrigation activity (1998) causing an improvement of the water budget: the average of overland flow decreased to 39 mm and the average groundwater value to 264 mm.

Figure 7 shows the evolution of the water level in the lake between 1953 and 2015; some isolated measurement data are from 1909.

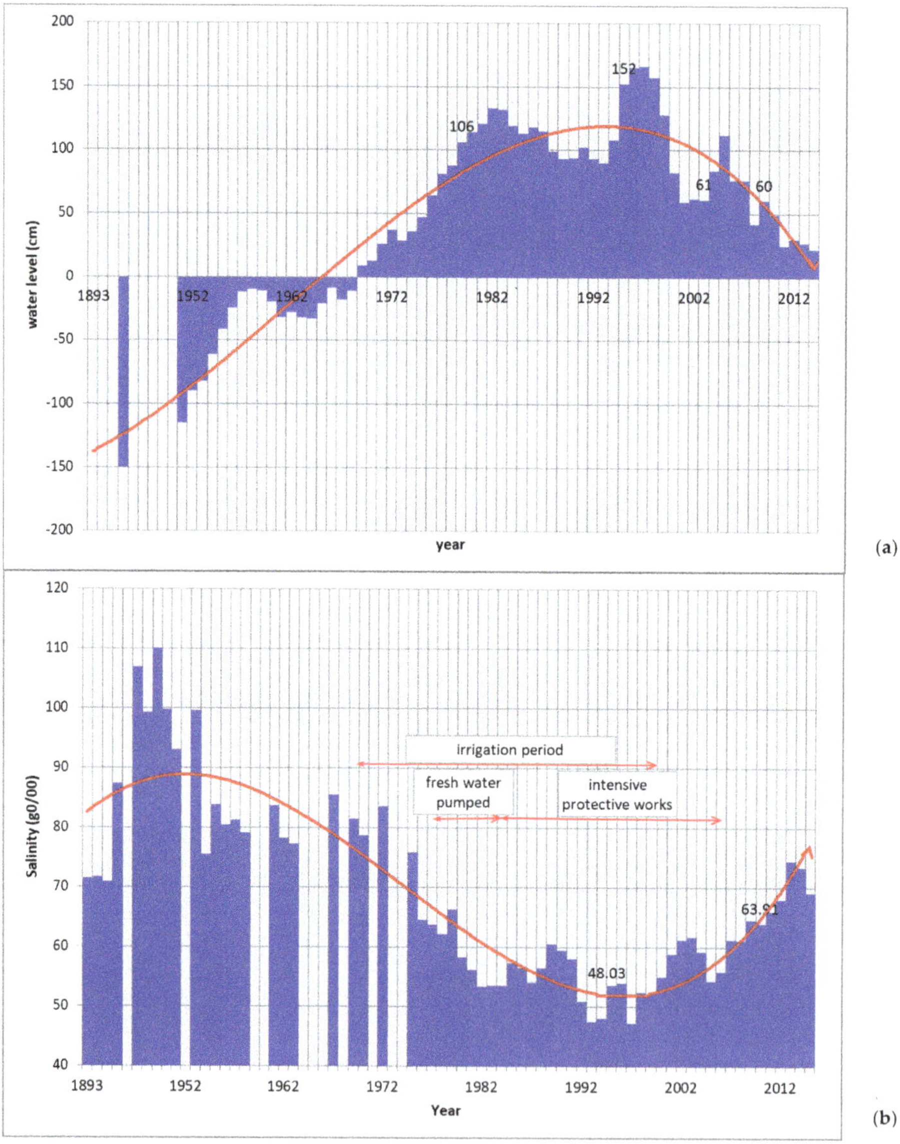

Figure 7. Water level and salinity evolution of Techiorghiol Lake under changes in climate and anthropic impact. (**a**) Salinity; (**b**) water level.

It can be seen that in the investigated period, water levels in the lake rose from −150 cm to +153 cm (the measurements are relative to the Black Sea level). Between 1909 and 1952 the water level in the lake increased by an average of 0.8 cm/year, between 1954 and 1966 the water level increased by 6 cm/year. Since 1970 the water level in the lake has become positive (+9 cm), relative to the Black Sea reference level (±0.00). The increase in water level in the lake was accentuated after 1970, the average value being 9 cm/year, as a result of land irrigation of the lake's catchment. From 1965 to 1989, the water level steadily increased to 133 cm, and in the period of 1996–1999 the level reached the highest values (+153 cm). As a result of the hydraulic works, there has been a trend of increasing the water level in the lake, combined with the decrease of the salinity of the water. Since the land irrigation was stopped (in 1998) the lake's water level has been slowly decreasing in the following years, in 2015 the level being 22 cm above Black Sea level (Figure 7).

Increases of water inputs led to the severe decrease of salinity around 47 g‰ in 1992 and 1997. In the irrigation period of 1970–1997, even if a number of protective measures were introduced, a substantial increase in salinity was not possible. After the irrigation was stopped (1998), the salinity began to increase, reaching the value of 70 g‰ (in 2015).

The changes caused by the increase of the freshwater inflow and the decrease of salinity caused quantitative changes in the lake's biotic community, especially in some organisms involved in the process of peloidogenesis. Some studies [21,36] reveal a decrease of green algae *Cladophora vagabunda* from 81.49 tons in 1978 to 42 tons in 1981. It is known [37] that the optimal salinity values in which this alga can develop are 73–83 g/L. This situation began to improve in 1987 when the second stage of protective works became operational. The completion of protective work (in 2005) and closure of the irrigation system (in 1998) has led to ecosystem regeneration.

The water of Techirghiol Lake is alkaline; the average value for pH is 8.3 (Figure 8). Normally, there is a direct relationship between water pH and salinity; a higher value of pH is given by the high content of mineralization. Even if the salinity of Techirghiol Lake waters decreased to a value of 47 g‰, the pH value of the waters was never under the value of 7.9. Figure 8 shows that the lake water pH is situated in the range 6.5–9 pH, more accurate under 8.5 units, except the values from 1999 and 2000.

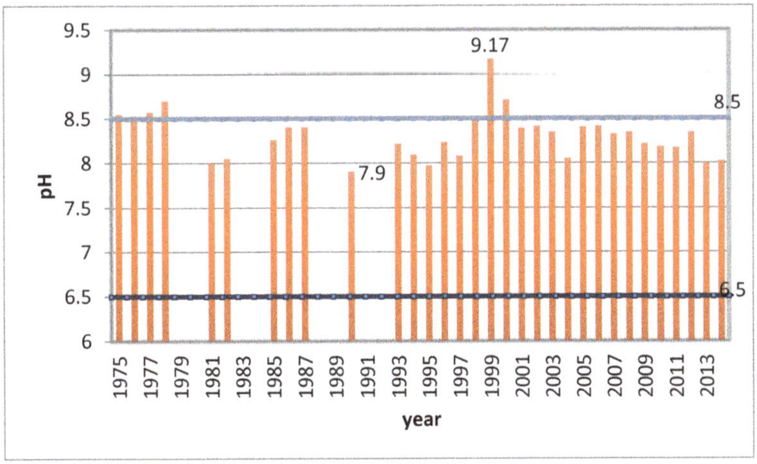

Figure 8. Variation of pH over time.

Figure 9 shows the variation of the dissolved oxygen (DO) time data series. It can be seen that the dissolved oxygen (DO) varies between 11 and 4.04 mg/L. In the period of 1975–1990 the DO values were over 8 mg/L and lake water could be included in the PEB category. The DO values have decreased after 1990 from an average of 8.91 mg/L to 6.24 mg/L. Correspondingly, the ecological potential has decreased, lake water could be included in the PEB/PEMo category. Three exceptions could be considered: 2009, 2013 and 2014, when water could be included in the PEM/PEB category.

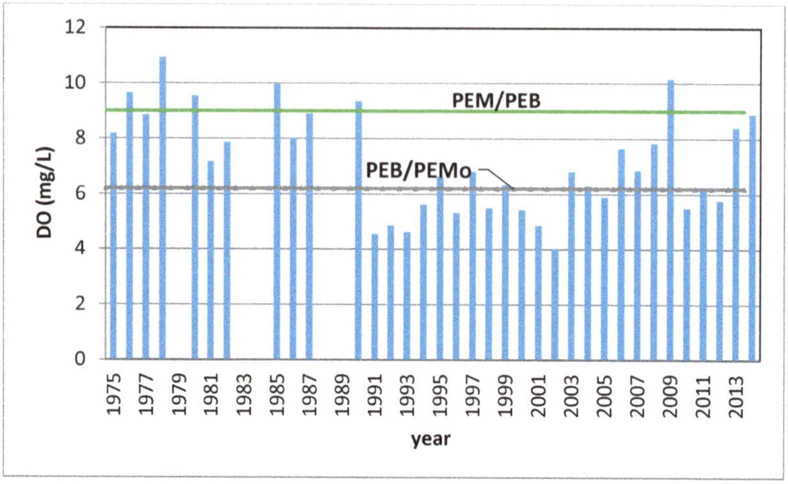

Figure 9. Variation of dissolved oxygen (DO).

The biologic oxygen demand (BOD) represents the mass concentration of dissolved oxygen consumed by microorganism or measures the chemical oxidation of inorganic matter in a given time (e.g., BOD5 stands for five days test). BOD5 affects the DO values. A greater BOD5 value means less oxygen for the microorganism's activity. The variation of this indicator is presented in Figure 10. Generally during the 1975–1993 period, the BOD5 values were between 3 mg/L and 6 mg/L and the lake water could be included in the PEM/PEB ecological potential category. After 1994 the BOD5 values have generally increased above 6 mg/L. In fact, during the period of 1994–2012, the average was three times higher than the average of the previous period.

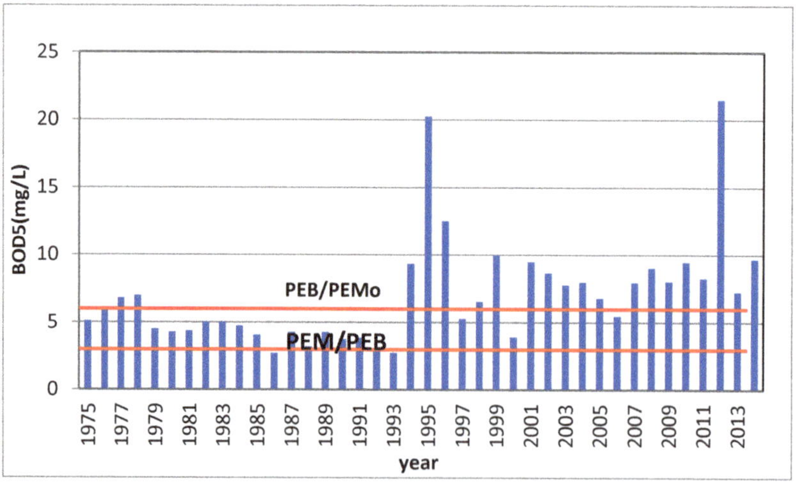

Figure 10. Variation of Biochemical oxygen demand (BOD).

The ecological potential of Techirghiol Lake from the point of view of the investigated elements varies throughout the period investigated (Table 1). It can be seen that Techirghiol Lake water could be included in the PEB/PEM category for the period of 1975–1990 (in this period, a larger quantity of freshwater was introduced via the irrigation system). After 1990 (1993 for BOD5) lake water was included in the PEB/PEMo category.

Table 1. Ecological Potential Variation.

Water Chemistry Elements	Ecological Potential	Period
pH	PEM	1975–2015
DO	PEB PEB/PEMo PEB	1975–1990 1990–2005 2005–2015
BOD5	PEM/PEB PEB/PEMo	1975–1993 1993–2015

It is very complex to explain the multiple factors that play a role in changing water chemistry variation. Barbulescu and Barbes [32] consider that one of the direct consequences of the decrease in water salinity was the modification of the lake biodiversity. Some studies appreciated that the phytoplankton structure was modified during the irrigation period: the number of species diminishing to 14–18 in relation to the previous period, when 38 species were found [21,22,37]. Another indicator is *Artemia* Salina. A drastic decreasing of *Artemia* Salina during the irrigation period, compared to the reference period (1952–1960), when densities above 100 g/L were recorded [22]. We can conclude that the cause of the increase of BOD5 in the last period (after 1995) could be an increase of aquatic life forms (phytoplankton or/and zooplankton).

4. Conclusions

An important aspect of Techirghiol Lake is its potential in the tourism industry, due to its unique properties: saline water and sapropelic mud. The malfunction of environmental protection measures and faulty or insufficient design (period of 1960–1987), in conjunction with the changes in climate has disturbed the normal functioning of the Techirghiol Lake ecosystem, finally resulting in a decrease in its capacity to yield economic values. It is therefore concluded that 1953 is considered as the last year in which the Techirghiol Lake system was under the influence of natural factors and the 1970 year is marked by the passage of the water level to positive values. Starting with 1971, the irrigation system became operational and the ecosystem degradation became aggressive. The rate of salinity decreased as a result of freshwater supply being 1.24‰ during the period of 1970–1987. The period of 1980–1987 is a critical one: overland flow increased (14 times the level of the period of 1953–1970 and the hydrological regime of Biruinta and Urlichioi tributary rivers became permanent); groundwater input increased by 7.2 mil.mc/year over the same period.

The most important challenge in the management of Techirghiol Lake basin is to integrate and balance the interest of the ecosystem and the economy. It is well known that the Dobrogea region is an arid area where crops cannot grow in optimum conditions without irrigation. New investigation in needed to provide the sound, scientific basis in order to find a balance between protecting the ecosystem, increasing the economy and designing hydrotechnical systems in the context of climate change.

Author Contributions: Conceptualization, C.M.; methodology, C.M., C.B. and I.C.P.; validation, C.M., C.B. and I.C.P.; investigation, C.M.; resources, I.C.P. and C.B.; data curation, C.B.; writing—original draft preparation, C.M.; writing—review and editing, C.M., C.B., I.C.P.; visualization, I.C.P.; supervision, C.M. All authors have read and agreed to the published version of the manuscript.

Funding: This research received no external funding.

Acknowledgments: The authors would like to thank the Dobrogea Littoral Water Basin Administration for technical support and National Meteorological Agency—Dobrogea meteorological Center which provided the climatic data.

Conflicts of Interest: The authors declare no conflict of interest.

References

1. Shadrin, N.; Zheng, M.; Oren, A. Past, present and future of saline lakes: Research for global sustainable development. *Chin. J. Ocean. Limnol.* **2015**, *33*, 1349–1353. [CrossRef]
2. Abbaspour, M.; Javid, A.H.; Mirbagheri, S.A.; Givi, F.A.; Moghimi, P. Investigation of lake drying attributed to climate change. *Int. J. Environ. Sci. Technol.* **2012**, *9*, 257–266. [CrossRef]
3. Delju, A.H.; Ceylan, A.; Piguet, E.; Rebetez, M. Observed climate variability and change in Urmia Lake Basin, Iran. *Theor. Appl. Climatol.* **2013**, *111*, 285–296. [CrossRef]
4. Valero-Garcés, B.L.; Navas, A.; Machin, J.; Stevenson, T.; Davis, B. Responses of a Saline Lake ecosystem in a semiarid region to irrigation and climate variability: The history of Salada Chiprana, Central Ebro Basin, Spain. *AMBIO A J. Hum. Environ.* **2000**, *29*, 344–350. [CrossRef]
5. Webster, K.E.; Kratz, T.K.; Bowser, C.J.; Magnuson, J.J.; Rose, W.J. The influence of landscape position on lake chemical responses to drought in northern Wisconsin. *Limnol. Oceanogr.* **1996**, *41*, 977–984. [CrossRef]
6. Ptak, M.; Sojka, M.; Choinski, A.; Nowak, B. Effect of Environmental Conditions and Morphometric Parameters on Surface Water Temperature in Polish Lakes. *Water* **2018**, *10*, 580. [CrossRef]
7. Stan, F.I.; Neculau, G.; Zaharia, L.; Ioana-Toroimac, G. The hydrological budget of lakes. Case studies Fântânele and Izvorul Muntelui Reservoirs (Romania). In *Water Resources and Wetlands, 4th International Conference Water Resources and Wetlands, Tulcea, Romania, 5–9 September 2018*; Gastescu, P., Bretcan, P., Eds.; Romanian Limnogeographical Association: Targoviste, Romania, 2018; pp. 56–63.
8. Salameh, E.; El-Naser, H. Changes in the Dead Sea Level and their Impacts on the Surrounding Groundwater Bodies. *Acta Hydrochim. Hydrobiol.* **2000**, *28*, 24–33. [CrossRef]
9. Carrasco, N.K.; Perissinotto, R. Development of a Halotolerant Community in the St. Lucia Estuary (South Africa) during a Hypersaline Phase. *PLoS ONE* **2012**, *7*, e29927. [CrossRef]
10. Piovano, E.L.; Ariztegui, D.; Moreira, S.D. Recent environmental changes in Laguna Mar Chiquita (central Argentina): A sedimentary model for a highly variable saline lake. *Sedimentology* **2002**, *49*, 1371–1384. [CrossRef]
11. Sojka, M.; Choinski, A.; Ptak, M.; Siepak, M. The Variability of Lake Water Chemistry in the Bory Tucholskie National Park (Northern Poland). *Water* **2020**, *12*, 394. [CrossRef]
12. Posea, G. *Geomorfologia României—Relief, Tipuri, Geneză, Regionare*; (Ediția a 2-a); Fundatia "România de Mâine": Bucuresti, Romania, 2005. (In Romanian)
13. Gastescu, P.; Bretca, P. Aspecte privind starea actuală a lacurilor Siutghiol şi Techirghiol. *Analele Universităţii Valahia Târgovişte Seria Geografie* **2003**, *3*, 134–138. (In Romanian)
14. Panin, N. Black Sea coast line changes in the last 10,000 years. A new attempt at identifying the Danube mouths as described by the ancients. *Dacia Revue d'Archéologie et d'Histoire Ancienne Bucuresti* **1983**, *27*, 175–184.
15. Gâştescu, P.; Breier, A. *Le complexe lacustre Razim-Sinoe (Roumanie). Genèse, Morphométrie et Régime Hydrique*; Scritti Geografici in onore di Riccardo Riccardi (Parte I); Società Geografica Italiana: Roma, Italia, 1974; pp. 247–269, (In Italian with French abstract).
16. Maftei, C.; Bărbulescu, A.; Hubert, P.; Serban, C.; Dobrica, G. Statistical Analysis of the Precipitation from Constanța (Romania) meteorological station. In *Recent Researches in Applied Computers and Computational Science*; Niola, V., Bojkovic, Z., Garcia-Planas, M.I., Eds.; WSEAS Press: Athens, Greece, 2012; p. 52.
17. Zamfirescu, F.; Moldoveanu, V.; Dinu, C.; Pitu, N.; Albu, M.; Danchiv, A.; Nash, F. Vulnerability to pollution of karst system in Southern Dobrogea. In *Impact of Industrial Activities on Groundwater—Proceedings of the International Hydrogeological Symposium*; Bucharest University Press: Constanţa, Romania, 1994; pp. 591–602.
18. Pascu, R. Cercetari preliminare asupra Lacului Techirghiol (jud. Constanta). *Anuarul Institutului Geologic al Romaniei* **1910**, *IV Fasc. I*, 1–20. (In Romanian)
19. Planul Naţional de Management Actualizat Aferent Porţiunii din Bazinul Hidrografic International al Fluviului Dunărea Care este Cuprinsă în teritoriul României. Available online: http://www.rowater.ro/dadobrogea/Planul%20de%20Management%20Bazinal/Plan%20de%20Management%20actualizat%20al%20Fluviului%20Dunarea,%20Deltei%20Dunarii,%20SH%20Dobrogea%20si%20Apelor%20Costiere%2020162021/ABADL_Planul%20de%20Management%20actualizat%20TEXT.pdf (accessed on 15 May 2019). (In Romanian).

20. ORDIN nr. 161 din 16 Februarie 2006 Pentru Aprobarea Normativului Privind Clasificarea Calitatii Apelor de Suprafata in Vederea Stabilirii Starii Ecologice a Corpurilor de apa. Available online: http://www.rowater.ro/dacrisuri/Documente%20Repository/Legislatie/gospodarirea%20apelor/ORD.%20161_16.02.2006.pdf (accessed on 15 May 2019). (In Romanian).
21. Research report MEN Grant 410B/1996. Studiu privind urmarirea parametrilor de calitate si a factorilor care il influenteaza in vederea protectiei resurselor natural ale lacului Techirghiol. 1996; (unpublished work).
22. Research report Grant nr. 587/1986 Consiliul National al Apelor—ICPGA—Cercetari privind evolutia si protectia calitatii apelor lacurilor Techirghiol si Amara in conditiile amenajarilor in curs. 1986; (unpublished work).
23. Bujor, P. Nouvelle contribution á l'étude de la biologie du lac salé du Tékirghiol. In Proceedings of the 5th Congr. Intern. Thalassiothérapie, Iasi, Romania, 23–30 May 1928; pp. 1–85, (In Romanian with French abstract).
24. ABADL Plan de Management—Aria Protejata Lacul Techirghiol 2008–2013 +X Anexe. Available online: http://www.rowater.ro/dadobrogea/Aria%20protejata%20Lacul%20Techirghiol/Forms/AllItems.aspx (accessed on 15 May 2019).
25. Solomon, S. (Ed.) *Intergovernmental Panel on Climate Change (IPCC), Climate Change 2007: The Scientific Basis. Contribution of Working Group I to the Fourth Assessment Report of the Intergovernmental Panel on Climate Change*; Cambridge University Press: New York, NY, USA, 2001.
26. Kundzewicz, Z.W.; Robsson, A. (Eds.) *Water Detecting Trend and Other Changes in Hydrological Data, WCDMP-45, WMO/TD-No*; World Meteorological Organization: Geneva, Switzerland, 2000; p. 1013.
27. Maftei, C.; Barbulescu, A.; Buta, C.; Serban, C. Change points detection and variability analysis of some precipitation series. In *Recent Researches in Computational Techniques, Non-Linear Systems and Control*; WSEAS Press: Athens, Greece, 2007; p. 232.
28. Maftei, C.; Barbulescu, A. Statistical analysis of precipitation time series in Dobrudja region. *Mausam* **2012**, *63*, 553–564.
29. Barbulescu, A.; Maftei, C. Modeling the climate in the area of Techirghiol Lake (Romania). *Rom. J. Phys.* **2015**, *60*, 1163–1170.
30. Bărbulescu, A.; Serban, C.; Maftei, C. Statistical analysis and evaluation of Hurst coefficient for annual and monthly precipitation time series. *WSEAS Trans. Math.* **2010**, *9*, 791–800.
31. Romanescu, G.; Iosub, M.; Sandu, I.; Minea, I.; Enea, A.; Dascalita, D.; Hapciuc, O.-E. Spatio-temporal Analysis of the Water Quality of the Ozana River. *Rev. Chim. (Bucharest)* **2016**, *67*, 42–47.
32. Barbulescu, A.; Barbes, L. Assessment of surface water quality Techirghiol Lake using statistical analysis. *Rev. Chim. (Bucharest)* **2013**, *64*, 868–874.
33. *ECSN Climate of Europe: Recent Variation, Present State and Future Prospects*; KNMI: de Bilt, The Netherlands, 1995; p. 72.
34. Houghton, J.T.; Ding, Y.; Griggs, D.J.; Noguer, M.; Van der Linden, P.J.; Xiaosu, D. (Eds.) *Intergovernmental Panel on Climate Change (IPCC) Climate Change 2001: The Scientific Basis. Contribution of Working Group I to the Third Assessment Report of the Intergovernmental Panel on Climate Change*; Cambridge University Press: New York, NY, USA, 2001.
35. Maftei, C.; Buta, C.; Draghici, G.; Filip, C. Analysis of changes in hydrometeorological variables of Techirghiol Lake. *IOP Conf. Ser. Earth Environ. Sci.* **2019**, *344*, 012025. [CrossRef]
36. Romanescu, G.; Tirnovan, A.; Sandu, I.; Cojoc, G.M.; Breaban, I.; Mihu-Pintilie, A. Water Chemism Within the Settling Pond of Valea Straja and the Quality of the Suha Water Body (Eastern Carpathians). *Rev. Chim. (Bucharest)* **2015**, *66*, 1700–1706.
37. Gheorghievici, L.M.; Pompei, I.; Gheorghievici, G.; Tanase, I. The influence of abiotic factors on suppliers of organic matter in the peloidogenesis process from Lake Techirghiol, Romania. *Aquac. Aquar. Conserv. Legis.* **2012**, *5*, 69–78.

water

MDPI

Article

Multivariate Statistical Analysis of Water Quality and Trophic State in an Artificial Dam Reservoir

Md Mamun, Ji Yoon Kim and Kwang-Guk An *

Department of Bioscience and Biotechnology, Chungnam National University, Daejeon 34134, Korea; mamun1006001@gmail.com (M.M.); jiyoonn20@naver.com (J.Y.K.)
* Correspondence: kgan@cnu.ac.kr; Tel.: +82-010-6404-9844; Fax: +82-42-882-9690

Abstract: Paldang Reservoir, located in the Han River basin in South Korea, is used for drinking water, fishing, irrigation, recreation, and hydroelectric power. Therefore, the water quality of the reservoir is of great importance. The main objectives of this study were to evaluate spatial and seasonal variations of surface water quality in the reservoir using multivariate statistical techniques (MSTs) along with the Trophic State Index (TSI) and Trophic State Index deviation (TSID). The empirical relationships among nutrients (total phosphorus, TP; total nitrogen, TN), chlorophyll-a (CHL-a), and annual variations of water quality parameters were also determined. To this end, 12 water quality parameters were monitored monthly at five sites along the reservoir from 1996 to 2019. Most of the parameters (all except pH, dissolved oxygen (DO), and total coliform bacteria (TCB)) showed significant spatial variations, indicating an influence of anthropogenic activities. Principal component analysis combined with factor analysis (PCA/FA) suggested that the parameters responsible for water quality variations were primarily correlated with nutrients and organic matter (anthropogenic), suspended solids (both natural and anthropogenic), and ionic concentrations (both natural and anthropogenic). Stepwise spatial discriminant analysis (DA) identified water temperature (WT), DO, electrical conductivity (EC), chemical oxygen demand (COD), the ratio of biological oxygen demand (BOD) to COD (BOD/COD), TN, TN:TP, and TCB as the parameters responsible for variations among sites, and seasonal stepwise DA identified WT, BOD, and total suspended solids (TSS) as the parameters responsible for variations among seasons. COD has increased ($R^2 = 0.63$, $p < 0.01$) in the reservoir since 1996, suggesting that nonbiodegradable organic loading to the water body is rising. The empirical regression models of CHL-a-TP ($R^2 = 0.45$) and CHL-a-TN ($R^2 = 0.27$) indicated that TP better explained algal growth than TN. The mean TSI values for TP, CHL-a, and Secchi depth (SD) indicated a eutrophic state of the reservoir for all seasons and sites. Analysis of TSID suggested that blue-green algae dominated the algal community in the reservoir. The present results show that a significant increase in algal chlorophyll occurs during spring in the reservoir. Our findings may facilitate the management of Paldang Reservoir.

Keywords: multivariate statistical methods; Trophic State Index; water quality; empirical model; Paldang Reservoir

Citation: Mamun, M.; Kim, J.Y.; An, K.-G. Multivariate Statistical Analysis of Water Quality and Trophic State in an Artificial Dam Reservoir. *Water* **2021**, *13*, 186. https://doi.org/10.3390/w13020186

Received: 9 November 2020
Accepted: 10 January 2021
Published: 14 January 2021

Publisher's Note: MDPI stays neutral with regard to jurisdictional claims in published maps and institutional affiliations.

1. Introduction

Although water is indispensable to life, it is one of the most threatened resources worldwide [1]. Clean and safe freshwater is a basic need for human health and economic development, but anthropogenic activities like industrialization, urbanization, and intensive agricultural farming have negatively impacted freshwater sources, hindering their use for drinking, irrigation, fishing, recreational, domestic, and industrial purposes [2–5]. Therefore, serious attention should be paid to protect freshwater resources. Among these, reservoirs are the most vulnerable due to high loads of pollutants, nutrients, organic matter, and suspended solids from the watershed [6,7]. For effective water management, gathering reliable information on reservoir water quality, evaluating spatial and seasonal water quality changes, detecting pollution sources, determining water quality status, and controlling water pollution in reservoirs are essential [1,3,8–11].

To assess the water quality of surface water resources, MSTs, TSI, and TSID have been widely used, and therefore have played a significant role in water resource management [2,9,11,12]. Multivariate statistical methods, such as discriminant analysis (DA), principal component analysis (PCA), factor analysis (FA), correlation analysis, and analysis of variance (ANOVA) facilitate the interpretation of complex water quality datasets [1,13,14]. These methods are also used to identify factors that influence surface water quality, serving as a valuable tool for effective surface water quality management [2,11]. These approaches can be used to evaluate temporal and spatial changes in surface water quality caused by natural and anthropogenic factors [2].

However, MSTs have some limitations when used alone [2]. Therefore, applying MSTs, TSI, and TSID in combination can be advantageous for assessing the water quality of reservoirs. To date, a few studies have used MSTs, TSI, and TSID together for surface water quality assessment of reservoirs [2,15]. The TSI and TSID were used to quantify the degree of eutrophication of a water body. Carlson [12] proposed a quantitative index to calculate the degree of eutrophication in lakes and reservoirs based on total phosphorus (TP), chlorophyll-a (CHL-a), and Secchi depth (SD). According to Carlson and others, TP is the best forecaster of algal growth, while CHL-a is the most reliable algal biomass indicator, and SD is the best proxy for water clarity in water bodies [4,16–18]. Moreover, TSI and TSID are used to evaluate spatial and seasonal changes in the water quality of reservoirs, and thereby provide useful information for reservoir management [19,20].

Seasonal rainfall patterns, hydrology, and watershed morphology are the major factors known to regulate water quality within a watershed [21]. These factors are closely related to the ecosystem's nutrient regime, water clarity, and algal growth. Rainfall is directly linked to inflow, outflow, depth, and water residence time (WRT), which control nutrient and suspended solids loads to the water body [22,23]. Empirical evidence suggests that phosphorus (P) is the key factor limiting CHL-a growth in freshwater systems [4,17,19]. Excessive concentrations of nutrients, especially P, may accelerate algal growth and cause eutrophication in reservoirs [24]. Total suspended solids are a potential source of P and play an essential role in the P cycle in reservoirs [2].

Paldang Reservoir is one of the largest reservoirs in South Korea, with a water volume of 255×10^6 m^3 and a surface area of 28.9 km^2 [25,26]. The maximum depth of water at full supply level is 21 m, and the mean depth is 8.3 m [25]. It is a manmade lake formed after the construction of a hydroelectric dam in 1973 and is located in the central Korean Peninsula [25]. Paldang Reservoir has been used for fishing, irrigation, recreation, hydroelectric power, and drinking water purposes. Additionally, it serves as an essential water resource for people living in the Seoul metropolitan area and surrounding cities [26]. More than 24 million people (48% of the Korean population) rely on the Paldang Reservoir for drinking water [27]. Therefore, the water quality of the reservoir is of great importance to Korea. However, human activities in the watershed have increased, resulting in significant pollution problems in the reservoir. Urbanization, domestic and industrial wastewater discharge, intensive agricultural activities, waste from animal farms, and inflowing rivers are all major sources of water pollution in the reservoir [27–29].

For these reasons, comprehensive water quality assessments of the reservoir are needed. The purposes of the present study are to (1) determine the spatial and seasonal variations of water quality parameters and identify the key factors affecting water quality in the reservoir using MSTs, (2) assess the trophic status of the reservoir using TSI and TSID, (3) determine how water quality parameters are correlated with hydrology, and (4) develop empirical models of the CHL-a-TP, CHL-a-TN, TSS-TP, and TSS-TN in the reservoir. Thus, this study will assess the current status of water quality and aid the development of effective management and conservation strategies to protect water quality in Paldang Reservoir.

2. Materials and Methods

2.1. Study Sites and Water Quality Parameters

Paldang Reservoir is the most downstream reservoir in the Han River system, and is situated at the confluence of the North Han River, South Han River, and Kyoungan Stream (Figure 1; [30]). In this study, five reservoir sampling sites (S1–S5) were selected. Sites 1 and 2 were located in the South Han River part of the reservoir. In contrast, sites 3–5 were located at the North Han River, Kyoungan Stream, and dam, respectively. The water intake tower for Paldang Reservoir is located at S5 (Figure 1). Based on their hydrological characteristics, reservoirs can be divided into two types, namely, lake- and river-type reservoirs. Lake-type reservoirs are generally characterized by high depth and long water WRT, while river-type reservoirs have shallower depths and shorter WRTs [31]. Paldang Reservoir is considered a river-type reservoir due to its shallow depth (mean depth: 8.3 m) and short WRT (3–10 days) [25,32]. Paldang Reservoir does not fully stratify throughout the year [30]. The overall inflow and outflow rates of the reservoir are almost equal, resulting in very small annual water level fluctuations. The annual amount of rainfall and water inflow from the upstream watershed directly influence WRT in Paldang Reservoir [30].

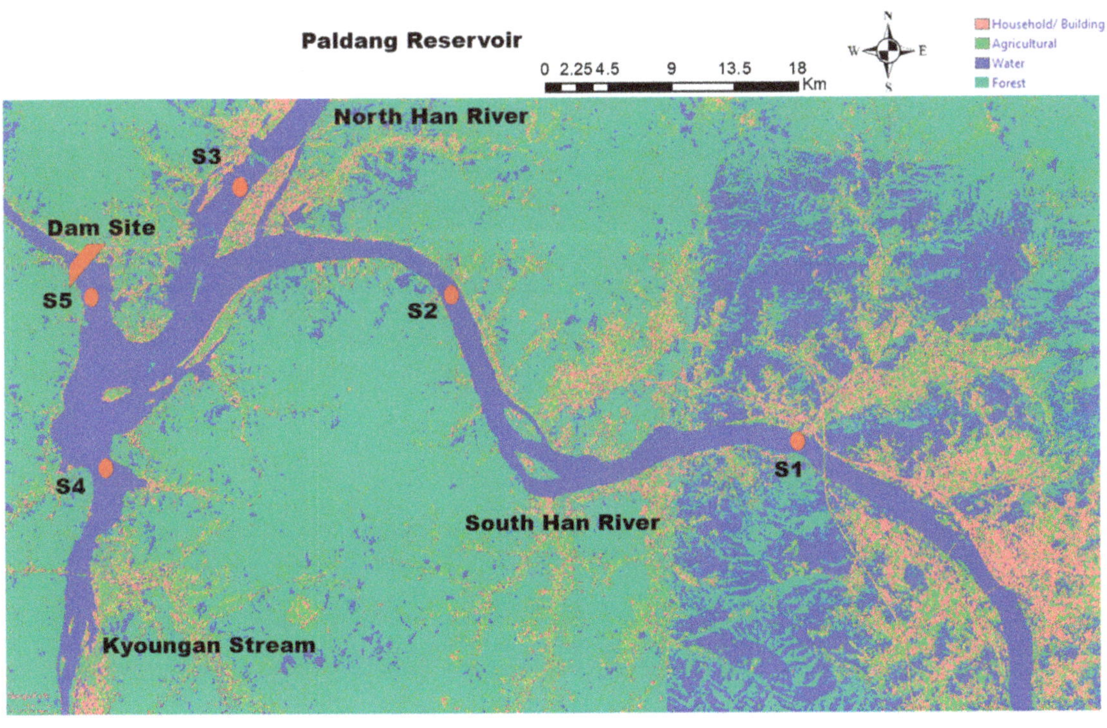

Figure 1. The map showing the sampling sites of Paldang Reservoir.

Monthly surface water quality data for the Paldang Reservoir from 1996 to 2019 were obtained from the Ministry of Environment's national water quality measurement network (http://water.nier.go.kr). Monthly rainfall and inflow and outflow data were collected from the Korean Meteorological Administration and the Korean Water Resource Corporation, respectively. WRT was defined as the reservoir water volume divided by the inflow rate [33]. The loading data for TP, TN, TSS, BOD, and COD were calculated using a conversion factor derived from the corresponding concentrations.

2.2. Trophic State Index and Trophic State Index Deviation

The trophic status of the Paldang Reservoir was determined using Carlson's TSI. The range of average TSI values designated Oligotrophic is 30–40, Mesotrophic is 40–50, Eutrophic is 50–70, and Hypereutrophic is >70 [18]. The following equations were used to calculate TSI values for the Paldang Reservoir [12]:

$$\text{TSI (CHL-a, } \mu\text{g L}^{-1}) = 10 \times [6 - (2.04 - 0.68\ln(\text{CHL-a}))/\ln 2] \quad (1)$$

$$\text{TSI (TP, } \mu\text{g L}^{-1}) = 10 \times [6 - \ln(48/\text{TP})/\ln 2] \quad (2)$$

$$\text{TSI (SD, m)} = 10 \times [6 - \ln(\text{SD})/\ln 2] \quad (3)$$

Using two-dimensional approaches, the TSID was defined using the relationships TSI (CHL-a)-TSI (SD) and TSI (CHL-a)-TSI (TP). This method has also been used frequently to quantify the degree of eutrophication and identify the limiting nutrient in reservoirs [18].

2.3. Statistical Analysis

The Kolmogorov–Smirnov single-sample test was used to examine the distribution of water quality data prior to statistical analyses [1]. One-way ANOVA was performed to determine whether there were significant spatial and seasonal variations in the reservoir's water quality parameter values. Pearson correlation analysis was used to analyze the relationships between various water quality variables. PCA/FA was conducted to determine the factors and pollution sources affecting the surface water quality [34]. Bartlett's sphericity test and the Kaiser–Meyer–Olkin (KMO) test were conducted first to determine the suitability of the data for PCA/FA [2]. DA was performed to assess both spatial and temporal variations in water quality and to identify water quality variables that could best distinguish among sites and seasons [11,34]. Standard and stepwise DA was applied to raw data. PCA/FA was applied to experimental data, standardized through Z-scale transformation, to avoid misclassification [2]. SPSS software (version 22.0; SPSS Inc., Chicago, IL, USA) was used for all statistical analyses. Bar, box, and scatter plots were prepared using SigmaPlot 14.0 software (Systat Software, Inc., San Jose, CA, USA). Interpolation of TSI values was conducted using QGIS 3.14 (QGIS Development Team, Gossau, Switzerland). Conditional plotting analysis was carried out with R 3.5.2 (R Development Core Team, Vienna, Austria).

3. Results and Discussion

3.1. Spatial and Seasonal Variations

The mean values of 12 water quality parameters recorded at five sampling sites in Paldang Reservoir are presented in Table 1. In this study, all variables except pH, dissolved oxygen (DO), and total coliform bacteria (TCB) showed significant spatial differences among sites ($p < 0.05$, Table 1). The spatial variations of these parameters indicate impacts of anthropogenic activities in the reservoir [25,32]. For example, BOD, COD, TSS, TN, TP, and CHL-a concentrations were significantly higher at site S4 than any other site; this site receives inputs from industrial and domestic wastewater [30]. Site S4 in Paldang Reservoir is affected by Kyoungan Stream. The water quality of this tributary stream is worse than that of the South Han River (Sites S1 and S2) and North Han River (Site S3), and thus, it may significantly impact the reservoir's water quality [25]. Water clarity (SD) was higher at Site S3 than other sites, indicating that the North Han River input is cleaner than the South Han River and Kyoungan Stream inflows [25]. The highest mean electrical conductivity (EC) was recorded at Site S1 due to agricultural activities and untreated household wastewater effluent.

Table 1. Mean value of water quality parameters at different sites and seasons (units mg L^{-1}, except pH, WT (°C), EC (μS cm^{-1}), TP (μ gL^{-1}), CHL-a (μg L^{-1}), SD (m), and TCB ($MPNML^{-100}$). pH—hydrogen ion concentration, WT—water temperature, DO-dissolved oxygen, EC—electrical conductivity, BOD—biological oxygen demand, COD—chemical oxygen demand, TSS—total suspended solids, TN—total nitrogen, TP—total phosphorus, CHL—chlorophyll-a, SD—Secchi depth, TCB—total coliform bacteria.

Sites	pH	WT	DO	EC	BOD	COD	BOD/COD	TSS	TN	TP	TN:TP	CHL-a	SD	TCB
S1	8.19 ± 0.27	13.78 ± 0.82 ^{a}	11.18 ± 0.61	217.38 ± 31.78 ^{a}	1.62 ± 0.32 ^{a}	3.71 ± 0.50 ^{a}	0.43 ± 0.12 ^{abc}	8.22 ± 4.88 ^{a}	2.56 ± 0.21 ^{a}	50.38 ± 12.26 ^{a}	67.24 ± 15.24 ^{a}	17.46 ± 6.42 ^{a}	1.31 ± 0.25 ^{a}	2225.12 ± 2579.35
S2	8.27 ± 0.29	14.01 ± 0.73 ^{abcd}	11.28 ± 0.70	211.72 ± 27.12 ^{b}	1.73 ± 0.34 ^{b}	3.82 ± 0.53 ^{b}	0.45 ± 0.11 ^{abc}	7.8 ± 5.44 ^{b}	2.51 ± 0.19 ^{b}	49.45 ± 12.37 ^{b}	67.87 ± 16.59 ^{b}	18.64 ± 4.16 ^{b}	1.32 ± 0.20 ^{b}	2357.05 ± 4172.17
S3	8.02 ± 0.23	13.54 ± 0.74 ^{b}	11.04 ± 0.57	116.81 ± 28.96 ^{c}	1.19 ± 0.20 ^{c}	3.23 ± 0.41 ^{c}	0.38 ± 0.10 ^{a}	4.36 ± 1.43 ^{c}	1.88 ± 0.17 ^{c}	26.35 ± 7.34 ^{c}	99.65 ± 39.53 ^{c}	12.46 ± 2.79 ^{c}	1.73 ± 0.36 ^{c}	633.2 ± 650.72
S4	8.22 ± 0.37	14.59 ± 0.69 ^{c}	11.11 ± 0.78	214.53 ± 40.14 ^{d}	2.22 ± 0.44 ^{d}	4.65 ± 0.48 ^{d}	0.47 ± 0.10 ^{b}	8.5 ± 2.68 ^{d}	2.79 ± 0.35 ^{d}	69.99 ± 35.93 ^{d}	63.16 ± 22.64 ^{d}	26.24 ± 9.90 ^{d}	1.1 ± 0.18 ^{d}	1701.32 ± 2711.27
S5	8.14 ± 0.34	13.61 ± 0.76 ^{d}	10.79 ± 0.71	157.77 ± 24.11 ^{e}	1.27 ± 0.16 ^{e}	3.5 ± 0.35 ^{e}	0.37 ± 0.08 ^{c}	6.74 ± 3.10 ^{abcd}	2.16 ± 0.15 ^{e}	38.61 ± 10.63 ^{e}	77.1 ± 25.32 ^{e}	15.32 ± 3.67 ^{e}	1.49 ± 0.31 ^{e}	738.67 ± 697.12
Season														
Spring	8.41 ± 0.50 ^{a}	11.43 ± 4.77 ^{a}	12.16 ± 1.62 ^{a}	195.3 ± 33.66 ^{a}	2.09 ± 0.51 ^{a}	4.2 ± 0.72 ^{a}	0.49 ± 0.11 ^{a}	7.22 ± 4.02 ^{a}	2.57 ± 0.43 ^{a}	47.83 ± 23.86 ^{a}	79.55 ± 66.02 ^{a}	24.33 ± 11.90 ^{a}	1.3 ± 0.32 ^{a}	395.51 ± 732.87 ^{a}
Summer	8.13 ± 0.56 ^{b}	23.23 ± 2.51 ^{b}	9.09 ± 1.06 ^{b}	178.05 ± 36.10 ^{b}	1.76 ± 0.49 ^{b}	4.17 ± 0.80 ^{b}	0.42 ± 0.12 ^{b}	10.75 ± 12.49 ^{b}	2.24 ± 0.41 ^{b}	61.76 ± 29.20 ^{b}	48.83 ± 18.19 ^{b}	19.56 ± 10.84 ^{b}	1.21 ± 0.55 ^{b}	3421.38 ± 8492 ^{b}
Fall	8.04 ± 0.43 ^{c}	17.82 ± 4.71 ^{c}	9.9 ± 1.13 ^{c}	182.75 ± 40.21 ^{abc}	1.38 ± 0.41 ^{c}	3.63 ± 0.62 ^{c}	0.38 ± 0.13 ^{c}	7.07 ± 3.77 ^{c}	2.25 ± 0.31 ^{c}	48.46 ± 19.42 ^{c}	59.5 ± 21.58 ^{c}	16.14 ± 8.96 ^{c}	1.38 ± 0.39 ^{c}	1898.88 ± 7296 ^{abc}
Winter	8.09 ± 0.47 ^{d}	3.14 ± 1.83 ^{d}	13.18 ± 1.20 ^{d}	178.48 ± 39.83 ^{c}	1.2 ± 0.36 ^{d}	3.11 ± 0.43 ^{d}	0.38 ± 0.13 ^{d}	3.44 ± 1.42 ^{d}	2.46 ± 0.29 ^{d}	29.78 ± 16.08 ^{d}	112.13 ± 46.26 ^{d}	12.06 ± 8.02 ^{d}	1.67 ± 0.52 ^{d}	400.69 ± 927.97 ^{c}

^{a,b,c,d} Values with different superscript letters for each variable, indicate statistical differences among sites and seasons at $p < 0.05$ (Tukey HSD test).

The water quality parameters of the Paldang Reservoir showed significant heterogeneity ($p < 0.05$) among the four seasons (Table 1). Water temperature (WT), TSS, TP, and TCB exhibited significantly higher values in the summer, whereas pH, EC, BOD, COD, TN, and CHL-a were highest in the spring. TSS and TP concentrations were elevated during the summer due to intense precipitation (Supplementary Figure S1). The summer monsoon significantly influences the hydrology, nutrients, and suspended solids concentration in Korean reservoirs [35]. More intense rainfall contributes to increased TSS in Paldang Reservoir water. The daily loading data also showed that TP, TN, TSS, BOD, and COD were higher during the summer monsoon season than any other season (Supplementary Figure S2). This result supports the view that the summer monsoon is the main driver of high levels of nutrients, organic matter, and suspended solids in mid-latitude East Asian reservoirs, such as those in South Korea, Eastern China, and Japan [36]. The regression equation between TSS and TP indicates that TSS is associated with 45% of TP in Paldang Reservoir (Supplementary Figure S3). This result suggests that TSS may act as a nutrient carrier in the reservoir [37].

Organic matter (BOD and COD) in reservoirs can have either allochthonous or autochthonous origins. Allochthonous organic matter enters aquatic systems mainly via runoff derived from overland water flow during rainfall events, while autochthonous organic matter is produced through photosynthesis by phytoplankton and hydrophytes [32]. As Paldang Reservoir is a river-type reservoir, it experiences high flow rates during the summer season, and large amounts of allochthonous organic matter is introduced into the reservoir. Park et al. [32] showed that 69% of the total organic matter was allochthonous in Paldang Reservoir during the summer season. In contrast, a high autochthonous organic matter load was observed in the winter and spring due to low flow rates and increased WRT [32]. Park et al. [30] revealed that 73% of autochthonous organic matter loading occurs during the spring. The peak organic matter concentration coincides with the maximum production of algae (spring bloom). This finding suggests that autochthonous production by phytoplankton (CHL-a) during the spring period is critical to organic matter buildup in Paldang Reservoir; thus, the threat to the water quality of Paldang Reservoir is greatest in spring [30,32].

In the present study, the water quality status of each sampling site and season was assessed by comparing the mean values of water quality parameters with those listed in the Korean Lakes and Reservoirs Surface Water Quality Regulation, 2015 (Supplementary Table S1). As shown in Table S1, Site S4 had class III water quality (contaminated water), while all other sites had class II water quality (lightly contaminated water) in terms of COD. Based on TSS, all sites fell into the class III water quality category except site S3. Sites S1 and S4 were in the class IV water quality (somewhat poor; contaminated water) in terms of TP. Site S4 had class IV water quality in terms of CHL-a. During spring, algal growth increased in the reservoir, and water quality was somewhat poor (class IV; contaminated water). TP concentrations were higher during summer due to surface runoff, and the water body was in class IV. All sites and seasons had class Ia water quality in terms of pH and DO. Notably, Site S5 (near the water intake tower) was in class III (average; contaminated water) in terms of CHL-a, TSS, and TP.

3.2. Correlation Analysis

Pearson correlation analysis was used to evaluate the relationships among 12 water quality parameters (Supplementary Table S2). As anticipated, DO was negatively associated with WT, as oxygen is more soluble in cold water [1]. High BOD and COD levels indicate organic pollution in the reservoir [34,35]. Increasing nutrient concentrations (TP and TN) lead to elevated organic matter concentrations (BOD and COD) in the reservoir [1,15]. EC showed significant positive relationships with BOD ($r = 0.40$, $p < 0.01$), COD ($r = 0.59$, $p < 0.01$), and TN ($r = 0.55$, $p < 0.01$). TN, TP, and BOD showed strong positive relationships with each other, demonstrating that their sources were analogous. TSS showed a significant positive relationship with TP ($r = 0.47$, $p < 0.01$), indicating that suspended particles have a tendency to adsorb P [38]. During rainfall events and stream bank erosion in high-flow periods, agricultural and industrial runoff can contribute to high levels of TSS and TP in the reservoir [38]. CHL-a was positively correlated with TP ($r = 0.70$, $p < 0.01$) and TN ($r = 0.53$, $p < 0.01$), which are key factors affecting phytoplankton growth in this water body [39]. The reservoir's water clarity decreased with increase in the TP, TN, CHL-a, TSS, and BOD concentrations.

3.3. Annual Variations of Water Quality

Annual data can provide information about long-term water quality dynamics in Paldang Reservoir. The results showed that TP ($R^2 = 0.27$, $p < 0.01$), BOD ($R^2 = 0.26$, $p < 0.01$), and CHL-a ($R^2 = 0.33$, $p < 0.01$) have decreased significantly since 1996 (Figure 2). The loading data for TP ($R^2 = 0.21$, $p < 0.02$), TN ($R^2 = 0.19$, $p < 0.03$), and BOD ($R^2 = 0.35$, $p < 0.00$) also showed a decreasing pattern in Paldang Reservoir (Supplementary Figure S4). COD ($R^2 = 0.63$, $p < 0.01$) and SD ($R^2 = 0.37$, $p < 0.01$) have increased in Paldang Reservoir since 1996. Moreover, the loading pattern for COD has changed. BOD concentrations in most Korean lakes and reservoirs are continuously decreasing, while COD concentrations have been increasing in most cases, indicating that high concentrations of nonbiodegradable organic matter in the influent may be inefficiently degraded by the biological effluent treatment process [30,40]. COD represents both biodegradable and nonbiodegradable organic pollution in water systems. However, increases in the COD level suggest increased nonbiodegradable organic loading from wastewater treatment plants (WWTPs) and urban sewage systems to the water body [41]. A previous study conducted in the United States found an increase in the occurrence and persistence of inorganic solid loading from a WWTP to a water body [42]. Industries may not strictly comply with environmental regulations, and thus may contribute large amounts of nonbiodegradable compounds to aquatic systems [43]. Water quality management strategies in Korean reservoirs likely need to be re-evaluated with a focus on water pollutant management, especially for organic matter.

3.4. Hydrology, Nutrients, and Chlorophyll-a

Inflow, outflow, and WRT are major drivers of the distributions of nutrients, suspended solids, and CHL-a in aquatic environments. Compared to TN and CHL-a, inflow, outflow, and WRT were more significant determinants of the concentrations of TP ($R^2 = 0.30$, 0.29, 0.30, $p < 0.01$) and TSS ($R^2 = 0.39$, 0.36, 0.39, $p < 0.01$) (Figure 3). The present findings were similar to previous studies in Korean reservoirs. Previous research in various parts of the world has shown that external loadings of TP and TSS are highly correlated with inflow, outflow, and WRT in the watershed, and this conclusion was supported by the present study [23,44,45]. Many studies have reported effects of WRT on algal growth in aquatic systems [46,47]. However, the results of the present study did not concur with some previous studies. Thus, WRT may not always be linked to algal growth in the reservoir. This may indicate that release of autochthonous nutrients regulates algal growth in Paldang Reservoir. Lee et al. [36] suggested that algal chlorophyll growth was influenced by nutrients in Paldang Reservoir.

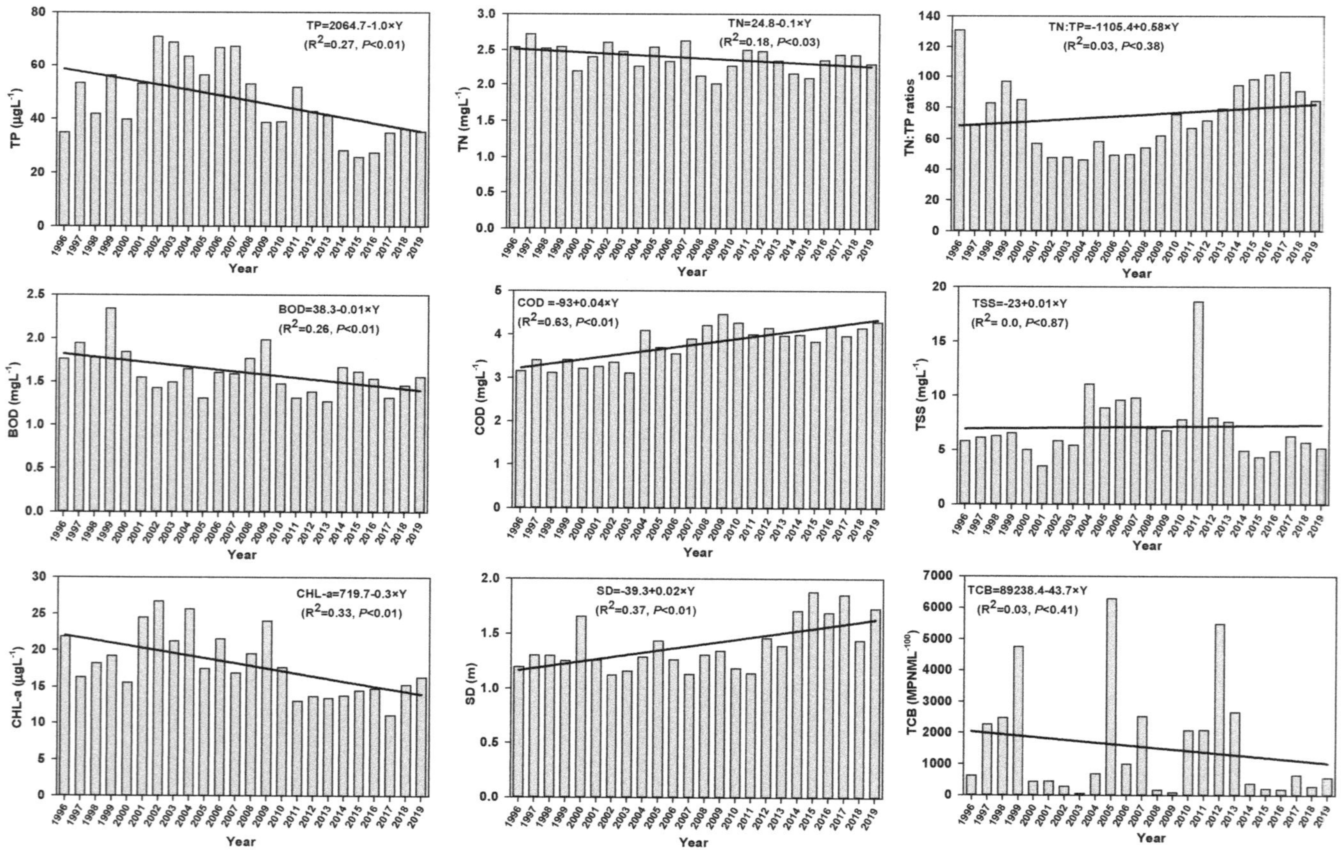

Figure 2. Yearly variation of water quality parameters in Paldang Reservoir. TP—total phosphorus, TN—total nitrogen, BOD—biological oxygen demand, COD—chemical oxygen demand, TSS—total suspended solids, CHL—chlorophyll-a, SD—Secchi depth, TCB—total coliform bacteria.

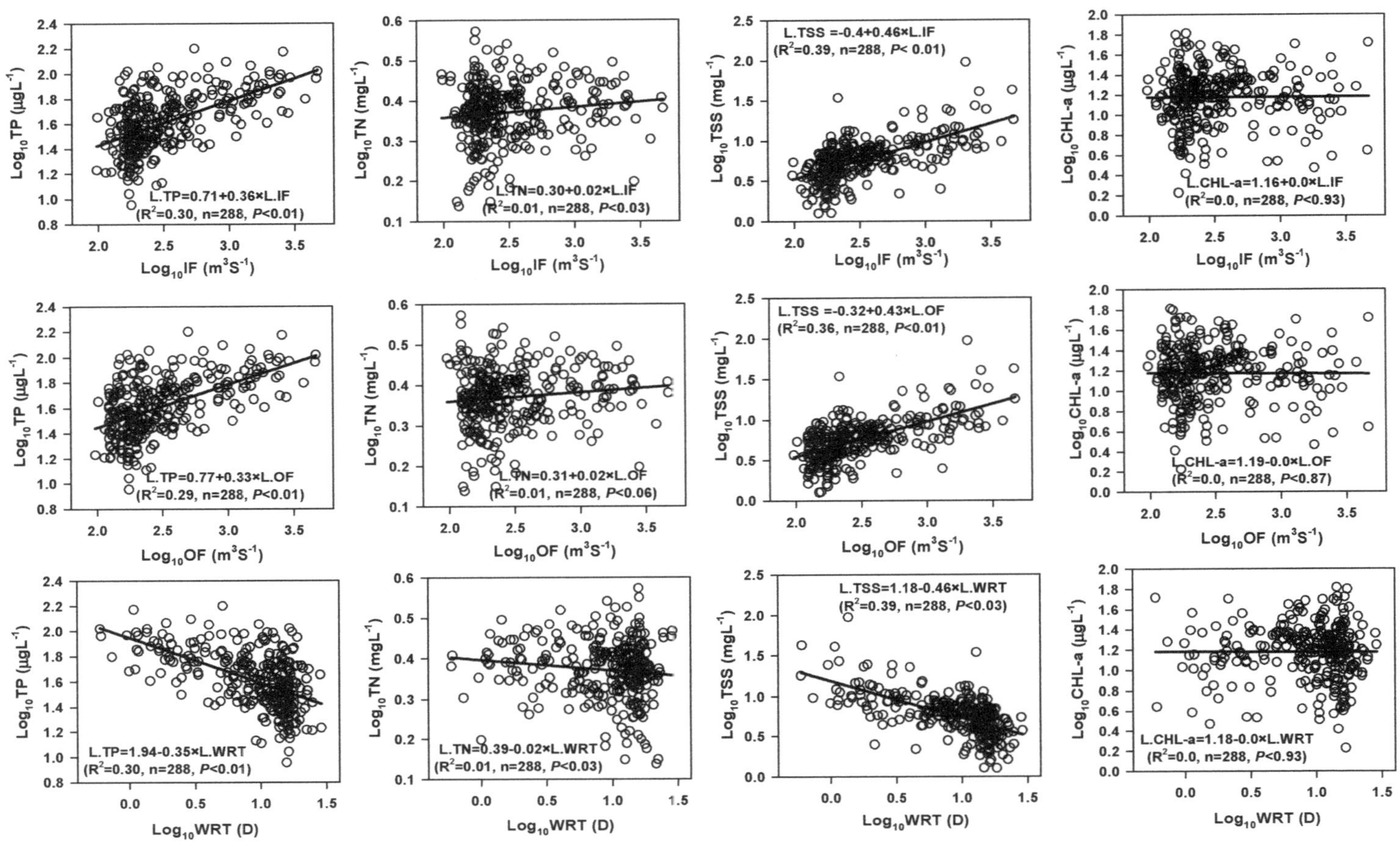

Figure 3. Influence of hydrology on TP, TN, TSS, and CHL-a. TP—total phosphorus, TN—total nitrogen, TSS—total suspended solids, CHL—chlorophyll-a.

The empirical models based on log-transformed CHL-a-TP and CHL-a-TN relationships are shown in Figure 4. Nutrients more strongly influenced chlorophyll growth in the reservoir's ambient water, and the concentration of TP (R^2 = 0.45, p < 0.01) better explained algal growth than that of TN (R^2 = 0.27, p < 0.01), indicating a P-limited system. When two predictors are strongly correlated (R^2 > 0.70), collinearity problems may arise that impede determination of the nutrient limiting algal growth. The present results showed that TP and TN (R^2 = 0.55) are moderately correlated in Paldang Reservoir. To avoid these problems, conditional plots have been used to identify limiting nutrients in aquatic systems [35,48]. Conditional plots showed that the association between CHL-a and TP was relatively steady in Paldang Reservoir, as indicated by the smooth lines on the four panels with similar slopes, which suggested that the effect of TP on CHL-a is consistent irrespective of the level of TN, in turn indicating a P-limited reservoir (Supplementary Figures S5 and S6). In addition, the conditional plot shows no interaction between TP and TN, further verifying that Paldang Reservoir is a P-limited system.

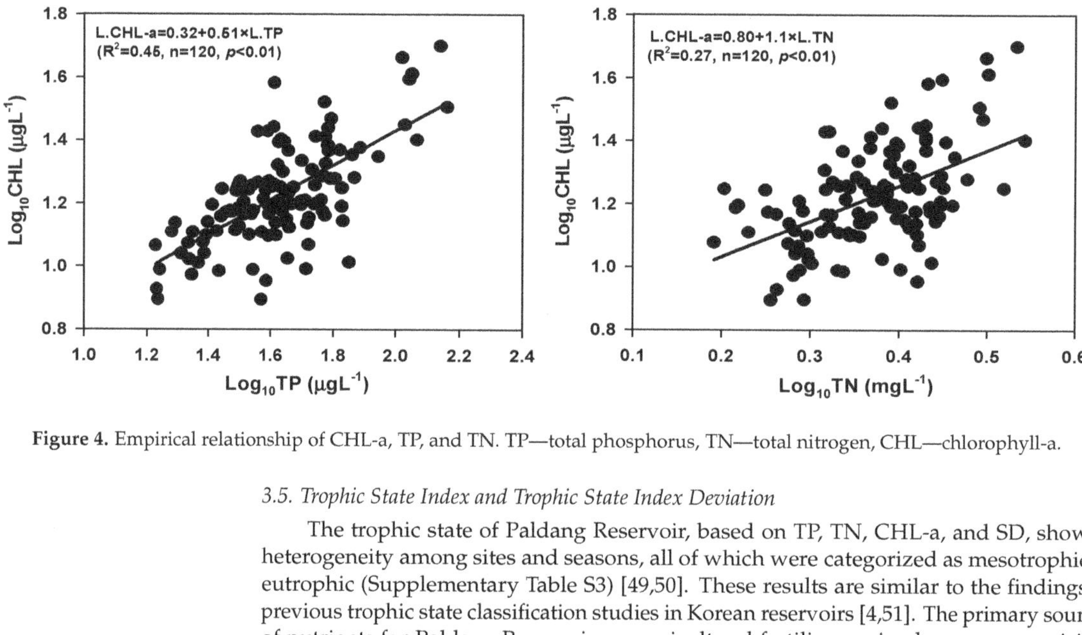

Figure 4. Empirical relationship of CHL-a, TP, and TN. TP—total phosphorus, TN—total nitrogen, CHL—chlorophyll-a.

3.5. Trophic State Index and Trophic State Index Deviation

The trophic state of Paldang Reservoir, based on TP, TN, CHL-a, and SD, showed heterogeneity among sites and seasons, all of which were categorized as mesotrophic to eutrophic (Supplementary Table S3) [49,50]. These results are similar to the findings of previous trophic state classification studies in Korean reservoirs [4,51]. The primary sources of nutrients for Paldang Reservoir are agricultural fertilizer, animal manure, municipal sewage, and industrial effluents [25]. Based on TP concentrations, all sites and seasons were under eutrophic conditions, except for Site S3 and the winter season. Notably, we found that Paldang Reservoir was in a eutrophic state in all sites and seasons, on the basis of TN, CHL-a, and SD. Considering the present results, measures should be taken to control eutrophication in Paldang Reservoir.

Assessing the potential of a water source to support cyanobacterial blooms or blue-green algae is essential for water resource management [52]. WT, TP, CHL-a, and SD are essential factors for determining potential cyanobacterial growth in a reservoir [53]. The concentrations of TP and CHL-a, along with SD, in Paldang Reservoir indicate a moderate level of risk for cyanobacterial exposure (Supplementary Table S4). CHL-a is a good indicator of overall phytoplankton biomass, and monitoring CHL-a is a direct method for semiquantitative estimation of cyanobacterial biomass in aquatic systems [20]. For South Korean reservoirs supplying drinking water, a cyanobacteria watch is issued when the concentration of CHL-a exceeds 15 µg L^{-1}. Furthermore, an alert is issued when the CHL-a concentration is greater than 25 µg L^{-1}. Once a watch or alert has been issued, additional treatment processes are required at water treatment plants until the watch or alert is cleared. Additionally, when an alert is issued, water intake below that at which algae can inhabit and analysis of cyanotoxin in the treated water, are required [54]. The results of the present study indicate that all sites and seasons (except site S3 and winter) were under watch conditions. Previous studies of Paldang Reservoir have suggested that cyanobacterial blooms occur during the spring season, which is in line with our findings [30,32].

Analysis of TSI and TSID provides valuable information on algal chlorophyll development, nutrient variability, and other parameters in lakes and reservoirs [4,12]. TSI and TSID were estimated based on TP, CHL-a, and SD in Paldang Reservoir, and their values showed spatial and seasonal variations (Figures 4 and 5). The mean TSI (TP), TSI (CHL-a), and TSI (SD) values indicate a eutrophic state during all seasons and at all sites (Figure 5, Supplementary Figure S7). These consistent eutrophic conditions may reduce DO and hamper ecosystem functions. The mean TSI (CHL-a) indicated more eutrophic conditions during spring and summer than the fall and winter. Water quality was worse in terms of TSI at Site S4 than other sites, and this site influenced water quality at the drinking water tower (Supplementary Figure S7; Site S5). Park et al. [32] revealed that Kyoungan Stream (Site S4) has a significant impact on the quality of drinking water in Paldang Reservoir.

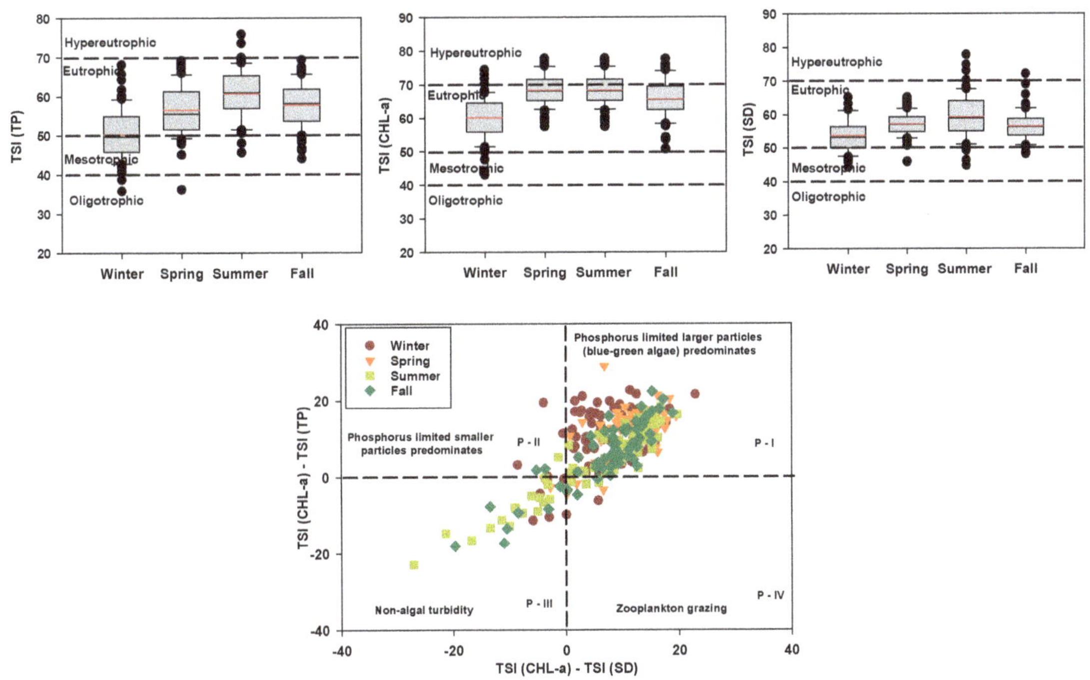

Figure 5. Seasonal Trophic State Index and its deviation.

Analysis of TSID showed that blue-green algae were predominant in the reservoir during all four seasons based on the relationships of TSI (CHL-a) with TSI (SD) and TSI (TP) (Figure 5). Blooms of blue-green algae are associated with eutrophic conditions [18]. Previous research identified the following genera of cyanobacteria in Paldang Reservoir: *Anabaena, Aphanocapsa, Chroococcus, Coelosphaerium, Dactylococcopsis, Microcystis, Merismopedia, Phormidium, Oscillatoria,* and *Pseudoanabaena* [26]. The occurrence of cyanobacteria is affected by light, temperature, pH, and nutrients. The concentration of TP is a major factor influencing the cyanobacterial contribution to total algal biomass [55]. Moreover, the biomass of cyanobacterial genera, such as *Aphanizomenon*, *Anabaena*, and *Microcystis*, is strongly influenced by the levels of TP and TN [55].

Nonalgal turbidity was observed during the summer and fall due to surface runoff from the watershed. Such turbidity is a common occurrence in Asian lakes and reservoirs during the monsoon period [35,56]. Small amounts of zooplankton grazing and P-limited small particles were observed in the reservoir. In addition, the TSID data indicated that TSI (CHL-a) was significantly higher than TSI (TP) during spring and winter, demonstrating that algal productivity was higher than expected and highlighting the controlling effect of P [4,18]. The water's trophic state must remain oligotrophic to mesotrophic for drinking water purposes according to the United States Environmental Protection Agency and Korean Ministry of Environment guidelines. The reservoir water intake towers face substantial bloom problems, impeding access to the water supply for local residents.

3.6. Discriminant Analysis

To study spatial and seasonal variations of water quality, DA was performed on the raw dataset. The spatial discriminant functions (DFs) and classification matrixes (CMs) used in this study are provided in Tables 2 and S5, respectively. Spatial standard and stepwise DFs with 14 and 8 discriminant variables, respectively, were used to generate CMs, which assigned 100% of cases correctly (Tables 2 and S5). The stepwise spatial DA results suggest that WT, DO, EC, COD, BOD/COD, TN, TN:TP, and TCB are the most important variables for explaining spatial variations in water quality in Paldang Reservoir among the five sites. The DFs indicated that WT, COD, BOD/COD, and TN were higher at Site S4 than other sites due to industrial and domestic wastewater effluents. These results are in accordance with previous findings in Paldang Reservoir [30]. Therefore, the spatial variations of these water quality parameters were mainly related to anthropogenic activities in the watershed.

Table 2. Classification functions for discriminant analysis of spatial variations in water quality of the reservoir. pH—hydrogen ion concentration, WT—water temperature, DO—dissolved oxygen, EC—electrical conductivity, BOD—biological oxygen demand, COD—chemical oxygen demand, TSS—total suspended solids, TN—total nitrogen, TP—total phosphorus, CHL—chlorophyll-a, SD—Secchi depth, TCB—total coliform bacteria.

Variables	Standard Mode					Stepwise Mode				
	Sites					Sites				
	S1	S2	S3	S4	S5	S1	S2	S3	S4	S5
pH	68.57	68.32	61.35	61.71	66.49					
WT	64.79	65.87	67.94	71.54	67.04	49.92	50.85	53.26	54.03	52.24
DO	76.22	76.18	73.18	75.920	72.93	41.12	40.93	36.90	39.34	37.81
EC	−0.30	−0.34	−0.72	−0.45	−0.56	0.008	−0.02	−0.37	−0.10	−0.23
BOD	−848.28	−851.76	−842.51	−878.11	−842.64					
COD	490.21	493.72	483.60	519.40	485.98	29.33	31.11	28.17	42.95	29.29
BOD/COD	3179.55	3193.22	3094.43	3308.40	3115.40	−0.84	0.36	−64.35	16.47	−42.13
TSS	−2.08	−2.07	−2.06	−2.23	−1.928					
TN	207.13	207.32	189.37	220.82	196.23	148.86	148.83	131.73	161.47	138.90
TP	1.37	1.37	1.37	1.42	1.35					
TN:TP	−1.35	−1.34	−1.10	−1.35	−1.20	−0.42	−0.41	−0.18	−0.41	−0.28
CHL-a	−4.95	−4.92	−4.45	−4.92	−4.67					
SD (m)	88.44	88.91	98.74	83.91	92.50					
TCB	−0.001	−0.001	−0.002	−0.001	−0.001	−0.0005	−0.0004	−0.001	−0.0008	−0.001
(Constant)	−2295.99	−2316.61	−2130.01	−2456.2	−2187.23	−807	−818.77	−691.97	−919.18	−724.82

Fisher's linear discriminant functions.

Seasonal DFs and CMs are shown in Tables 3 and S6, respectively, and were used to evaluate seasonal changes in water quality in Paldang Reservoir. Seasonal standard and stepwise mode DFs using 14 and 3 discriminant variables, respectively, generated CMs that assigned 100% of cases correctly (Tables 3 and S6). Temporal stepwise DA findings showed that WT, BOD, and TSS were the most important factors in temporal variations in the water quality of Paldang Reservoir among the four seasons. The DFs indicated that WT and TSS were higher during summer than other seasons. WT was highest in summer and lowest in winter due to the impact of climate seasonality [1]. TSS concentrations were higher during summer due to summer monsoon effects [35]. In contrast, BOD was highest in spring. Previous research revealed elevated organic matter concentrations during spring in Paldang Reservoir [32].

Table 3. Classification functions for discriminant analysis of seasonal variations in water quality of the reservoir. pH—hydrogen ion concentration, WT—water temperature, DO—dissolved oxygen, EC—electrical conductivity, BOD-biological oxygen demand, COD—chemical oxygen demand, TSS—total suspended solids, TN—total nitrogen, TP—total phosphorus, CHL—chlorophyll-a, SD—Secchi depth, TCB—total coliform bacteria.

Variables	Standard Mode				Stepwise Mode			
	Season				Season			
	Spring	Summer	Fall	Winter	Spring	Summer	Fall	Winter
pH	38.16	37.93	37.88	38.11				
WT	4.16	4.99	4.64	3.66	0.60	1.54	1.18	0.08
DO	17.83	17.41	17.42	17.79				
EC	0.18	0.17	0.17	0.16				
BOD	−207.80	−209.89	−210.07	−209.58	9.80	6.63	5.21	6.03
COD	102.58	102.18	101.04	101.13				
BOD/COD	713.31	711.79	708.23	708.02				
TSS	−1.00	−0.96	−1.00	−1.06	0.26	0.32	0.22	0.14
TN	47.09	45.11	45.15	45.83				
TP	0.52	0.54	0.51	0.50				
TN:TP	−0.09	−0.09	−0.09	−0.08				
CHL-a	−1.07	−1.07	−1.06	−1.07				
SD (m)	21.79	22.30	22.13	23.01				
TCB	−0.0002	−0.0001	−0.0001	−0.0001				
(Constant)	−553.79	−551.02	−535.15	−532.22	−16.02	−26.89	−16.30	−5.36

Fisher's linear discriminant functions.

Varol et al. [2] studied surface water quality variations in Keban Reservoir, Turkey, using the DA method, and found that eight and three variables successfully explained the temporal and spatial variations, respectively, among 19 water quality parameters. Chen et al. [14] studied surface water quality variations in Danjiangkou Reservoir, China, using the DA method, and their results indicated that six and four variables effectively explained spatial and temporal variations, respectively, among 11 water quality parameters. Mustapha et al. [57] studied surface water quality variations in the upper reach of the Kano River, Nigeria, using the DA method and successfully identified 7 variables, among 23 tested, having a statistically significant effect on the spatial variations. Singh et al. [9] showed that DA allows for data reduction, where only six and two variables were sufficient to discriminate spatial and temporal variations, respectively, in the Gomti River, India. Similarly, Zhang et al. [58] applied this method to evaluate spatial-temporal variations of water quality in the southwest New Territories and Kowloon, Hong Kong, and revealed that four and eight parameters could support 84.2% and 96.1% correct assignment in temporal and spatial analysis, respectively [58]. Furthermore, they suggested that the number of monitoring variables (and the associated cost) could be reduced, as their method allowed for considerable reduction of the dimensionality of the large dataset. Overall, DA led to a considerable reduction in the present research dataset and helped determine the parameters responsible for spatial and temporal variations.

3.7. Principal Component Analysis Combined with Factor Analysis (PCA/FA)

Urbanization, domestic sewage, industrial wastewater effluents, intensive agricultural activities, and waste from animal farms and inflowing rivers are the primary sources of water pollution in Paldang Reservoir. Bartlett's test and KMO were performed to examine the suitability of the data for PCA/FA. In the present study, the KMO value was 0.59, and Bartlett's test was significant ($p < 0.000$), indicating that the Paldang Reservoir data were suitable for PCA/FA and that meaningful relationships were present among the water quality variables. PCA/FA with varimax rotation identified five varifactors (VFs), which explained 82.32% of the total variance (Table 4). Varifactor 1 (VF1) represented 25.82% of the total variance and showed a strong positive loading (>0.70) for TP, strong negative loadings for TN:TP and SD, and moderate positive loadings (between 0.5 and 0.7) for TSS, TN, and CHL-a (Table 4). This VF represents inputs of nutrients and suspended matter from untreated domestic sewage, industrial effluents, and agricultural runoff. Nutrient inputs influence algal growth in Paldang Reservoir. The negative contribution of SD to this VF is related to high levels of nutrients, suspended solids, and algal growth [4,11]. VF2 showed strong positive loadings for pH and BOD/COD, and a moderate positive loading for BOD. This VF represents organic matter concentrations in the reservoir. VF3 (17.85% of the total variance) showed strong positive loadings for WT, EC, and COD and a moderate positive loading for BOD. This VF indicates the contributions of ions and organic matter input to the reservoir from untreated domestic sewage, industrial effluents, and agricultural runoff. VF4 (9.65 of the total variance) had a strong positive loading for DO, while VF5 (9.61% of the total variance) displayed a strong positive loading for TCB. The PCA/FA findings suggest that most of the variation in reservoir water quality can be attributed to nutrients and organic matter (anthropogenic), suspended solids (both natural and anthropogenic), and ionic concentrations (both natural and anthropogenic), which are regulated by both natural and anthropogenic activities.

Table 4. Varimax rotated component matrix for water quality parameters (Kaiser–Meyer–Olkin (KMO) = 0.59, Bartlett's test was significant (p = 0.000), extraction method: principal component analysis, and rotation method: varimax with Kaiser normalization, and bold and italic values represent strong and moderate loadings, respectively). pH—hydrogen ion concentration, WT—water temperature, DO—dissolved oxygen, EC—electrical conductivity, BOD-biological oxygen demand, COD—chemical oxygen demand, TSS—total suspended solids, TN—total nitrogen, TP-total phosphorus, CHL—chlorophyll-a, SD—Secchi depth, TCB—total coliform bacteria.

Variables	Components				
	VF1	VF2	VF3	VF4	VF5
pH	0.2	**0.81**	−0.08	0.07	−0.14
WT	−0.27	0.47	**0.71**	−0.17	−0.14
DO	0.001	0.11	0.02	**0.95**	0.03
EC	0.12	−0.02	**0.87**	−0.05	0.08
BOD	0.32	*0.64*	*0.53*	0.22	0.09
COD	0.34	−0.22	**0.78**	0.30	−0.04
BOD/COD	0.09	**0.92**	0.01	−0.02	0.16
TSS	*0.60*	−0.35	0.04	0.41	0.33
TN	*0.60*	0.24	0.46	−0.13	0.44
TP	**0.88**	0.20	0.15	−0.04	0.09

Table 4. *Cont.*

Variables	Components				
	VF1	VF2	VF3	VF4	VF5
TN:TP	**−0.84**	0.02	0.01	0.01	0.05
CHL-a	*0.68*	0.44	0.29	−0.001	−0.20
SD (m)	**−0.75**	−0.27	−0.06	−0.22	−0.24
TCB	0.06	−0.01	−0.01	0.06	0.91
Eigenvalues	3.61	2.71	2.50	1.35	1.34
Percentage of variance	25.82	19.37	17.85	9.65	9.61
Cumulative percentage	25.82	45.19	63.05	72.70	82.32

PCA/FA is a dimension-reduction technique that provides information about the most significant factors through simplification of the data. Therefore, this method has been utilized in various studies exploring the pollution sources affecting a water system. For example, PCA/FA was employed by Lim et al. [59] to identify sources of pollution in the Langat River, Malaysia. Four components were extracted in group 1, explaining 85% of the total variance, while six components were extracted in group 2, explaining 88% of the total variance. Based on these data, they determined that seawater intrusion, agricultural and industrial pollution, and geological weathering were mainly responsible for the river pollution. In addition, Tanriverdi et al. [60] used PCA/FA to analyze and assess the surface water quality of Ceyhan River and suggested that stations near cities were strongly affected by household wastewater, while other stations were influenced by agricultural facilities. Moreover, Jha et al. [61] identified major pollution sources influencing physicochemical variables in Aerial Bay, Andaman Islands, using the FA technique, which included rivulet flux into the bay, land run-off, prevailing biological processes, and tidal flow. Haji Gholizadeh et al. [11] identified five and four potential pollution sources to the Miami Canal in South Florida during the wet and dry seasons, respectively, which affected water quality variables. PCA/FA was used to distinguish four potential pollution types, namely, organic pollution, nutrient pollution, chemical pollution, and natural pollution, in Danjiangkou Reservoir, China, revealing that the study area was primarily influenced by industrial effluent and domestic sewage [14]

4. Conclusions

MSTs, TSI, and TSID were combined to assess the water quality of Paldang Reservoir. All variables except pH, DO, and TCB showed significant spatial variations due to the effects of anthropogenic activities. The mean values of TSI (TP), TSI (CHL-a), and TSI (SD) indicated a eutrophic state, and TSID showed that blue-green algae dominated the reservoir. PCA/FA results revealed that the concentrations of TP, TN, BOD, COD, TSS, and EC were generally linked to both anthropogenic activities and natural processes. Stepwise DA provided better results for both spatial and temporal analyses. Thus, this study demonstrated that MSTs, TSI, and TSID are effective approaches for evaluating reservoir water quality, and that these methods can be used in combination as useful water quality management tools. Relative to US EPA and MOE guidelines, the reservoir is in a eutrophic state in terms of CHL-a, which is unfavorable for drinking purposes. To improve the water quality of this reservoir, nutrient and organic matter loads from the watershed should be limited.

Supplementary Materials: The following are available online at https://www.mdpi.com/2073-4441/13/2/186/s1, Figure S1: Seasonal and Total rainfall pattern of Paldang watershed (Spring: March–May, Summer: June–August, Fall: September–November, Winter: December–February, and TRF: total rainfall), Figure S2: Loading Data of TP, TN, TSS, BOD, and COD in the Paldang Reservoir (TP—total phosphorus, TN—total nitrogen, TSS—total suspended solids, BOD—biological oxygen demand, COD—chemical oxygen demand, Spring: March–May, Summer: June–August, Fall: September–November, and Winter: December–February), Figure S3: Empirical relations among TSS, TP, and TN (TP—total phosphorus, TN—total nitrogen, TSS—total suspended solids), Figure S4: Yearly loading data of TP, TN, TSS, BOD, COD (TP—total phosphorus, TN—total nitrogen, TSS—total suspended solids, BOD—biological oxygen demand, COD—chemical oxygen demand), Figure S5: The relationship between CHL-a and TP is plotted conditional on the range of TN (TP—total phosphorus, TN—total nitrogen, CHL—chlorophyll-a), Figure S6: The relationship between CHL-a and TN is plotted conditional on the range of TP (TP—total phosphorus, TN—total nitrogen, CHL—chlorophyll-a), Figure S7: Trophic State Index of Paldang Reservoir at five different sites, Table S1: Water quality classes of Paldang Reservoir based on sites and seasons according to the Korean Ministry of Environment water quality standards for reservoirs and lakes (pH—hydrogen ion concentration, DO—dissolved oxygen, COD—chemical oxygen demand, TSS-total suspended solids, TN—total nitrogen, TP—total phosphorus, CHL—chlorophyll-a, TCB—total coliform bacteria, Ia: very good (high-quality water), Ib: good (high-quality water), II: somewhat good (lightly contaminated water), III: average (contaminated water), IV: somewhat poor (contaminated water), V: poor (highly contaminated water), VI: very poor (highly contaminated water)), Table S2: Pearson correlation analysis of water quality parameters (units mg L^{-1}, except pH, WT (°C), EC (μS cm^{-1}), TP (μg L^{-1}), CHL-a (μg L^{-1}), SD (m), and TCB ($MPNML^{-100}$)). pH—hydrogen ion concentration, WT—water temperature, DO—dissolved oxygen, EC—electrical conductivity, BOD—biological oxygen demand, COD—chemical oxygen demand, TSS—total suspended solids, TN—total nitrogen, TP—total phosphorus, CHL—chlorophyll-a, SD—Secchi depth, TCB—total coliform bacteria, Table S3: Trophic state criteria based on TP, TN, CHL-a, and SD from Nurnberg (1996) for Paldang Reservoir (TN—total nitrogen, TP—total phosphorus, CHL—chlorophyll-a, SD—Secchi depth, M: mesotrophic, E: eutrophic, and H: Hypereutrophic), Table S4: Thresholds of risk associated with potential exposure to cyanobacteria in Paldang Reservoir (adopted from WHO, 2015, LRE: lower risk of exposure, MRE: moderate risk of exposure and HRE: higher risk of exposure, TP—total phosphorus, CHL—chlorophyll-a, SD—Secchi depth), Table S5: Classification matrix for discriminant analysis of spatial variations in water quality of the reservoirs, Table S6: Classification matrix for discriminant analysis of seasonal variations in water quality of the reservoirs.

Author Contributions: Conceptualization, M.M.; methodology, M.M.; software, M.M. and J.Y.K.; formal analysis, M.M.; data curation, M.M. and J.Y.K.; writing—original draft preparation, M.M.; writing—review and editing, M.M. and K.-G.A.; visualization, M.M. and K.-G.A.; supervision, K.-G.A.; funding acquisition, K.-G.A. All authors have read and agreed to the published version of the manuscript.

Funding: This work was supported by the "Korea Environment Industry & Technology Institute (KEITI)" through the "Aquatic Ecosystem Conservation Research Program" funded by the Korean Ministry of Environment (Grant number: 2020003050004).

Institutional Review Board Statement: Not applicable.

Informed Consent Statement: Not applicable.

Data Availability Statement: The datasets presented in this study are available on reasonable request from the corresponding author.

Acknowledgments: The authors would like to acknowledge the Korean Ministry of Environment for their assistance.

Conflicts of Interest: The authors declare that they have no conflicts of interest.

References

1. Varol, M. Use of water quality index and multivariate statistical methods for the evaluation of water quality of a stream affected by multiple stressors: A case study. *Environ. Pollut.* **2020**, *266*, 115417. [CrossRef] [PubMed]
2. Varol, M. Spatio-temporal changes in surface water quality and sediment phosphorus content of a large reservoir in Turkey. *Environ. Pollut.* **2020**, *259*, 113860. [CrossRef] [PubMed]
3. Varol, M.; Gökot, B.; Bekleyen, A.; Şen, B. Spatial and temporal variations in surface water quality of the dam reservoirs in the Tigris River basin, Turkey. *Catena* **2012**, *92*, 11–21. [CrossRef]
4. Mamun, M.; Kwon, S.; Kim, J.E.; An, K.G. Evaluation of algal chlorophyll and nutrient relations and the N:P ratios along with trophic status and light regime in 60 Korea reservoirs. *Sci. Total Environ.* **2020**, *741*, 140451. [CrossRef] [PubMed]

5. Bhat, S.A.; Meraj, G.; Yaseen, S.; Pandit, A.K. Statistical Assessment of Water Quality Parameters for Pollution Source Identification in Sukhnag Stream: An Inflow Stream of Lake Wular (Ramsar Site), Kashmir Himalaya. *J. Ecosyst.* **2014**, *2014*, 1–18. [CrossRef]
6. Mamun, M.; An, K.G. Stream health assessment using chemical and biological multi-metric models and their relationships with fish trophic and tolerance indicators. *Ecol. Indic.* **2020**. [CrossRef]
7. Koçer, M.A.T.; Sevgili, H. Parameters selection for water quality index in the assessment of the environmental impacts of land-based trout farms. *Ecol. Indic.* **2014**, *36*, 672–681. [CrossRef]
8. Singh, K.P.; Malik, A.; Mohan, D.; Sinha, S. Multivariate statistical techniques for the evaluation of spatial and temporal variations in water quality of Gomti River (India)—A case study. *Water Res.* **2004**, *38*, 3980–3992. [CrossRef]
9. Singh, K.P.; Malik, A.; Sinha, S. Water quality assessment and apportionment of pollution sources of Gomti river (India) using multivariate statistical techniques—A case study. *Anal. Chim. Acta* **2005**, *538*, 355–374. [CrossRef]
10. Su, S.; Li, D.; Zhang, Q.; Xiao, R.; Huang, F.; Wu, J. Temporal trend and source apportionment of water pollution in different functional zones of Qiantang River, China. *Water Res.* **2011**, *45*, 1781–1795. [CrossRef]
11. Haji Gholizadeh, M.; Melesse, A.M.; Reddi, L. Water quality assessment and apportionment of pollution sources using APCS-MLR and PMF receptor modeling techniques in three major rivers of South Florida. *Sci. Total Environ.* **2016**, *566–567*, 1552–1567. [CrossRef] [PubMed]
12. Carlson, R.E. A trophic state index for lakes. *Limnol. Oceanogr.* **1977**, *22*, 361–369. [CrossRef]
13. Chen, J.; Li, F.; Fan, Z.; Wang, Y. Integrated application of multivariate statistical methods to source apportionment ofwatercourses in the liao river basin, northeast China. *Int. J. Environ. Res. Public Health* **2016**, *13*, 1035. [CrossRef] [PubMed]
14. Chen, P.; Li, L.; Zhang, H. Spatio-temporal variations and source apportionment of water pollution in Danjiangkou Reservoir Basin, Central China. *Water* **2015**, *7*, 2591–2611. [CrossRef]
15. Atique, U.; An, K.G. Landscape heterogeneity impacts water chemistry, nutrient regime, organic matter and chlorophyll dynamics in agricultural reservoirs. *Ecol. Indic.* **2020**, *110*, 105813. [CrossRef]
16. Markad, A.T.; Landge, A.T.; Nayak, B.B.; Inamdar, A.B.; Mishra, A.K. Trophic state modeling for shallow freshwater reservoir: A new approach. *Environ. Monit. Assess.* **2019**, *191*. [CrossRef]
17. Jones, J.R.; Knowlton, M.F.; An, K.G. Trophic state, seasonal patterns and empirical models in South Korean Reservoirs. *Lake Reserv. Manag.* **2003**, *19*, 64–78. [CrossRef]
18. Carlson, R.E. Expanding the Trophic State Concept to Identify Non-Nutrient Limited Lakes and Reservoirs. In *Proceedings of a National Conference on Enhancing the States' Lake Management Programs*; North American Lake Management Society: Chicago, IL, USA, 1991; pp. 59–71.
19. Mamun, M.; An, K.G. Major nutrients and chlorophyll dynamics in Korean agricultural reservoirs along with an analysis of trophic state index deviation. *J. Asia-Pac. Biodivers.* **2017**, *10*, 183–191. [CrossRef]
20. Mamun, M.; Kim, J.J.; Alam, M.A.; An, K.G. Prediction of algal chlorophyll-a and water clarity in monsoon-region reservoir using machine learning approaches. *Water* **2020**, *12*, 30. [CrossRef]
21. An, K.G.; Jones, J.R. Factors regulating bluegreen dominance in a reservoir directly influenced by the asian monsoon. *Hydrobiologia* **2000**, *432*, 37–48. [CrossRef]
22. Mamun, M.; Lee, S.J.; An, K.G. Temporal and spatial variation of nutrients, suspended solids, and chlorophyll in Yeongsan watershed. *J. Asia-Pac. Biodivers.* **2018**, *11*, 206–216. [CrossRef]
23. Nadarajah, S.; Wijenayake, W.M.H.K.; Amarasinghe, U.S. Influence of hydrology on water quality and trophic state of irrigation reservoirs in Sri Lanka. *Lakes Reserv. Res. Manag.* **2019**, *24*, 287–298. [CrossRef]
24. Bai, X.; Ding, S.; Fan, C.; Liu, T.; Shi, D.; Zhang, L. Organic phosphorus species in surface sediments of a large, shallow, eutrophic lake, Lake Taihu, China. *Environ. Pollut.* **2009**, *157*, 2507–2513. [CrossRef] [PubMed]
25. Eun, H.N.; Seok, S.P. A hydrodynamic modeling study to determine the optimum water intake location in Lake Paldang, Korea. *J. Am. Water Resour. Assoc.* **2005**, *41*, 1315–1332. [CrossRef]
26. Boopathi, T.; Wang, H.; Lee, M.-D.; Ki, J.-S. Seasonal Changes in Cyanobacterial Diversity of a Temperate Freshwater Paldang Reservoir (Korea) Explored by using Pyrosequencing. *Environ. Biol. Res.* **2018**, *36*, 424–437. [CrossRef]
27. Lee, J.E.; Youn, S.J.; Byeon, M.; Yu, S.J. Occurrence of cyanobacteria, actinomycetes, and geosmin in drinking water reservoir in Korea: A case study from an algal bloom in 2012. *Water Supply* **2020**, *20*, 1862–1870. [CrossRef]
28. Kim, D.W.; Min, J.H.; Yoo, M.; Kang, M.; Kim, K. Long-term effects of hydrometeorological and water quality conditions on algal dynamics in the Paldang dam watershed, Korea. *Water Sci. Technol. Water Supply* **2014**, *14*, 601–608. [CrossRef]
29. Youn, S.J.; Kim, H.N.; Yu, S.J.; Byeon, M.S. Cyanobacterial occurrence and geosmin dynamics in Paldang Lake watershed, South Korea. *Water Environ. J.* **2020**, 1–10. [CrossRef]
30. Park, H.K.; Cho, K.H.; Won, D.H.; Lee, J.; Kong, D.S.; Jung, D. Il Ecosystem responses to climate change in a large on-river reservoir, Lake Paldang, Korea. *Clim. Chang.* **2013**, *120*, 477–489. [CrossRef]
31. Straskraba, M.; Tundisi, G. *Guidelines of Lake Management: Volume 9 Reservoir Water Quality Management*; International Lake Environment Committee: Kusatsu, Japan, 1999; Volume 9, ISBN 4906356265.
32. Park, H.K.; Byeon, M.S.; Shin, Y.N.; Jung, D. Il Sources and spatial and temporal characteristics of organic carbon in two large reservoirs with contrasting hydrologic characteristics. *Water Resour. Res.* **2009**, *45*, 1–12. [CrossRef]
33. Kerneïs, A.; Nakache, F.; Deguin, A.; Feinberg, M. The effects of water residence time on the biological quality in a distribution network. *Water Res.* **1995**, *29*, 1719–1727. [CrossRef]

34. Varol, M.; Gökot, B.; Bekleyen, A.; Şen, B. Water quality assessment and apportionment of pollution sources in Tigris River (Turkey) using multivariate statistical techniques. *River Res. Applic* **2012**, *28*, 1428–1438. [CrossRef]
35. Mamun, M.; Kim, J.Y.; An, K.G. Trophic responses of the Asian reservoir to long-term seasonal and interannual dynamic monsoon. *Water* **2020**, *12*, 2066. [CrossRef]
36. Lee, Y.; Ha, S.Y.; Park, H.K.; Han, M.S.; Shin, K.H. Identification of key factors influencing primary productivity in two river-type reservoirs by using principal component regression analysis. *Environ. Monit. Assess.* **2015**, *187*, 1–12. [CrossRef] [PubMed]
37. Mallin, M.A.; Cahoon, L.B. The Hidden Impacts of Phosphorus Pollution to Streams and Rivers. *Bioscience* **2020**, *70*, 315–329. [CrossRef]
38. Fox, G.A.; Purvis, R.A.; Penn, C.J. Streambanks: A net source of sediment and phosphorus to streams and rivers. *J. Environ. Manag.* **2016**, *181*, 602–614. [CrossRef]
39. Mamun, M.; Lee, S.J.; An, K.G. Roles of nutrient regime and N:P ratios on algal growth in 182 Korean agricultural reservoirs. *Pol. J. Environ. Stud.* **2018**, *27*, 1175–1185. [CrossRef]
40. Iloms, E.; Ololade, O.O.; Ogola, H.J.O.; Selvarajan, R. Investigating industrial effluent impact on municipal wastewater treatment plant in vaal, South Africa. *Int. J. Environ. Res. Public Health* **2020**, *17*, 1096. [CrossRef]
41. Lee, J.; Lee, S.; Yu, S.; Rhew, D. Relationships between water quality parameters in rivers and lakes: BOD5, COD, NBOPs, and TOC. *Environ. Monit. Assess.* **2016**, *188*, 1–8. [CrossRef]
42. Kolpin, D.W.; Furlong, E.T.; Meyer, M.T.; Thurman, E.M.; Zaugg, S.D.; Barber, L.B.; Buxton, H.T. Pharmaceuticals, hormones, and other organic wastewater contaminants in U.S. streams, 1999–2000: A national reconnaissance. *Environ. Sci. Technol.* **2002**, *36*, 1202–1211. [CrossRef]
43. Vizcaino, I.P.; Carrera, E.V.; Sanromán-Junquera, M.; Muñoz-Romero, S.; Rojo-Álvarez, J.L.; Cumbal, L.H. Spatio-temporal analysis of water quality parameters in machángara river with nonuniform interpolation methods. *Water* **2016**, *8*, 507. [CrossRef]
44. Mallin, M.A.; Johnson, V.L.; Ensign, S.H. Comparative impacts of stormwater runoff on water quality of an urban, a suburban, and a rural stream. *Environ. Monit. Assess.* **2009**, *159*, 475–491. [CrossRef] [PubMed]
45. Abell, J.M.; Özkundakci, D.; Hamilton, D.P.; Jones, J.R. Latitudinal variation in nutrient stoichiometry and chlorophyll-nutrient relationships in lakes: A global study. *Fundam. Appl. Limnol.* **2012**, *181*, 1–14. [CrossRef]
46. Kawara, O.; Yura, E.; Fujii, S.; Matsumoto, T. A study on the role of hydraulic retention time in eutrophication of the Asahi River Dam reservoir. *Water Sci. Technol.* **1998**, *37*, 245–252. [CrossRef]
47. Lee, S.; Lee, S.; Kim, S.H.; Park, H.; Park, S.; Yum, K. Examination of critical factors related to summer chlorophyll a concentration in the Sueo dam reservoir, Republic of Korea. *Environ. Eng. Sci.* **2012**, *29*, 502–510. [CrossRef] [PubMed]
48. Qian, S.S. *Environmental and Ecological Statistics with R*, 2nd ed.; Chapman and Hall: London, UK; CRC: Boca Raton, FL, USA, 2016; ISBN 9781498728720.
49. Nürnberg, G.K. Trophic state of clear and colored, soft- and hardwater lakes with special consideration of nutrients, anoxia, phytoplankton and fish. *Lake Reserv. Manag.* **1996**, *12*, 432–447. [CrossRef]
50. United States Environmental Protection Agency. *Guideline for Data Quality Assessment*; USEPA: Washington, DC, USA, 2007.
51. Jung, S.; Shin, M.; Kim, J.; Eum, J.; Lee, Y.; Lee, J.; Choi, Y.; You, K.; Owen, J.; Kim, B. The effects of Asian summer monsoons on algal blooms in reservoirs. *Inland Waters* **2016**, *6*, 406–413. [CrossRef]
52. World Health Organization. *Guidelines for Drinking Water Quality: Management of Cyanobacteria in Drinking Water Suppliers Information for Regulators and Water Suppliers*, 4th ed.; WHO: Geneva, Switzerland, 2015.
53. Miller, A.W. Trophic state evaluation for selected lakes in Yellowstone National Park, USA. *WIT Trans. Ecol. Environ.* **2010**, *135*, 143–155. [CrossRef]
54. Cha, Y.; Park, S.S.; Kim, K.; Byeon, M.; Stow, C.A. Probablistic prediction of cyanobacteria abundance in a Korean reservoir using a Bayesian Poisson model. *Water Resour. Res.* **2014**, *50*, 2518–2532. [CrossRef]
55. Wagner, C.; Adrian, R. Cyanobacteria dominance: Quantifying the effects of climate change. *Limnol. Oceanogr.* **2009**, *54*, 2460–2468. [CrossRef]
56. Zou, W.; Zhu, G.; Cai, Y.; Vilmi, A.; Xu, H.; Zhu, M.; Gong, Z.; Zhang, Y.; Qin, B. Relationships between nutrient, chlorophyll a and Secchi depth in lakes of the Chinese Eastern Plains ecoregion: Implications for eutrophication management. *J. Environ. Manag.* **2020**, *260*, 109923. [CrossRef] [PubMed]
57. Mustapha, A.; Aris, A.Z.; Yusoff, F.M.; Zakaria, M.P.; Ramli, M.F.; Abdullah, A.M.; Kura, N.U.; Narany, T.S. Statistical Approach in Determining the Spatial Changes of Surface Water Quality at the Upper Course of Kano River, Nigeria. *Water Qual. Expo. Health* **2014**, *6*, 127–142. [CrossRef]
58. Zhang, X.; Wang, Q.; Liu, Y.; Wu, J.; Yu, M. Application of multivariate statistical techniques in the assessment of water quality in the Southwest New Territories and Kowloon, Hong Kong. *Environ. Monit. Assess.* **2011**, *173*, 17–27. [CrossRef]
59. Lim, W.Y.; Aris, A.Z.; Praveena, S.M. Application of the chemometric approach to evaluate the spatial variation of water chemistry and the identification of the sources of pollution in Langat River, Malaysia. *Arab. J. Geosci.* **2013**, *6*, 4891–4901. [CrossRef]
60. Tanriverdi, Ç.; Alp, A.; Demirkiran, A.R.; Üçkardeş, F. Assessment of surface water quality of the Ceyhan River basin, Turkey. *Environ. Monit. Assess.* **2010**, *167*, 175–184. [CrossRef]
61. Jha, D.K.; Vinithkumar, N.V.; Sahu, B.K.; Das, A.K.; Dheenan, P.S.; Venkateshwaran, P.; Begum, M.; Ganesh, T.; Prashanthi Devi, M.; Kirubagaran, R. Multivariate statistical approach to identify significant sources influencing the physico-chemical variables in Aerial Bay, North Andaman, India. *Mar. Pollut. Bull.* **2014**, *85*, 261–267. [CrossRef]

water

MDPI

Article

Evaluation of the Gulf of Aqaba Coastal Water, Jordan

Ahmed A. Al-Taani [1,2,*], Maen Rashdan [2], Yousef Nazzal [1], Fares Howari [1], Jibran Iqbal [1], Abdulla Al-Rawabdeh [2], Abeer Al Bsoul [3] and Safaa Khashashneh [2]

1 College of Natural and Health Sciences, Zayed University, Abu Dhabi 144534, UAE; Yousef.Nazzal@zu.ac.ae (Y.N.); Fares.Howari@zu.ac.ae (F.H.); Jibran.Iqbal@zu.ac.ae (J.I.)

2 Department of Earth and Environmental Sciences, Yarmouk University, Irbid 21163, Jordan; maen.rashdan@gmail.com (M.R.); abd_rawabdeh@yu.edu.jo (A.A.-R.); safaa@yu.edu.jo (S.K.)

3 Department of Chemical Engineering, Al-Balqa Applied University, As-Salt 19117, Jordan; dr.abeeralbsoul@bau.edu.jo

* Correspondence: Ahmed.Al-Taani@zu.ac.ae

Received: 7 July 2020; Accepted: 21 July 2020; Published: 27 July 2020

Abstract: (1) Background: The Gulf of Aqaba (GoA) supports unique and diverse marine ecosystems. It is one of the highest anthropogenically impacted coasts in the Middle East region, where rapid human activities are likely to degrade these naturally diverse but stressed ecosystems. (2) Methods: Various water quality parameters were measured to assess the current status and conditions of GoA seawater including pH, total dissolved solids (TDS), total alkalinity (TA), Cl^-, NO_3^-, SO_4^{2-}, PO_4^{3-}, NH_4^+, Ca^{2+}, Mg^{2+}, Na^+, K^+, Sr, Cd, Co, Cr, Cu, Fe, Mn, Pb, and Zn. (3) Results: The pH values indicated basic coastal waters. The elevated levels of TDS with an average of about 42 g/L indicated highly saline conditions. Relatively low levels of inorganic nutrients were observed consistent with the prevalence of oligotrophic conditions in GoA seawater. The concentrations of Ca^{2+}, Mg^{2+}, Na^+, K^+, Sr, Cl^-, and SO_4^{2-} in surface layer varied spatially from about 423–487, 2246–2356, 9542–12,647, 513–713, 9.2–10.4, 22,173–25,992, and 317–407 mg/L, respectively. The average levels of Cd, Co, Cr, Cu, Fe, Mn, Pb and Zn ranged from 0.51, 0.38, 1.44, 1.29, 0.88, 0.38, and 6.05 µg/L, respectively. (4) Conclusions: The prevailing saline conditions of high temperatures, high evaporation rates, the water stratification and intense dust storms are major contributing factors to the observed seawater chemistry. The surface distribution of water quality variables showed spatial variations with no specific patterns, except for metal contents which exhibited southward increasing trends, closed to the industrial complex. The vast majority of these quality parameters showed relatively higher values compared to those of other regions.

Keywords: water quality; coastal area; metals; pollution source; Gulf of Aqaba; Jordan; Red Sea

1. Introduction

The Gulf of Aqaba (GoA) is the upper northeastern segment of the Red Sea [1]. It is a partially-isolated, narrow and deep coastal water body. The Strait of Tiran connects GoA with the Red Sea (Figure 1). Despite the extreme environmental conditions, the GoA supports unique aquatic ecosystems and biodiversity, and is a habitat for one of the world's richest coral communities [2,3].

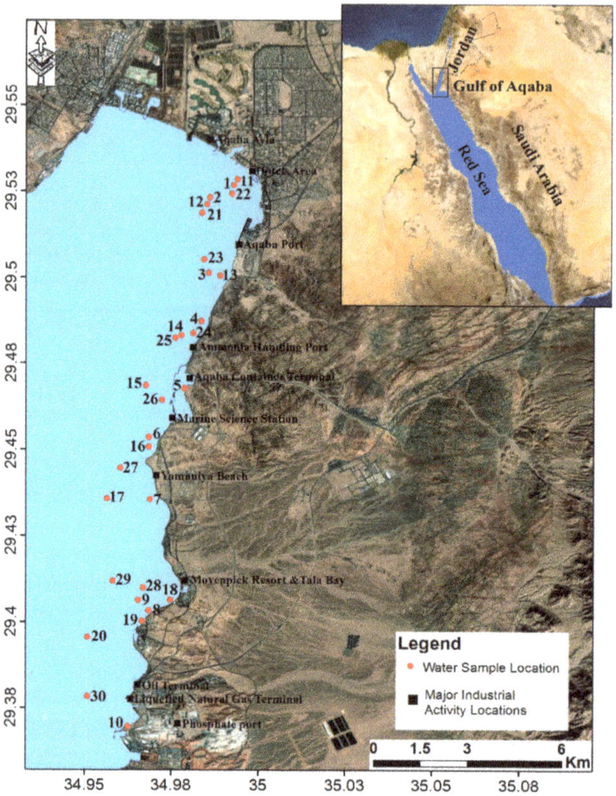

Figure 1. Location map of the Gulf of Aqaba (GoA) and sampling sites.

The GoA is one of the high anthropogenically impacted coasts in the Middle East region [4]. The expansion in economic and industrial activities in the Gulf's bordering countries have contributed to the degradation of naturally stressed coastal and marine ecosystems. They are subjected to various impacts and sources of pollution including dredging and reclamation activities, coastal construction development, industrial waste, ports, oil spills, and domestic sewage, among others [5].

GoA is the only marine port for Jordan, and is highly vulnerable to pollution, where all marine-related activities are concentrated within a few kilometers of the coast (27 km). In addition, many economic, industrial, and recreational activities are taking place along the Jordanian coastline, many of which are of potential environmental impacts [6–9]. Additionally, the region plans to have a number of large coastal projects (such as the Red-Dead Sea conduit, new resorts, and ports relocation), which will certainly accelerate the degradation cycle of existing environmental conditions and threaten these unique marine communities [7]. Signs of human impacts were reported [10–12].

In addition to human impacts, the GoA is subject to regular dust storm events that contribute metals and other chemicals to the GoA coastal water [8,9]. Aeolian dust flux to GoA is likely to influence seawater chemistry [13], where atmospheric dry deposition in the GoA is considered an important external source of trace metals [8,9,14–16]. The mineral dust rate on GoA region is one of the highest on Earth [8,9,13,17]. It is believed that the frequency of dust storm events will become more common in the GoA, due to increase in regional aridity and dust fluxes [18,19].

The relatively small volume and absence of significant wave action along with the low rate of water circulation and renewal (between GoA and the Red Sea), render the Gulf particularly vulnerable to pollution. The residence time of water in the Gulf averaged 1–3 years [20,21].

The growing concern over the sustainability of these unique aquatic ecosystems of GoA has recently gained momentum and became a priority issue in Jordan. The impact of intense and widespread human activities from the Gulf's bordering countries poses imminent threats to GoA coast, which requires a proper monitoring plan. The objective of the present study is to assess the current status of surface water quality along the coastal region of the Jordanian GoA coast. It also intends to evaluate the spatial distribution of a variety heavy metals and to identify potential sources of contamination. This assessment will help develop a sustainable management plan for coastal water resources.

2. Materials and Methods

2.1. Description of Study Area

The GoA is the Red Sea's northeastern extension. It is a partially-enclosed, narrow and deep coastal water body. The Strait of Tiran connects the Red Sea to the GoA (Figure 1). The GoA extends approximately 180 km southward with a width ranging from 5–25 km (the average of 16 km maximum), and a maximum depth of about 1800 m (the average is about 800 m). Only 27 km of the eastern coast belong in Jordan, and the remaining coastline, unpopulated, and largely underdeveloped, lie in the Saudi territory.

The GoA is influenced by prevailing subtropical conditions with extremely high temperatures, high evaporation rate (about 400 cm/year) and negligible rainfall (of less than 2.2 cm/year) [22]. Surface water flow in the Gulf is nonexistent or limited solely during rare intense rainstorms occurring as flash floods in winter. The average water temperature in the upper 200 m varies seasonally from 20 °C in winter to 28 °C in summer, whereas the average air temperature ranges between 32.20 ± 3.16 °C in summer and 17.60 ± 3.46 °C in winter [23]. The maximum sea level of 154.30 cm was recorded in 2013 [23].

These conditions result in a high salinity in surface water layer, ranging from 40.3 to 40.8‰ in winter and from 40.5 to 46.6‰ in summer [9,24–26]. The surface coastal water of the GoA is extremely oligotrophic, because of its nutrient-poor water originating from the Red Sea surface waters through the Straits of Tiran. The surface water is shallow with stable thermocline throughout the year, except in wintertime, when a wind-driven convective mixing occurs between the deep (nutrient-rich) and surface waters. Water stratification occurs in spring. However, the oceanographic characteristics of extensive solar irradiance, high transparency, deep sunlight penetration, and warm water created unique aquatic ecosystems and biodiversity, with one of the world's richest coral communities [2,3].

The northerly wind, with a high speed and activity during summertime, is the prevailing wind direction and is responsible for the majority of aeolian deposition events in the region. However, Khamaseen winds blowing in springtime account for most sand and dust storms in southern Jordan and the adjacent areas [27]. They deliver dust from the interior of the Sahara Desert in north Africa.

2.2. Sampling and Analysis

Surface water samples were collected in September 2017 from 30 different locations along the coastal areas of GoA, Jordan (from north to south), sampling sites are presented in Figure 1. Coastal water samples were collected in 1-L precleaned polyethylene containers pre-rinsed with 10% HCl and 2 mL of HNO_3, Samples were labeled and measured in the field for pH, electrical conductivity (EC, mS/cm at 25 °C), and total dissolved solids (TDS) using pH-meter (Sensions 5, HACH portable case), and EC/TDS-meter (ECOSCAN-hand held series, EUTECH instruments). Water samples were kept refrigerated at 4 °C and transported to a water laboratory (Yarmouk University, Jordan) for subsequent chemical analyses. Sample preparation and analysis followed APHA [28] procedures.

In the laboratory, all samples were filtered by Whatman filter paper (No. 42) and analyzed for total alkalinity (TA), Cl^-, NO_3^-, SO_4^{2-}, PO_4^{3-}, NH_4^+, Ca^{2+}, Mg^{2+}, Na^+, K^+, Cd, Co, Cr, Cu, Fe, Mn, Pb, Sr, and Zn, as follows: 50 mL of filtered samples were used to determine the concentrations of Na^+, K^+,

Ca^{2+}, Mg^{2+}, and Cd, Co, Cr, Fe, Mn, Pb, Sr, Zn using flame atomic absorption spectrophotometer FAAS (NOVAA 300 Analytica JENA AJ with detection limits varying from 0.001–0.02 μg/L. Each sample was analyzed in duplicate. The accuracy and precision of the analytical method was evaluated by the analysis of a reference material (NASS-5), with recoveries ranging between 98.02–104.01%. A total of 5 mL of filtered samples was used to measure NH_4^+, Cl^-, NO_3^-, SO_4^{2-}, PO_4^{3-} by ion chromatography (IC) (Dionex ICS 1600, Thermoscientific). A total of 25 mL of filtered samples were titrated with 0.02 N H_2SO_4, using phenolphthalein and methyl orange as indicators to determine total alkalinity (TA) of the water samples. There are two pH endpoints corresponding to the above indicators at 8.3 and at 4.3. TA was calculated using the following equation:

$$TA = (\text{volume of acid used} * \text{normality of acid} * 50{,}000)/\text{volume of sample}.$$

The average ionic mass balance for water quality data was −0.02% indicating a high level of accuracy.

3. Results and Discussion

The results of dissolved metals and physicochemical properties of coastal water are tabulated in Table 1, and presented in Figures 2–5. The pH values of surface water layer ranged between 8 and 8.49, with a mean value of 8.26 (Table 1 and Figure 2). They indicate a slightly basic coastal water. They showed a slight spatial variability with no distinct trends. This is likely related to the calcium carbonate buffering capacity of water [23].

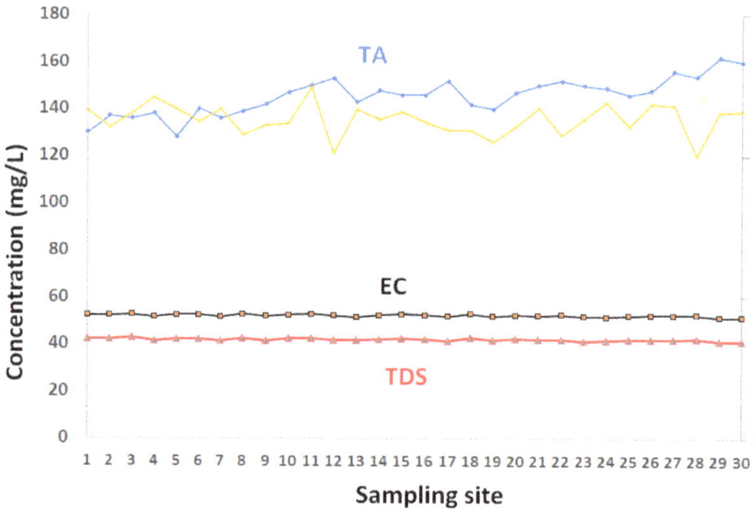

Figure 2. The electrical conductivity (EC), total dissolved solids (TDS), and pH of GoA coastal water.

Table 1. Characteristics of surface seawater layer, GoA, Jordan.

S.N.	pH	EC	TDS	TA	Cl^-	NO_3^-	SO_4^{2-}	PO_4^{3-}	NH_4^+	Ca^{2+}	Mg^{2+}	Na^+	K^+	Sr	Cd	Co	Cr	Cu	Fe	Mn	Pb	Zn
		mS/cm	g/L						mg/L									µg/L				
1	8.32	52.27	41.97	130	24,592.68	15.29	382.26	0.27	15.57	442.45	2295.70	11,165.80	591.78	9.88	0.20	0.38	1.01	0.91	0.94	0.76	0.19	3.93
2	8.20	52.12	41.89	137	24,992.25	13.80	382.12	0.27	15.19	443.91	2283.90	10,759.90	587.75	9.98	0.28	0.31	0.97	0.76	1.23	0.74	0.23	5.10
3	8.30	52.48	42.74	136	25,392.13	14.26	381.29	0.28	15.35	486.99	2355.90	11,736.90	602.31	10.40	0.40	0.31	0.96	1.18	0.75	0.76	0.26	5.23
4	8.41	51.56	41.34	138	23,72.94	13.50	407.45	0.26	14.15	433.9	2253.90	9891.20	547.82	9.44	0.49	0.33	1.32	0.87	1.40	0.70	0.17	4.45
5	8.33	52.40	42.10	128	25,632.05	12.65	396.85	0.29	13.08	444.76	2312.40	11,571.90	579.97	9.93	0.42	0.37	1.38	0.69	1.29	0.70	0.22	5.70
6	8.24	52.41	42.15	140	25,192.19	14.01	401.12	0.26	15.46	468.79	2320.90	11,610.50	570.23	10.03	0.38	0.38	1.38	1.31	0.84	0.86	0.34	5.51
7	8.33	51.42	41.22	136	23,792.62	12.78	386.94	0.27	14.83	423.32	2246.20	9541.50	581.90	9.19	0.26	0.34	1.28	1.71	1.11	0.73	0.23	6.41
8	8.15	52.74	42.34	139	24,792.31	13.88	379.78	0.25	14.06	470.41	2330.20	11,309.00	594.68	10.02	0.50	0.37	1.35	1.11	0.97	0.87	0.27	6.44
9	8.22	51.90	41.41	142	22,253.10	12.35	380.60	0.25	13.67	432.60	2258.90	10,065.40	558.09	9.40	0.46	0.29	1.29	1.30	1.28	0.87	0.21	5.14
10	8.23	52.42	42.34	147	24,792.31	13.80	392.86	0.24	14.55	480.36	2350.30	10,520.90	711.29	10.04	0.44	0.34	1.38	1.52	1.41	0.88	0.24	4.87
11	8.49	52.68	42.34	150	25,392.13	14.48	381.29	0.22	15.02	473.00	2349.10	11,358.90	580.34	10.03	0.48	0.39	1.25	0.91	1.16	0.84	0.28	5.40
12	8.02	52.13	41.78	153	23,992.56	13.75	359.81	0.17	14.98	444.55	2284.60	10,394.50	619.86	9.97	0.48	0.29	1.30	0.81	1.00	0.77	0.27	4.02
13	8.33	51.48	41.54	143	22,172.88	14.82	395.20	0.20	15.93	438.58	2261.00	9784.60	604.60	9.55	0.50	0.38	1.19	0.79	0.78	0.76	0.30	3.63
14	8.26	52.29	42.10	148	22,593.46	13.41	400.98	0.19	14.72	459.73	2315.50	10,456.20	567.31	10.37	0.44	0.31	1.39	1.50	0.94	0.83	0.26	6.90
15	8.31	52.69	42.42	146	25,591.75	15.33	391.07	0.20	16.44	429.90	2251.70	12,646.90	555.55	9.62	0.52	0.36	1.30	1.30	1.26	0.79	0.26	7.05
16	8.24	52.40	42.10	146	25,792.00	15.29	396.85	0.22	16.91	458.40	2313.50	12,437.90	712.91	10.08	0.40	0.34	1.40	0.99	1.12	0.84	0.28	5.52
17	8.18	51.89	41.34	152	24,292.47	12.73	387.35	0.18	13.65	429.54	2251.60	9724.70	598.42	9.64	0.43	0.34	1.40	1.30	1.05	0.68	0.28	3.78
18	8.18	52.87	42.70	142	25,991.94	13.58	390.52	0.19	14.24	483.88	2354.80	12,420.90	579.69	10.15	0.44	0.28	1.22	1.31	0.80	0.79	0.28	4.54
19	8.10	51.91	41.62	140	24,592.37	12.22	402.91	0.16	14.44	439.57	2270.70	9745.90	572.09	9.62	0.44	0.36	1.41	1.30	1.37	0.76	0.26	5.40
20	8.21	52.45	42.34	147	25,991.32	12.99	405.53	0.16	14.44	484.87	2350.90	10,592.90	512.84	10.42	0.42	0.38	1.30	1.12	1.22	0.81	0.32	5.37
21	8.34	52.15	41.92	150	24,592.37	13.20	351.27	0.16	15.00	441.12	2280.10	10,608.90	559.62	9.17	0.44	0.39	1.80	0.99	0.82	1.08	0.40	6.11
22	8.15	52.51	42.24	152	24,792.31	15.24	316.99	0.22	13.33	468.72	2327.10	10,740.70	600.64	10.27	0.62	0.41	1.75	1.01	1.36	1.03	0.49	7.26
23	8.26	51.83	41.48	150	22,253.41	12.78	398.64	0.16	13.92	438.41	2259.00	9889.90	612.91	9.56	0.67	0.46	1.70	1.06	1.62	0.96	0.47	7.63

Table 1. *Cont.*

S.N.	pH	EC	TDS	TA	Cl^-	NO_3^-	SO_4^{2-}	PO_4^{3-}	NH_4^+	Ca^{2+}	Mg^{2+}	Na^+	K^+	Sr	Cd	Co	Cr	Cu	Fe	Mn	Pb	Zn
		mS/cm	g/L	mg/L											µg/L							
24	8.38	51.69	41.62	149	22,293.09	14.26	391.34	0.22	16.48	438.21	2269.30	10,147.90	695.98	9.58	0.75	0.50	1.90	1.38	1.79	0.99	0.66	7.93
25	8.21	52.14	41.99	146	24,592.37	13.63	393.82	0.16	15.29	444.22	2295.40	11,180.10	606.13	9.99	0.68	0.51	1.73	1.50	1.66	1.02	0.59	8.22
26	8.37	52.44	42.22	148	25,191.57	13.46	364.49	0.16	16.08	463.92	2325.70	11,720.90	597.54	10.26	0.75	0.46	1.74	1.69	1.87	1.05	0.73	8.23
27	8.36	52.40	42.12	156	24,991.01	15.5	386.11	0.22	15.38	460.75	2317.40	11,536.90	577.79	10.18	0.76	0.48	1.66	1.70	1.87	1.04	0.79	7.58
28	8.00	52.57	42.46	154	23,792.62	13.16	381.84	0.17	14.01	483.45	2352.20	10,865.90	596.38	10.40	0.72	0.49	1.75	1.51	1.82	1.14	0.70	7.86
29	8.31	51.55	41.34	162	22,533.01	15.12	395.61	0.21	16.40	430.51	2252.60	9934.20	564.46	9.37	0.71	0.46	1.72	1.89	1.88	1.12	0.76	8.17
30	8.32	51.43	41.28	160	23,152.82	14.09	390.24	0.19	15.86	425.14	2249.70	9681.40	581.45	9.21	0.72	0.50	1.91	1.91	1.94	1.15	0.76	8.28
Mean	8.26	52.17	41.95	146	24,326.40	13.85	385.77	0.21	14.98	452.13	2298.01	10,801.44	594.08	9.86	0.51	0.38	1.44	1.24	1.29	0.88	0.38	6.05
Min	8.00	51.42	41.22	128	22,172.88	12.22	316.99	0.16	13.08	423.32	2246.20	9541.50	512.84	9.17	0.20	0.28	0.96	0.69	0.75	0.68	0.17	3.63
Max	8.49	52.87	42.74	162	25,991.94	15.50	407.45	0.29	16.91	486.99	2355.90	12,646.90	712.91	10.42	0.76	0.51	1.91	1.91	1.94	1.15	0.79	8.28
STD	0.11	0.42	0.44	8.06	1211.19	0.96	18.22	0.04	1.034	20.03	38.05	891.41	44.03	0.38	0.15	0.07	0.27	0.34	0.37	0.14	0.20	1.49
Median	8.26	52.28	42.05	146.50	24,592.53	13.78	390.38	0.21	14.99	444.39	2295.55	10,674.80	584.83	9.98	0.47	0.38	1.38	1.30	1.25	0.84	0.28	5.61
Co. Var	0.01	0.01	0.01	0.05	0.05	0.068	0.05	0.27	0.07	0.04	0.02	0.08	0.07	9.86	0.51	0.38	1.44	1.24	1.29	0.88	0.38	6.05

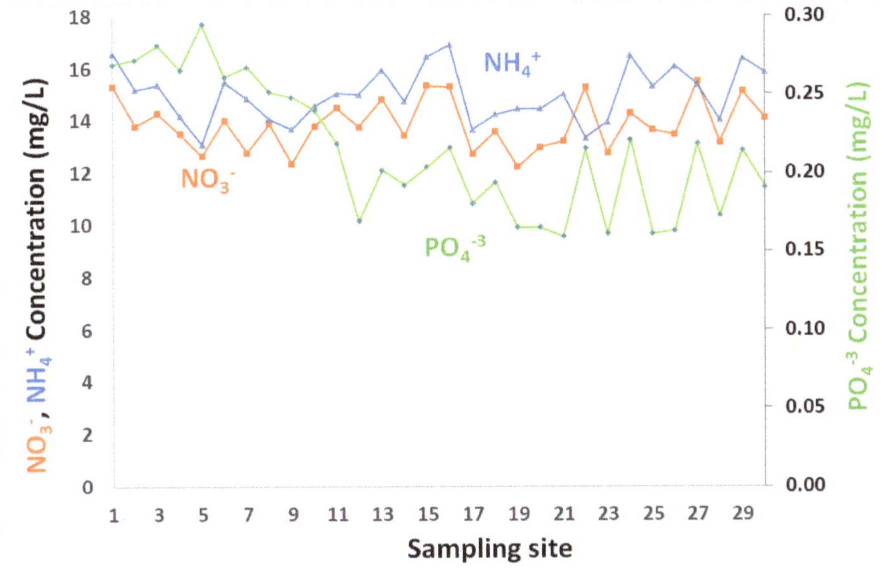

Figure 3. NO_3^-, NH_4^+ and PO_4^{3-} in coastal water of the GoA, Jordan.

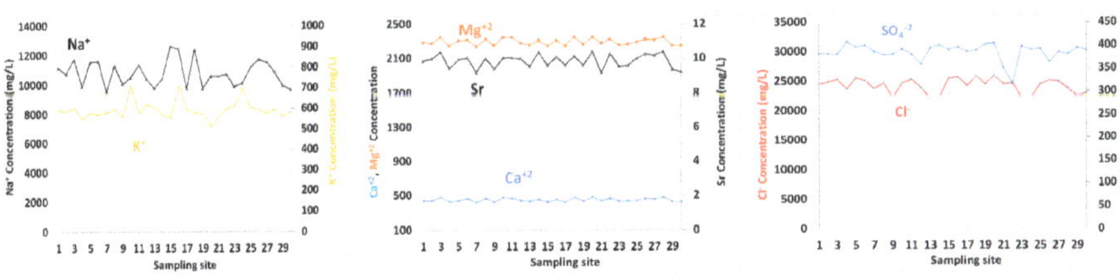

Figure 4. The spatial distributions of Na^+, K^+, Mg^{2+}, Sr, Ca^{2+}, SO_4^{2-} and Cl^- concentrations in surface seawater of GoA, Jordan.

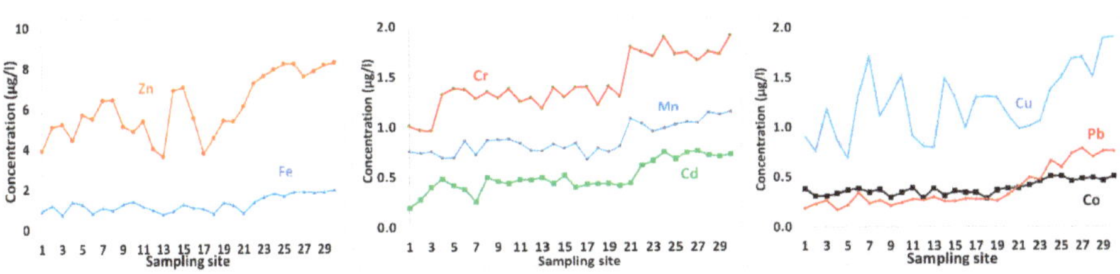

Figure 5. Metal contents in surface seawater in GoA, Jordan.

In addition, these pH values are probably attributable to low growth levels, and production of algal biomass that would contribute organic acids to coastal water when decomposed. While the weather conditions of high temperatures and abundant sunlight allow phytoplankton to grow in abundance, our sampling campaign coincided with a period of nutrient-depleted and stratified water, where photosynthetic activity was at its lowest levels. Manasreh et al. [23] recorded the lowest chlorophyll-a levels in summertime and highest during winter. Microbial decomposition of dead phytoplankton, algae and other flora, produces humic substances, organic acids and amino acids that raise the seawater acidity. Additionally, higher temperature during summertime will reduce dissolved CO_2 levels in coastal water and increases the pH value.

TA (total alkalinity) concentrations ranged between 128 and 162 mg/L, with an average value of 146 mg/L. These high values are consistent with the coastal water's buffering capacity (due to high contents of calcium carbonates) of the GoA's water. The TDS varied from about 41.22–42.74 g/L with an average of 41.95 g/L. EC varied from 51.42 to 52.87 mS/cm with average of 52.17 mS/cm. Similar values of TDS were reported along the Saudi GoA with an average of 41.4 g/L [9].

The spatial pattens of TDS in water showed insignificant variations (Figure 2). The elevated levels of TDS indicate highly saline conditions in the GoA and are primarily attributable to its geographic location in a subtropical desert region, with very high evaporation rates, very low precipitation, and negligible freshwater input. Manasreh et al. [23] reported an evaporation of 2 m/year in the GoA with an increasing salinity toward the north. In addition to these salinity raising factors, the high TDS values are linked to water stratification and poor water circulation during the sampling period. These factors created unique environmental conditions of higher temperature, evaporation, and salinity than average, compared to the average range for oceans. Lack of input of freshwater into the coastal water contributes to high salinity water. A negligible supply of terrigenous sediments into the water results in clear water conditions with high transparency.

Higher water density is often observed in summertime in response to high salinity (due to poor water mixing). From July–August, a stratified water column dominates with thermocline and pyncnocline, occurring at about 250 m in 2013 and near 350 m in 2014 and 2015 [23]. The northwards currents drag warm and saline waters to GoA from the Red Sea [25,29]. The flow of surface water from the Red Sea to the GoA is triggered by the high evaporation rate, where the flowing water offsets the evaporation loss [30]. The GoA surface seawater temperature is 2 °C lower than that of the Red Sea, where the flowing water brings heat that increases temperature, evaporation rates and salinity of surface seawater layer.

In addition, the TDS values become higher in August, corresponding to the summer season, a period of high dust storm events. The atmospheric dust input is an significant source of salts (and metals) to GoA water [8,9]. The GoA is located in a desert-belt region with frequent dust storms, where Negev and Sinai Deserts are in the west and the Arabian Desert is in east. It is believed that large quantities of dust aerosols delivered to GoA is originated from adjacent deserts [31]. Dust deposition will significantly influence the composition of GoA seawater, where the deposition rate of dust in GoA is one of the highest on Earth [13] ranging between 28 g/m^2/year in the northwestern part [13] to about 34.68 g/m^2/year in Aqaba city at the northeastern corner of GoA [8]. TDS was significantly correlated with Ca^{2+}, Mg^{2+}, Na^+, Sr and Cl^- with r = 0.86, 0.88, 0.81 and 0.70, respectively (Table 2). These ions are major contributors to seawater salinity.

Inorganic nutrients (nitrate, ammonium, phosphate) are minor constituents of seawater, but are essential for marine ecosystem productivity and growth. Relatively low levels of inorganic nutrients (NH_4^+, PO_4^{3-}, NO_3^-) (Table 1 and Figure 3) were observed in surface water layer, consistent with the findings of others [23,32–35]. The coastal water in GoA is in extremely oligotrophic conditions, with very limited nutrients supplied to Gulf's water through terrestrial runoff.

Nitrates are present in all water samples, where the concentrations increased slightly in some locations, although not all. NO_3^- concentrations ranged between 12.22 mg/L up to 15.50 mg/L, with overall mean and median levels of about 13.85 and 13.78 mg/L, respectively (Table 1 and Figure 3). Ammonium levels fluctuated from 13.08 and 16.91 mg/L, with mean and median values of about 15 mg/L (Table 1 and Figure 3). While the nutrient levels generally varied, their variations showed no spatial trends. Nitrate and ammonium showed relatively similar ups and downs and were significantly correlated with r = 0.65 (Table 2).

Table 2. Correlation matrices for surface water quality parameters.

Cl^-	NO_3^-	SO_4^{2-}	PO_4^{3-}	NH_4^+	Ca^{2+}	Mg^{2+}	Na^+	K^+	Cd	Co	Cr	Cu	Fe	Mn	Pb	Sr	Zn	pH	EC	TDS	TA	
1.00	0.14	−0.11	0.14	−0.03	0.54	0.60	0.74	−0.07	−0.33	−0.25	−0.32	−0.23	−0.27	−0.26	−0.24	0.52	−0.21	−0.05	0.74	0.70	−0.31	**Cl^-**
	1.00	−0.20	0.18	0.65	0.10	0.11	0.42	0.25	0.16	0.21	−0.01	0.01	0.04	0.22	0.24	0.17	0.12	0.29	0.19	0.25	0.23	**NO_3^-**
		1.00	0.10	0.16	−0.13	−0.15	−0.08	−0.05	−0.12	−0.03	−0.22	0.15	0.06	−0.32	−0.16	−0.12	−0.11	0.18	−0.24	−0.18	−0.21	**SO_4^{2-}**
			1.00	−0.02	0.01	0.06	0.15	0.06	−0.49	−0.36	−0.53	−0.24	−0.27	−0.43	−0.42	−0.01	−0.30	0.30	0.02	0.05	−0.65	**PO_4^{3-}**
				1.00	−0.18	−0.18	0.24	0.28	0.09	0.26	0.03	0.23	0.11	0.20	0.28	−0.12	0.18	0.38	−0.14	−0.03	0.16	**NH_4^+**
					1.00	0.97	0.52	0.11	−0.01	−0.10	−0.16	−0.06	−0.19	0.07	−0.01	0.85	−0.05	−0.24	0.77	0.86	−0.03	**Ca**
						1.00	0.59	0.12	−0.05	−0.09	−0.16	−0.08	−0.18	0.07	−0.02	0.87	−0.04	−0.18	0.82	0.88	−0.10	**Mg**
							1.00	0.09	−0.07	−0.13	−0.21	−0.14	−0.20	−0.05	−0.06	0.57	0.04	0.05	0.83	0.81	−0.22	**Na**
								1.00	0.11	0.10	0.15	−0.01	0.11	0.08	0.08	0.11	0.01	−0.07	0.01	0.06	0.08	**K**
									1.00	0.79	0.83	0.50	0.80	0.78	0.88	0.01	0.76	0.05	−0.11	−0.07	0.72	**Cd**
										1.00	0.80	0.48	0.77	0.78	0.88	−0.09	0.78	0.18	−0.20	−0.14	0.51	**Co**
											1.00	0.52	0.72	0.84	0.82	−0.20	0.79	0.03	−0.22	−0.23	0.69	**Cr**
												1.00	0.57	0.57	0.65	−0.11	0.64	0.04	−0.19	−0.14	0.51	**Cu**
													1.00	0.66	0.80	−0.10	0.75	0.09	−0.26	−0.26	0.53	**Fe**
														1.00	0.88	−0.03	0.80	0.00	−0.04	0.01	0.71	**Mn**
															1.00	0.01	0.81	0.07	−0.15	−0.08	0.71	**Pb**
																1.00	0.00	−0.32	0.76	0.81	−0.07	**Sr**
																	1.00	0.11	−0.04	0.00	0.50	**Zn**
																		1.00	−0.22	−0.16	−0.09	**pH**
																			1.00	0.94	−0.14	**EC**
																				1.00	−0.16	**TDS**
																					1.00	**TA**

Nitrate is the major nitrogen species in the oxic zone, while ammonium dominates in the anoxic zone. The nitrification is a kinetic reaction and is dependent on several water conditions such as salinity, pH, and Eh [36,37]. However, the coexistence of nitrate and ammonium can trigger or slow down nitrogen conversion like nitrification or denitrification [38]. The coexistence of NH_4^+ and NO_3^- may result from poor mixing in groundwater, especially in locations where both nitrogen species are released from active pollution sources [39,40]. The GoA waters are well oxygenated with redox indicators of oxidizing conditions [9]. Manasreh et al. [23] observed that the GoA water is well-ventilated due to the annual water mixing with complete saturation (100%).

This suggests that the presence of high NH_4^+ levels could be associated with leaks from sewer system and/or because of water discharged from fish farm or fertilizer plume. Phosphate varied from 0.16–0.29 mg/L, with an average of 0.21 mg/L (Table 1 and Figure 3). Relatively higher concentrations were observed in the northern GoA, in close proximity to the phosphate terminal, where deposition of dust containing phosphate during loading/unloading activities may contribute phosphates to seawater. While phosphate showed low levels, it is the limiting nutrient for phytoplankton growth. Weak correlations between phosphate and both ammonium and nitrate were observed with $r = -0.02$ and 0.18, respectively (Table 2).

Aeolian deposits can provide important nutrients which stimulate the primary productivity in marine ecosystems [41–44], especially in oligotrophic water [45], like the GoA [10,20]. However, they can also deliver various contaminants that negatively impact the aquatic biodiversity.

Similar observations were reported by Badran [46], where phosphate and nitrate levels in surface water varied seasonally, with the lowest in summer and the highest in winter. The sampling period was concurrent with water stratification, and the concentrations of nutrients in surface water layer were low. In the winter season, winds drive convective mixing of deep (nutrient-rich) and surface waters, where nitrate and other nutrients are injected into the euphotic zone, resulting in seasonal plankton blooms [47]. The highest productivity (chlorophyll-a) is expected during the winter season, which declines to minimum levels in summertime [23].

Water stratification and high sunlight irradiation during summertime further draw down the inorganic nutrients in the surface water by enhancing primary productivity at the subsurface water layer (50–75 m) [34]. During photosynthesis, phytoplankton assimilate nutrients, and it is the availability of inorganic nitrogen that often limits the rate of primary production in the sea [32]. Nutrients uptake within the euphotic zone in oligotrophic water body results in a considerable depletion of their levels. Phytoplankton communities in oligotrophic waters are likely to survive by utilizing recycled nutrients [48,49].

Nutrient levels in the southern Red Sea are greater than its northern and central regions. The inflow of surface water from the Red Sea to GoA (to compensate for the high evaporation loss) is a contributing factor to lower levels of nutrients observed in GoA seawater (oligotrophic water). In late summer, an increase of 25% in nutrient levels is observed in the southern Red Sea compared to the central region, due to the inflow of nutrient-rich waters from the Gulf of Aden to the southern area of the Red Sea [50]. The highest levels of phosphate in the southern Red Sea are usually observed in October, following upwellings in the Arabian Sea.

In addition to water mixing, it is likely that nitrate and other nutrients are associated with atmospheric deposition, as this area experiences frequent dust storms. Rare flash floods carrying terrigenous sediments can be a minor contributor to nitrate and phosphate in coastal water. Dust from the phosphate terminal in the GoA provides further evidence of contribution of aeolian dust to coastal water. Fish farming and wastewater discharges may be important sources of nitrate and phosphate, as water samples were collected adjacent to the coastlines closer to touristic, industrial and other human facilities.

Sr content in seawater ranged between 9.17 and 10.42 mg/L (Table 1). The concentrations of Ca^{2+}, Mg^{2+}, Na^+ and K^+ in surface seawater layer varied from 423.32–486.99, 2246.2–2355.9, 9541.5–12,646.9 and 512.84–712.91 mg/L, respectively (Table 1). High temperature and evaporation rates are main contributors to high levels of ions, among others. The spatial distributions of Ca^{2+}, Mg^{2+} and Sr in surface water exhibited relatively similar patterns with no trends. Ca^{2+} was positively correlated with Mg^{2+}, Sr, and Na^+ with correlation coefficients of 0.97, 0.85, and 0.52, respectively. The correlation coefficients between Mg^{2+} and Sr was 0.87, and between Mg^{2+} and Na^+ is 0.59. The K^+ levels in seawater was not significantly correlated with any cation tested.

Cl^- exhibited spatial changes in surface water, with concentrations ranging from 22,172.88–25,991.94 mg/L. The SO_4^{-2} values in seawater samples varied between 316.99 and 407.45 mg/L. Cl was correlated with Ca^{2+}, Mg^{2+} and Na^+ with correlation coefficients of 0.54, 0.60 and 0.74, respectively. Similar to K^+, the SO_4^{2-} content showed no significant correlations with any other ions of seawater. TDS values were well correlated with Cl^-, Ca^{2+}, Mg^{2+}, Na^+ and Sr, with r = 0.70, 0.86, 0.88, 0.81 and 0.81, respectively. Whereas salinity was neither correlated with SO_4^{2-} nor with K^+. Dust deposition to GoA is also an important contributor to TDS and other ions.

Table 3 compares seawater chemistry of the GoA (northernmost Red Sea) analyzed in this study relative to other regions of the Red Sea. Relatively elevated levels of pH, TA, Cl^-, NO_3^-, PO_4^{3-}, NH_4^+, Mg^{2+} and K were observed for GoA, compared to the central and northern Red Sea. Whereas SO_4^{2-}, TDS, Ca, and Na showed values that are comparable to or lower than those for other parts of the Red Sea water.

Table 3. Comparison of water quality parameters in different regions of the Red Sea and GoA.

	pH	EC	TDS	TA	Cl^-	NO_3^-	SO_4^{2-}	PO_4^{3-}	NH_4^+	Ca	Mg	Na	K
		mS/cm	g/L	mg/L						mg/L			
Present study (northernmost Red Sea)	8.26	52.17	41.95	146	24,326.4	13.85	385.77	0.21	14.98	452.13	2298.01	10,801.44	594.08
Northern Red Sea (Egypt) [51]	7.7	60.3	42	98	23,607	12.4	1260		0.39	738	1570	12,339	287
Central Red Sea (Jeddah, Saudi) [52]	8.1	72.55	43.55	128	22,336	1	2440	<0.13	-	496	1512	11,920	-

The concentrations of metals in surface seawater layer are shown in Table 1 and presented in Figure 5. Zn showed the highest concentration, with an average concentration of 6.05 μg/L. Other metals that followed were in the order Cr > Fe > Cu > Mn > Cd > Pb = Co (Table 1, Figure 5).

Spatial variability of metals contents in seawater samples exhibited increasing trends toward the south (Figure 5), where the industrial complex is located. Potential impacts from heavy metals are commonly confined to areas in the vicinity of urban or industrialized regions on the coastal edge. However, these levels of metals also suggest that they have probably been derived from multiple sources, including a geogenic origin.

Al-Taani et al. [9] reported high levels of dissolved oxygen, with redox values indicating oxidizing conditions in the coastal water GoA, which may favor immobilization of some metals with relatively low levels in seawater samples. Zn in seawater varied from 3.63–8.28 μg/L with an average of 6.05 μg/L (Table 1 and Figure 5). These values are higher than those reported for the Saudi GoA [9], the offshore surface seawater of Red Sea [53], the average oceanic concentration [54], and the Mediterranean surface seawater [55] (Table 4).

In addition, atmospheric dry deposition is the primary external source of trace metals to GoA [14]. Aeolian dry fluxes of certain trace elements (e.g., Cd, Pb, Cu and Zn) to the ocean water may surpass those of riverine sources [56,57]. Aeolian dust of Zn to GoA ranges between 1.68 mg/m^2/year (in Eilat city at the northwestern corner; [31]) and 4.02 mg/m^2/year (in Aqaba city in northeastern region; [8]). High concentrations of Zn were observed GoA seawater ranging from 5.71–11.55 μg/L [58] in the vicinity of Industrial Complex.

Table 4. Comparison of selected metals (μg/L) in surface seawater of GoA relative to other regions.

	Cd	Co	Cr	Cu	Fe	Mn	Pb	Zn
Present study	0.51	0.38	1.44	1.24	1.29	0.88	0.38	6.05
Saudi GoA [9]	0.03	0.236	0.957	6.183	15.255	0.259	0.202	3.323
Offshore surface seawater, northern Red Sea [53]	0.53	0.15	-	0.11	1.58	0.11	0.36	0.22
Oceanic concentration [54]	0.07	0.005 [a]	0.33	0.12	0.04	0.01	0.001	0.4
Mediterranean surface seawater [55]	0.42 [b]	1 [c]	3.4 [d]	0.2 [e]	0.056–0.336 [f]	0.11–0.19 [f]	0.05 [g]	0.17

[a]: [59,60]. [b]: [61]. [c]: [62]. [d]: [63,64]. [e]: [65,66]. [f]: [67]. [g]: [62,65].

Fe contents of seawater varied from 0.75–1.94 μg/L with a mean value of 1.29 μg/L. These concentrations of Fe showed spatial variability in surface water layer with generally greater values in the southern GoA (Figure 5). During the stratified summer, surface water becomes enriched in Fe [13], but the winter mixing of surface and deep water layers, decreases these Fe levels. The average concentration of Fe measured in the present study is relatively comparable to that for the Red sea offshore seawater [53], but higher than the averages for oceanic concentration [54] and the Mediterranean seawater [68] (Table 4). Higher concentrations of Fe were observed in the Saudi GoA with about 15.25 μg/L, suggesting that atmospheric dry deposition in this area is more intense. It is believed that Fe is probably derived from crustal sources. The average dry flux of Fe to the GoA waters varied from about 216 mg/m^2/year [31] to 440 mg/m^2/year [8].

Cr levels varied from 0.96–1.91 μg/L with average value about 1.44 μg/L (Table 1). These values are higher than those observed in the Saudi GoA seawater [9], the mean oceanic level [54], but less than those for the Mediterranean Sea [63,64] (Table 3). The spatial pattern of Cr suggests that these high levels of Cr at the southern end of GoA are most likely related to discharge of brine water from desalination plant [68]. In addition to the industrial wastewater, mineral dust from fertilizer and cement factories remain potential sources of Cr to GoA seawater. atmospheric aerosol deposition to GoA fluctuated between 0.96 mg/m^2/year in Eilat city [31], and about 1.42 mg/m^2/year in Aqaba city [8].

Mn content in southern water samples exhibited higher values relative to the northern part of GoA (Table 1 and Figure 5), and it is likely to originate form anthropogenic emissions [31]. The Mn contents in surface seawater ranged between 0.68 and 1.15 μg/L, with an average of 0.88 μg/L, which is three times as much as that measured for the Saudi GoA [9]. These values are also greater than those reported for the offshore water of Red Sea [53], the average oceanic concentration, and those for the Mediterranean Sea [67] (Table 4). Windblown dust of Mn to the uppermost eastern GoA (Elat city) averaged 5.28 mg/m^2/year [31], whereas, in the northernmost extension, a mean value of 10.29 mg/m^2/year was reported [8]. In addition, Mn is probably related to desalination plants in the neighboring cities (Eilat, Taba and Haql) [3,9], where various heavy metals, including Mn, may be released with the discharged water of thermal desalination plants, depending on the metal alloys used [68,69].

Cd concentrations in seawater ranged between 0.2 and 0.76 μg/L, with an average of 0.51 μg/L (Table 1). Similarly, elevated levels of Cd were found in the southern GoA (Figure 5), where industrial activities are concentrated. Comparable levels of Cd have been reported by Shriadah et al. [53] for the northern Red Sea (offshore seawater), but higher than those for the average oceanic levels of about 0.07 μg/L [54]. Additionally, the average Cd content in surface seawater measured in the present study is greater than those for Saudi GoA and the Mediterranean Sea (Table 4).

In addition to desalination plants, the anthropogenically derived Cd (and other metals such as Pb and Co) is likely related to the discharge of cooling water and sewage in the southern GoA [26,70,71]. The average concentration of Cd from atmospheric dust varied from 0.012 mg/m^2/year in Eilat city [31] to 0.04 mg/m^2/year in Aqaba city [8].

Co content in surface seawater exhibited little spatial variations, with relatively higher levels were observed in the southern portion of GoA (Figure 5). Co ranged from 0.28–0.51 μg/L, with average Co values of about 0.38 μg/L (Table 1). Lower average Co contents were reported for the northern Red Sea offshore water [53] and for the Saudi GoA [9]. Co occurs in seawater at concentrations below 0.005 μg/L [59,60]. However, higher average value has been detected in the surface water of Mediterranean Sea [62]. Potential sources of Co in GoA seawater are likely similar to those of Co [9]. Dust particles collected from GoA region showed an average of 0.1 mg/m^2/year [31].

Relatively elevated levels of Cu were detected in seawater samples, varying from 0.69–1.91 μg/L, with greater values observed for the southern sampling sites (Figure 5). These values are higher than the average ocean level [54] and the offshore water of Red Sea [53] (Table 3). An average Cu value of 0.2 μg/L was reported for the Mediterranean Sea [67,68]. However, elevated levels of Cu were detected in the Saudi GoA (at the Jordan–Saudi border). The atmospheric dry deposition flux of Cu to GoA region ranged from 0.38 mg/m^2/year in Eilat [31] to 0.68 mg/m^2/year in Aqaba city [8]. Elevated concentrations of Cu have been reported in Jordanian GoA water, in the range of 0.74–2.28 μg/L [58], and are higher than those measured in this study.

Pb levels in seawater samples varied from 0.17–0.79 μg/L, with a spatial pattern of increasing levels in the southern part of GoA. The average Pb value of 0.38 μg/L is comparable to those measured in offshore surface water sites of Red Sea [53], but higher than those reported for the Saudi GoA [9], the average ocean, and the Mediterranean surface seawater (Table 3). Pb is of anthropogenic origin, mainly from fossil fuel burning [72]. Elevated concentrations of Pb were found in the Jordanian GoA water ranging between 0.73 and 1.43 μg/L [56]. GoA receives high dry flux of Pb varying from 0.8 mg/m^2/year in Eilat [31] to about 1.42 mg/m^2/year in Aqaba city [8]. All metals tested were significantly correlated (Table 2) indicating that they may have been derived from similar sources.

4. Conclusions

GoA is a place for rich and diverse marine ecosystems. It is highly vulnerable to pollution, wherehuman activities in the bordering countries are intense, with high potential for water contamination. This requires cross-border collaboration to protect these naturally diverse but stressed ecosystems. This study intended to ascertain the seawater quality conditions along the eastern coast of GoA, Jordan. The sampling campaign coincided with a period of low levels of inorganic nutrients, low rates of algal growth with reduced microbial decomposition of dead algal cells. In addition to prevailing saline conditions of high temperatures and high evaporation rates, the water stratification and intense dust storms are the major contributing factors to the observed seawater chemistry. The surface distribution of water quality variables showed spatial variations with no specific patterns, except for metal contents, which exhibited southward increasing trends, closed to the industrial complex. The vast majority of these quality parameters showed relatively higher values compared to those of other regions.

Author Contributions: Conceptualization, A.A.A.-T. and M.R.; methodology, J.I.; software, A.A.-R.; validation, F.H., Y.N.; formal analysis, S.K.; investigation, A.A.A.-T.; resources, A.A.B.; data curation, Y.N.; writing—original draft preparation, A.A.A.-T.; writing—review and editing, F.H.; visualization, A.A.-R.; supervision, A.A.A.-T.; project administration, M.R. All authors have read and agreed to the published version of the manuscript.

Funding: This research received no external funding.

Conflicts of Interest: The authors declare no conflict of interest.

References

1. Batayneh, A.T.; Al-Taani, A. Integrated resistivity and water chemistry for evaluation of groundwater quality of the Gulf of Aqaba coastal area in Saudi Arabia. *Geosci. J.* **2015**, *20*, 403–413. [CrossRef]
2. Dullo, W.-C.; Gektidis, M.; Golubic, S.; Heiss, G.A.; Kampmann, H.; Kiene, W.; Kroll, D.K.; Kuhrau, M.L.; Radtke, G.; Reijmer, J.G.; et al. Factors controlling holocene reef growth: An interdisciplinary approach. *Facies* **1995**, *32*, 145–188. [CrossRef]

3. Al-Rousan, S.A.; Al-Shloul, R.N.; Al-Horani, F.A.; Abu-Hilal, A.H. Heavy metal contents in growth bands of Porites corals: Record of anthropogenic and human developments from the Jordanian Gulf of Aqaba. *Mar. Pollut. Bull.* **2007**, *54*, 1912–1922. [CrossRef]
4. Al-Taani, A.; Batayneh, A.; Mogren, S.; Nazzal, Y.; Ghrefat, H.; Zaman, H.; Elawadi, E. Groundwater Quality of Coastal Aquifer Systems in the Eastern Coast of the Gulf of Aqaba, Saudi Arabia. *J. Appl. Sci. Agric.* **2013**, *8*, 768–778.
5. Batayneh, A.T.; Ghrefat, H.; Zumlot, T.T.; Elawadi, E.; Mogren, S.; Zaman, H.; Al-Taani, A.A.; Nazzal, Y.; Elwahaidi, M.; Elwaheidi, M. Assessing of Metals and Metalloids in Surface Sediments along the Gulf of Aqaba Coast, Northwestern Saudi Arabia. *J. Coast. Res.* **2014**, *31*, 163–176. [CrossRef]
6. Al-Rousan, S.; Al-Moghrabi, S.; Pätzold, J.; Wefer, G. Environmental and biological effects on the stable oxygen isotope records of corals in the northern Gulf of Aqaba, Red Sea. *Mar. Ecol. Prog. Ser.* **2002**, *239*, 301–310. [CrossRef]
7. Al-Rousan, S.; Al-Taani, A.A.; Rashdan, M. Effects of pollution on the geochemical properties of marine sediments across the fringing reef of Aqaba, Red Sea. *Mar. Pollut. Bull.* **2016**, *110*, 546–554. [CrossRef]
8. Al-Taani, A.A.; Rashdan, M.; Khashashneh, S. Atmospheric dry deposition of mineral dust to the Gulf of Aqaba, Red Sea: Rate and trace elements. *Mar. Pollut. Bull.* **2015**, *92*, 252–258. [CrossRef]
9. Al-Taani, A.A.; Batayneh, A.; Nazzal, Y.; Ghrefat, H.; Elawadi, E.; Zaman, H. Status of trace metals in surface seawater of the Gulf of Aqaba, Saudi Arabia. *Mar. Pollut. Bull.* **2014**, *86*, 582–590. [CrossRef]
10. Batayneh, A.; Elawadi, E.; Zaman, H.; Al-Taani, A.A.; Nazzal, Y.; Ghrefat, H. Environmental Assessment of the Gulf of Aqaba Coastal Surface Waters, Saudi Arabia. *J. Coast. Res.* **2014**, *30*, 283–290. [CrossRef]
11. Batayneh, A.; Zumlot, T.; Ghrefat, H.; Nazzal, N.; Qaisy, S.; Zaman, H.; Elawadi, E.; Mogren, S.; Bahkaly, I.; Al-Taani, A. Hydro chemical Facies and Ionic Ratios of the Coastal Groundwater Aquifer of Saudi Gulf of Aqaba: Implication for Seawater Intrusion. *J. Coast. Res.* **2014**, *30*, 75–87. [CrossRef]
12. Al-Trabulsy, H.; Khater, A.; Habbani, F. Heavy elements concentrations, physiochemical characteristics and natural radionuclides levels along the Saudi coastline of the Gulf of Aqaba. *Arab. J. Chem.* **2013**, *6*. [CrossRef]
13. Chase, Z.; Paytan, A.; Johnson, K.S.; Street, J.; Chen, Y. Input and cycling of iron in the Gulf of Aqaba, Red Sea. *Glob. Biogeochem. Cycles* **2006**, *20*. [CrossRef]
14. Chase, Z.; Beck, A.; Biller, D.; Bruland, K.; Measures, C.; Sañudo-Wilhelmy, S. Evaluating the impact of atmospheric deposition on dissolved trace-metals in the Gulf of Aqaba, Red Sea. *Mar. Chem.* **2011**, *126*, 256–268. [CrossRef]
15. El-Radaideh, N.; Al-Taani, A.A.; Al Khateeb, W.M. Status of sedimentation in King Talal Dam, case study from Jordan. *Environ. Earth Sci.* **2017**, *76*, 132. [CrossRef]
16. El-Radaideh, N.; Al-Taani, A.A.; Al Khateeb, W.M. Characteristics and quality of reservoir sediments, Mujib Dam, Central Jordan, as a case study. *Environ. Monit. Assess.* **2017**, *189*, 143. [CrossRef]
17. Yusuf, N.E.; Al-Taani, A.A.; Al-Fukaha, F.W.; Al-Shereideh, S.A. Sediments Transport across the Fringing Reef in the Gulf of Aqaba, Red Sea and the Implications on Live Corals. *Basic Sci. Eng.* **2011**, *20*, 35–52.
18. Tegen, I.; Werner, M.; Harrison, S.P.; Kohfeld, K.E. Relative importance of climate and land use in determining present and future global soil dust emission. *Geophys. Res. Lett.* **2004**, *31*. [CrossRef]
19. Woodward, S.; Roberts, D.L.; Betts, R.A. A simulation of the effect of climate change–induced desertification on mineral dust aerosol. *Geophys. Res. Lett.* **2005**, *32*. [CrossRef]
20. Klinker, J.; Reiss, Z.; Kropach, C.; Levanon, I.; Harpaz, H.; Halicz, E.; Assaf, G. Observation on the circulation pattern in the Gulf of Aqaba, Red Sea. *Isr. J. Earth Sci.* **1976**, *25*, 85–103.
21. Hulings, N.C. Currents in the Jordan Gulf of Aqaba. *Dirasat* **1979**, *6*, 21–33.
22. Reiss, Z.; Hottinger, L. *The Gulf of Aqaba, Ecological Micropaleontology*, 1st ed.; Springer: Berlin/Heidelberg, Germany, 1984; ISBN 978-3-642-69787-6.
23. Manasrah, R.; Abu-Hilal, A.; Rasheed, M. *Physical and Chemical Properties of Seawater in the Gulf of Aqaba and Red Sea BT-Oceanographic and Biological Aspects of the Red Sea*; Rasul, N.M.A., Stewart, I.C.F., Eds.; Springer: Cham, Switzerland, 2019; pp. 41–73. ISBN 978-3-319-99417-8.
24. Manasrah, R.; Al-Zibdah, M.; Al-Ougaily, F.; Yusuf, N.; Al-Najjar, T. Seasonal changes of water properties and current in the northernmost Gulf of Aqaba, Red Sea. *Ocean Sci. J.* **2007**, *42*, 103–116. [CrossRef]
25. Manasrah, R.; Mohammad, B.; Lass, H.; Fennel, W. Circulation and winter deep-water formation in the northern Red Sea. *Oceanologia* **2004**, *46*, 5–23.

26. Abu-Hilal, A.; Badran, M.; de Vaugelas, J. Distribution of trace elements in Callichirus laurae burrows and nearby sediments in the gulf of Aqaba, Jordan (Red Sea). *Mar. Environ. Res.* **1988**, *25*, 233–248. [CrossRef]
27. Abed, A.; Al Kuisi, M.; Abul Khair, H. Characterization of the Khamaseen (Spring) Dust in Jordan. *Atmos. Environ.* **2009**, *43*, 2868–2876. [CrossRef]
28. *APHA Standard Methods for Examination of Water and Wastewater*, 18th ed.; American Public Health Association: Washington, DC, USA, 1998; pp. 45–60.
29. Genin, A.; Lazar, B.; Brenner, S. Vertical mixing and coral death in the Red Sea following the eruption of Mount Pinatubo. *Nature* **1995**, *377*, 507–510. [CrossRef]
30. Murray Stephen, P.; Hecht, A.; Babcock, A. Marine research. *J. Mar. Res.* **1984**, *42*, 265–287.
31. Chen, Y.; Paytan, A.; Chase, Z.; Measures, C.; Beck, A.J.; Sañudo-Wilhelmy, S.A.; Post, A.F. Sources and fluxes of atmospheric trace elements to the Gulf of Aqaba, Red Sea. *J. Geophys. Res. Atmos.* **2008**, *113*. [CrossRef]
32. Ryther, J.H.; Dunstan, W.M. Nitrogen, Phosphorus, and Eutrophication in the Coastal Marine Environment. *Science* **1971**, *171*, 1008–1013. [CrossRef]
33. Badran, M.I.; Al Zibdah, M.K. Quality standard codes of reference of Jordanian coastal waters of the Gulf of Aqaba, Red Sea. *Chem. Ecol.* **2005**, *21*, 337–350. [CrossRef]
34. Manasrah, R.; Alsaad, L.; Trabeen, K.; Rasheed, M.; Al-Absi, E.; Dixon, L.K.; Al-Sawalmih, A. Physical and chemical properties of seawater during 2013–2015 in the 400 m water column in the northern Gulf of Aqaba, Red Sea. *Environ. Monit. Assess.* **2020**, *192*, 188. [CrossRef] [PubMed]
35. Al-Rshaidat, M.M.D.; Segonds-Pichon, A.; Salem, M. Chlorophyll-nutrient relationships of an artificial inland lagoon equipped with seawater replenishment system in the Northern Red sea (Gulf Of Aqaba). *J. Mar. Sci. Eng.* **2020**, *8*, 147. [CrossRef]
36. Chen, S.; Ling, J.; Jean-paul, B. Nitrification kinetics of biofilm as affected by water quality factors. *Aquac. Eng.* **2006**, *34*, 179–197. [CrossRef]
37. Al-Taani, A.A.; Al-Qudah, K.A. Investigation of desert subsoil nitrate in Northeastern Badia of Jordan. *Sci. Total Environ.* **2013**, *442*, 111–115. [CrossRef] [PubMed]
38. Masserini, R.T.J. *Ammonium, Nitrate, and Nitrite in the Oligotrophic Ocean: Detection Methods and Usefulness as Tracers*; University of South Florida: Tampa, FL, USA, 2005.
39. Clark, I.; Timlin, R.; Bourbonnais, A.; Jones, K.; Lafleur, D.; Wickens, K. Origin and Fate of Industrial Ammonium in Anoxic Ground Water—15N Evidence for Anaerobic Oxidation (Anammox). *Groundw. Monit. Remediat.* **2008**, *28*, 73–82. [CrossRef]
40. Sbarbati, C.; Colombani, N.; Mastrocicco, M.; Aravena, R.; Petitta, M. Performance of different assessment methods to evaluate contaminant sources and fate in a coastal aquifer. *Environ. Sci. Pollut. Res.* **2015**, *22*. [CrossRef]
41. Jickells, T.; An, Z.; Andersen, K.; Baker, A.R.; Bergametti, G.; Brooks, N.; Cao, J.; Boyd, P.; Duce, R.; Hunter, K.; et al. Global Iron Connections Between Desert Dust, Ocean Biogeochemistry, and Climate. *Science* **2005**, *308*, 67–71. [CrossRef]
42. Dulac, F.; Moulin, C.; Lambert, C.E.; Guillard, F.; Poitou, J.; Guelle, W.; Quetel, C.R.; Schneider, X.; Ezat, U. Quantitative Remote Sensing of African Dust Transport to the Mediterranean. In *The Impact of Desert Dust Across the Mediterranean*; Springer: Berlin/Heidelberg, Germany, 1996; pp. 25–49. ISBN 978-94-017-3354-0.
43. Bergametti, G.; Remoudaki, E.; Losno, R.; Steiner, E.; Chatenet, B.; Buat-Menard, P. Source, transport and deposition of atmospheric phosphorus over the Northwestern Mediterranean. *J. Atmos. Chem.* **1992**, *14*, 501–513. [CrossRef]
44. Bishop, J.K.B.; Davis, R.E.; Sherman, J.T. Robotic observations of dust storm enhancement of carbon biomass in the North Pacific. *Science* **2002**, *298*, 817–821. [CrossRef]
45. Duce, R.A.; LaRoche, J.; Altieri, K.; Arrigo, K.R.; Baker, A.R.; Capone, D.G.; Cornell, S.; Dentener, F.; Galloway, J.; Ganeshram, R.S.; et al. Impacts of Atmospheric Anthropogenic Nitrogen on the Open Ocean. *Science* **2008**, *320*, 893–897. [CrossRef]
46. Badran, M.I. Dissolved Oxygen, Chlorophyll a and Nutrients: Seasonal Cycles in Waters of the Gulf of Aquaba, Red Sea. *Aquat. Ecosyst. Health Manag.* **2001**, *4*, 139–150. [CrossRef]
47. Al-Qutob, M.; Häse, C.; Tilzer, M.; Lazar, B. Phytoplankton drive nitrite dynamics in the Gulf of Aqaba, Red Sea. Marine Ecology Progress Series. *Mar. Ecol. Prog. Ser.* **2002**, *239*, 233–239. [CrossRef]
48. Eppley, R.W.; Renger, E.H.; Venrick, E.L.; Mullin, M.M. A Study Of Plankton Dynamics and Nutrient Cycling in the Central Gyre of the North Pacific Ocean. *Limnol. Oceanogr.* **1973**, *18*, 534–551. [CrossRef]

49. Kiefer, D.A.; Atkinson, A.C. Cycling of nitrogen by plankton: A hypothetical description based upon efficiency of energy conversion. *J. Mar. Res.* **1984**, *42*, 655–675. [CrossRef]
50. Thiel, H.; Weikert, H.; Karbe, L. Risk assessment for mining metalliferous muds in the deep Red Sea. *Ambio* **1986**, *15*, 34–41.
51. Abdel-Aal, E.A.; Farid, M.E.; Hassan, F.S.M.; Adila, E.M. Desalination of Red Sea water using both electrodialysis and reverse osmosis as complementary methods. *Egypt. J. Pet.* **2015**, *24*, 71–75. [CrossRef]
52. Al-Moubaraki, A.; Al-Judaibi, A.; Asiri, M. Corrosion of C-Steel in the Red Sea: Effect of Immersion Time and Inhibitor Concentration. *Int. J. Electrochem. Sci.* **2015**, *10*, 4252–4278.
53. Shriadah, M.A.; Okbah, M.A.; El-Deek, M.S. Trace Metals in the Water Columns of the Red Sea and the Gulf of Aqaba, Egypt. *Water Air Soil Pollut.* **2004**, *153*, 115–124. [CrossRef]
54. Broecker, W.S.; Peng, T.H. Tracers in the Sea. In *The Lamont-Doherty Geological Observatory*; Columbia University: New York, NY, USA, 1982; pp. 45–60.
55. Tankere, S.P.C.; Statham, P.J. Distribution of dissolved Cd, Cu, Ni and Zn in the Adriatic Sea. *Mar. Pollut. Bull.* **1996**, *32*, 623–630. [CrossRef]
56. Guerzoni, S.; Chester, R.; Dulac, F.; Herut, B.; Loÿe-Pilot, M.-D.; Measures, C.; Migon, C.; Molinaroli, E.; Moulin, C.; Rossini, P.; et al. The role of atmospheric deposition in the biogeochemistry of the Mediterranean Sea. *Prog. Oceanogr.* **1999**, *44*, 147–190. [CrossRef]
57. Koçak, M.; Kubilay, N.; Herut, B.; Nimmo, M. Dry atmospheric fluxes of trace metals (Al, Fe, Mn, Pb, Cd, Zn, Cu) over the Levantine Basin: A refined assessment. *Atmos. Environ.* **2005**, *39*, 7330–7341. [CrossRef]
58. Marine Science Station (MSS). *Environmental Appraisal of the Jordanian Coast of the Gulf of Aqaba*; Unpublished Report; MSS: Aqaba, Jordan, 1999; p. 88.
59. Danielsson, L.-G. Cadmium, cobalt, copper, iron, lead, nickel and zinc in Indian Ocean water. *Mar. Chem.* **1980**, *8*, 199–215. [CrossRef]
60. Millero, F.J. *Chemical Oceanography*, 2nd ed.; CRC Press: New York, NY, USA, 1996; ISBN1 0849384230. ISBN2 9780849384233.
61. El-Moselhy, K.; Hamed, M. Impact of land-based activities on hydrographic conditions and levels of heavy metals in water and sediments along the Mediterranean coast of Egypt. *Egypt. J. Aquat. Res.* **2006**, *32*, 63–82.
62. Migon, C.; Nicolas, E. The trace metal recycling component in the North-western Mediterranean. *Mar. Pollut. Bull.* **1998**, *36*, 273–277. [CrossRef]
63. Emelyanov, E.M.; Shimkus, K.M. Geochemistry and Sedimentology of the Mediterranean Sea. *Geochem. Sedimentol. Mediterr. Sea* **1986**. [CrossRef]
64. Furness, R.W.; Rainbow, P.S. (Eds.) *Heavy Metals in the Marine Environment*; CRC Press: Boca Raton, FL, USA, 1990.
65. Laumond, F.; Copin-Montegut, G.; Courau, P.; Nicolas, E. Cadmium, copper and lead in the western Mediterranean Sea. *Mar. Chem.* **1984**, *15*, 251–261. [CrossRef]
66. Boyle, E.A.; Chapnick, S.D.; Bai, X.X.; Spivack, A. Trace metal enrichments in the Mediterranean Sea. *Earth Planet. Sci. Lett.* **1985**, *74*, 405–419. [CrossRef]
67. Saager, P.M.; Schijf, J.; de Baar, H.J.W. Trace-metal distributions in seawater and anoxic brines in the eastern Mediterranean Sea. *Geochim. Cosmochim. Acta* **1993**, *57*, 1419–1432. [CrossRef]
68. Höpner, T.; Lattemann, S.; Hoepner, T.; Lattemann, S. Chemical Impacts from Seawater Desalination Plants—A Case Study of the Northern Red Sea. *Desalination* **2002**, *152*, 133–140. [CrossRef]
69. Höpner, T. A procedure for environmental impact assessments (EIA) for seawater desalination plants. *Desalination* **1999**, *124*. [CrossRef]
70. Abu-Hilal, A.H.; Badran, M.M. Effect of pollution sources on metal concentration in sediment cores from the Gulf of Aqaba (red sea). *Mar. Pollut. Bull.* **1990**, *21*, 190–197. [CrossRef]
71. Abu-Hilal, A. Observations on heavy metal geochemical association in marine sediments of the Jordan Gulf. of Aqaba. *Mar. Pollut. Bull.* **1993**, *26*, 85–90. [CrossRef]
72. El-Radaideh, N.; Al-Taani, A. Geo-environmental study of heavy metals of the agricultural highway soils, NW Jordan. *Arab. J. Geosci.* **2018**, *11*. [CrossRef]

water

MDPI

Article

Characteristics and Causes of Long-Term Water Quality Variation in Lixiahe Abdominal Area, China

Chenjuan Jiang [1], Jia'nan Zhou [1], Jingcai Wang [1], Guosheng Fu [2] and Jiren Zhou [1,*]

[1] College of Hydraulic Science and Engineering, Yangzhou University, Yangzhou 225000, China; jiangchenjuan001@163.com (C.J.); zhoujianan0410@163.com (J.Z.); wangjingcai@yzu.edu.cn (J.W.)

[2] Taizhou Branch of Jiangsu Hydrological and Water Resources Survey Bureau, Taizhou 225300, China; fgs37@126.com

* Correspondence: zhoujiren001@163.com

Received: 20 May 2020; Accepted: 11 June 2020; Published: 13 June 2020

Abstract: The Lixiahe abdominal area is a representative plain river network in the lower reaches of the Huai River, being an upstream section of south-to-north water diversion from the Yangtze River in Jiangsu Province, China. The assessment of long-term water quality variation and the identification of probable causes can provide references for sustainable water resources management. Based on the monthly water quality data of 15 monitoring stations in the Lixiahe abdominal area, the periodic characteristics and tendency of water quality variation were studied by combining wavelet analysis, the Mann–Kendall trend test, and Sen's slope estimator, and the correlation between water quality variation, water level, and water diversion was discussed with cross wavelet transform and wavelet coherence. The results show that the comprehensive water quality index (CWQI) included periodic fluctuations on multiple scales from 0.25 to 5 years. The CWQI of 7 out of 15 monitoring stations has a significant decreasing trend, indicating regional water quality improvement. The trend slope ranges from −0.071/yr to 0.007/yr, where −0.071/yr indicates the water quality improvement by one grade in 15 years. The spatial variation of water quality in the Lixiahe abdominal area was significant. The water quality of the main water diversion channels and its nearby rivers was significantly improved, while the improvement of other areas was not significant or even became worse due to the increasing discharge of pollutants. The CWQI of the main water diversion channels and its nearby rivers was inversely correlated with the amount of water diversion. The greater the amount of water diversion, the better the water quality. The water diversion from the Yangtze River has played an important role in improving the regional water environment.

Keywords: water quality variation; wavelet analysis; low-pass filtering; linear regression

1. Introduction

Water environment deterioration is a prominent issue in river basin management throughout the world, which has become a serious threat to water security [1]. Surface water and groundwater is affected by geological, climatic, and other natural conditions as well as anthropogenic activities [2] such as precipitation, the pumping of groundwater, and regional droughts [3,4], which is of great significance to the ecological environment of the basin and the production and life of residents in the surrounding areas [5]. The assessment of long-term water quality variation and identification of probable causes can provide information supports and references for sustainable water resources management.

The single factor index method [6,7], comprehensive pollution index method [8,9], Canadian Council of Ministers of the Environment Water Quality Index (CCME CWQI) [10,11], multivariate statistical techniques [12,13], such as cluster analysis [14], discriminant analysis [12], factor analysis [13], principal component analysis [15], and artificial neural network [16,17] are widely used in river water quality evaluation. DRASTIC is a widely used indexing method to assess groundwater vulnerability to a wide range of potential contaminants [18,19]. All these methods are used to comprehensively assess the water quality as well as identify spatial and temporal variations in water quality and main sources of contamination.

Wavelet analysis is becoming a common tool for analyzing localized variations of power within a time series [20,21], which is widely used in hydrology in the study of noise elimination, filtering of time series, monitoring of abrupt points, identification of periodic components [22,23], and assessing the long-term variation of water quality [24,25]. Besides, wavelet analysis is also combined with Artificial Neural Network (wavelet-ANN), Adaptive Neuro-Fuzzy Inference System (wavelet-ANFIS), or extreme learning machine to predict monthly water quality, which is successfully used in the Aji-Chay River in Northwestern Iran [26,27], the Yamuna river in India [28], and the Johor River in Malaysia [29].

In previous studies on temporal variation of water quality, multivariate statistical techniques and continuous wavelet transform are used to present a significant and validated picture of the seasonal periodic behavior of water quality, but they do not directly explore long-term periods, variation tendency, and the coherence of the periodic behavior of water quality variables with influencing factors. This present study aims to remedy this shortcoming by investigating long-term periods, variation tendency, and the coherence of water quality with water level and water diversion from outside the basin with combined methods of continuous wavelet transform, cross wavelet transform, wavelet coherence, Mann–Kendall trend test, and Sen's slope estimator.

The Lixiahe abdominal area is a representative plain river network in the lower reaches of Huai River, where the water used for industry and agriculture is mainly from water diversion from the Yangtze River. The water diversion project is widely constructed to solve the problems of regional water shortage and water pollution in many countries, including Australia, China, Canada, India, the United States, and others [30]. Hence, assessment of the effects of the water diversion on the regional water resource and water quality is significant for sustainable water resources management [31,32]. Since the water diversion project in the Lixiahe abdominal area has been executed for decades, it is valuable to investigate the long-term water quality variation and its possible causes to give scientific guidelines for water resources management and the optimization of water diversion operation. Based on the monthly water quality data of 15 monitoring stations from 2003 to 2017, the comprehensive water quality index (CWQI) was used to evaluate the water quality of the river, and the methods of wavelet analysis, Mann–Kendall trend test, and Sen's slope estimator were used to study periodic characteristics and tendency of water quality variation. Furthermore, the possible causes of water quality variation were discussed.

2. Materials and Methods

2.1. Study Area Description

The Lixiahe abdominal area is a relatively closed plain river network in the lower reaches of the Huai River in Northern Jiangsu Province, China, with an area of 11,722 km^2, which is located to the east of Li Canal, the south of Subei Main Irrigation Canal, the west of Tongyu River, the north of 328 National Highway from Yangzhou to Nantong, and the Rutai canal (Figure 1). There are many rivers, polder networks, lakes, marshes, and wetlands in the area. The terrain is high around, low in the middle, with an altitude ranging from 0 to 10 m above the old-yellow river datum plane. The area belongs to a subtropical monsoon climate and it is affected by a marine climate. The annual average precipitation is 1025 mm, and the precipitation in flood season accounts for about 70% of the annual rainfall.

Since it is located in the lower reaches of the Huai River, the water quality of the water from the upper reaches is generally poor. The water used in the area is mainly from the Jiangdu water conservancy project and the Gaogang water conservancy project, by which the high-quality Yangtze River water enters the area through the Xintongyang canal and the Taizhouyingjiang River to meet the demand for water for life, industry, and agriculture (Figure 1). Point source pollutant is a major pollution source, while the contribution of agricultural non-point source pollution has been growing. The mean annual load of four main water quality indicators (permanganate index: COD_{Mn}, ammonium nitrogen: NH_3-N, total nitrogen: TN, total phosphorus: TP) was 29,000, 10,000, 23,000, and 7000 tons in 2003–2017.

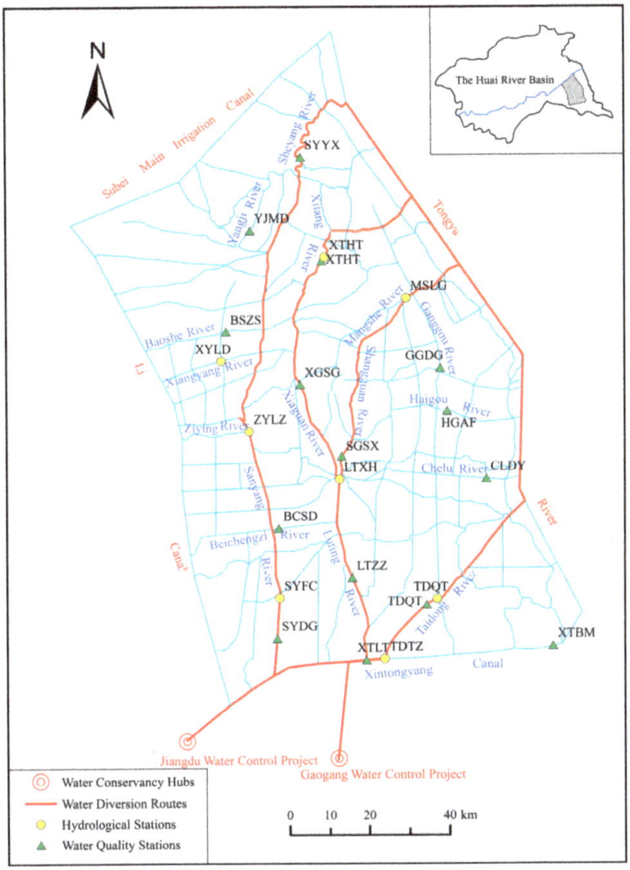

Figure 1. Sketch of river networks and monitoring stations in the Lixiahe abdominal area.

2.2. Hydrological and Water Quality Monitoring

In this study, the monthly water quality data of 15 monitoring stations and the daily water level data of 8 hydrological stations were collected from 2003 to 2017. The monitoring stations cover Taizhou, Yancheng, and Yangzhou, which are three important areas in the Lixiahe abdominal area (Figure 1). The data were collected by the Taizhou Branch, Yancheng branch, and Yangzhou Branch of the Jiangsu Bureau of the hydrological and water resources survey. In addition, daily water diversion data of Gaogang and Jiangdu station from 2003 to 2017 were collected from the Taizhou and Yangzhou Branch of the Jiangsu hydrological and Water Resources Survey Bureau.

2.3. Research Methods

Two main water quality indicators, permanganate index (COD_{Mn}) and ammonia nitrogen (NH_3-N), are selected for analysis, and the comprehensive water quality index (CWQI) determined by these two factors is calculated. The long-term trend of water quality is studied by using methods of continuous wavelet transform, Mann–Kendall trend test, and Sen's slope estimator. The correlation between water quality and water level as well as water diversion volume are investigated by cross wavelet transform and wavelet coherence analysis.

2.3.1. Comprehensive Water Quality Identification Index (CWQI)

The comprehensive water quality index comprehensively takes into account a variety of pollution indexes, which can fully express the overall comprehensive water quality information of the river. The CWQI is computed using the formula of Xu [33].

$$I_{wq} = X_1 \cdot X_2 = \frac{1}{m}\sum (P_1 + P_2 + \cdots + P_m) \tag{1}$$

where X_1 represents the overall water quality grade in the water quality category, X_2 represents the position of the water quality within the same grade, m is the number of water quality indexes participating in the comprehensive water quality evaluation, and P_1, P_2 and P_m represent the single factor water quality identification index (SWQI) of different monitoring items. The calculation of SWQI follows the method of Xu [7],

$$P_i = X_1 \cdot X_2 \tag{2}$$

where X_1 equals grade value of water quality, and X_2 reads

$$X_2 = \frac{\rho_i - \rho_{ikl}}{\rho_{iku} - \rho_{ikl}} \tag{3}$$

where ρ_i is the measured mass concentration, k equals X_1, and ρ_{ikl} and ρ_{iku} are the lower and upper limit value of grade k for monitoring item i. The advantage of SWQI compared to pollutant concentration is that the measured values of different indicators can be converted into the values corresponding to the water quality category.

The water quality category used in this study is the Standard for surface water environmental quality assessment of China (GB3838-2002) [34]. The detailed range values of water quality indicators and SWQI/CWQI corresponding to different water quality grades as well as its description are presented in Table 1, where the range values of single indicators are according to GB3838-2002 [34], the range values of SWQI/CWQI are according to the calculation method introduced by Xu [7,33], and the water quality description is subjectively given according to the regional water quality objective of Grade III.

Table 1. The range values of water quality category.

Indicator	Range Values (mg/L)					
	Grade I	**Grade II**	**Grade III**	**Grade IV**	**Grade V**	**Inferior to Grade V**
COD_{Mn}	[0, 2)	[2, 4)	[4, 6)	[6, 10)	[10, 15)	[15, infinity)
NH_3-N	[0, 0.15)	[0.15, 0.5)	[0.5, 1)	[1, 1.5)	[1.5, 2)	[2, infinity)
SWQI/CWQI	[1, 2)	[2, 3)	[3, 4)	[4, 5)	[5, 6)	[6, infinity)
Description	Extremely good	Very good	Good	Poor	Very poor	Extremely poor

Note: The range values of COD_{Mn} and NH_3-N are cited from (GB3838-2002) [34].

2.3.2. Continuous Wavelet Transform

Continuous wavelet analysis can clearly reveal a variety of variation periods hidden in the time series by decomposing a time series into time–frequency space. The time series of CWQI is standardized, and the significant period CWQI variation is analyzed by using the continuous wavelet transform (CWT) method. The CWT used in this work followed the method of Torrence and Compo [20,35], and the Morlet wavelet was employed as the mother function for the analysis. The prototype formula of Morlet is

$$\Psi_0(\eta) = \pi^{-1/4} e^{i\omega_0\eta} e^{-\eta^2/2} \tag{4}$$

where η is dimensionless time, and ω_0 is a dimensionless frequency, which can be set to 6 to satisfy the admissible condition. The CWT of the discrete time series x_n is defined as

$$W_n^X(s) = \sum_{n'=0}^{N-1} x_{n'} \Psi^* \left[(n' - n) \frac{\delta t}{s} \right] \tag{5}$$

where $W_n^X(s)$ is the wavelet coefficients, N is the length of the time series x_n, * is the complex conjugate, and δt is the temporal sampling interval. $|W_n(s)|^2$ represents the wavelet power spectrum. By varying the wavelet scale s and translating along the localized time index n, the wavelet power spectrum reveals the fluctuating energy of different periodicities defined by s versus time.

Red noise is used as background spectrum to test the wavelet spectrum. The first order autoregressive (AR1) process is used in the red noise test. The power spectrum of background red noise is defined as

$$P_k = (1 - \alpha_2) / \left(\left| 1 - \alpha_2 e^{-2i\pi k} \right| \right) \tag{6}$$

where α is the correlation coefficient of autoregressive equation in red noise power spectrum, and k is the Fourier frequency index. In general, values outside the wavelet influence cone (COI) at various scales are estimated at the significance level of 5%. For detailed information, please refer to references [20,21]. In this study, cross wavelet transform (XWT) and wavelet coherence (WTC) were done in Matlab software.

2.3.3. Trend Analysis Methods

Tests for the detection of significant trends in a hydro-meteorological time series can be classified as parametric and non-parametric methods. Parametric trend tests require data to be independent and normally distributed, while non-parametric trend tests require only that the data be independent [36]. In this study, two non-parametric methods (Mann–Kendall trend test and Sen's slope estimator) were used to detect the trends of water quality indicators. The Mann–Kendall statistical test is able to quantify the significance of trends in time series, and the Sen's slope estimator is used for estimating the slope of trend, both of which are widely used in hydro-meteorological time series [37,38].

In the Mann–Kendall statistical test, the standard normal test statistic Z_S is computed according to reference [36]. Positive values of Z_S indicate increasing trends, while negative Z_S values show decreasing trends. Testing trends is done at the significance level $\alpha = 0.05$.

When $|Z_s| > Z_{1-\alpha}$, α significant trend exists in the time series, where $Z_{1-\alpha} = 1.96$. In the Sen's slope estimator, Q_{med} is computed, the sign of which reflects data trend reflection, while the value of which indicates the steepness of the trend. For detailed information on the Mann–Kendall and Sen's slope estimator, please refer to references [36,39].

2.3.4. Cross Wavelet Transform and Wavelet Coherence

Cross wavelet transform (XWT) and wavelet coherence (WTC) can reveal the multi-scale relationship between two time series in the time-frequency domain, and they can analyze the resonance period and phase relationship in high-energy and low-energy regions, respectively [21]. In this study, XWT and WTC methods are used to analyze the relationship between water quality variation and water level and water diversion volume, and the methods follow Grinsted et al. [21].

The result of the XWT of two series X_n and Y_n is defined as

$$W_n^{XY} = W_n^X W_n^{Y^*} \tag{7}$$

where $\left|W^{XY}\right|$ is the cross wavelet power, and * is the complex conjugate.

The WTC is defined as

$$R_n^2(s) = \frac{\left|S\left(s^{-1}W_n^{XY}(s)\right)\right|^2}{S\left(s^{-1}\left|W_n^X(s)\right|^2\right)\cdot S\left(s^{-1}\left|W_n^Y(s)\right|^2\right)} \quad (8)$$

where s is a smoothing operator. The level of statistical significance of WTC was estimated using the Monte Carlo method. For detailed information, please refer to references [20,21]. In this study, XWT and WTC were done in Matlab software.

3. Characteristics of Water Quality Variation

3.1. General Description of Water Quality

During 2003–2017, the annual mean COD_{Mn} and NH_3-N in the Lixiahe abdominal area are 4.69 mg/L and 0.68 mg/L, and the CWQI is 3.24, which is Grade III according to GB3838-2002 [34]. According to the results of the Mann–Kendall trend test and Sen's slope estimator on water quality indicators, NH_3-N and CWQI show a significant decreasing trend (Table 2). Hence, the water quality improves (Figure 2).

The regional water quality shows significant seasonal changes (Figure 2). From January to March, COD_{Mn} and NH_3-N are both high; from April to June, COD_{Mn} and NH_3-N are gradually decreasing, and the water quality is gradually improving; in July, COD_{Mn} and NH_3-N are both soaring, and the water quality is the worst in the year; from August to October, COD_{Mn} and NH_3-N are gradually decreasing, and the water quality is gradually improving; from November to December, the NH_3-N increases, and the water quality is slightly worse (Table 3). The seasonal variation of water quality is mainly affected by the rainfall and the amount of water diversion. From January to May, the water quality gradually improves with the increase of water diversion volume; from June to September, the water diversion volume gradually decreases, and the precipitation rises to the maximum value in the year, which brings a large amount of non-point source pollutants to rivers, resulting in the deterioration of the water body; from October to December, the precipitation is low, but the water diversion volume increases significantly, and the water quality improves [40].

There is significant spatial variation in the water quality. The COD_{Mn} and NH_3-N of YJMD in the northwest and SYYX, LTZZ, SYDG, TDQT, XTLT in the main river channels of water diversion are relatively low, and the water quality is very good, with CWQI smaller than 3. The NH_3-N of SGSX is high, and the COD_{Mn} of GGDG, XTHT, CLDY, HGAF, XGSG is high, and the water quality is good, with CWQI in the range of 3–4. At BCSD, BSZS and XTBM, both COD_{Mn} and NH_3-N are high, with a CWQI larger than 4; therefore, the water quality is poor (Table 3). The regional water quality target is Grade III of GB3838-2002, but the COD_{Mn} and NH_3-N of each station exceed the standard by a certain proportion, with regional mean overproof rates of 15% and 21%, respectively. Here, the overproof rate means the portion of observed water quality data inferior to Grade III, whose upper limit value of COD_{Mn} and NH_3-N is 6 mg/L and 1 mg/L, respectively [19]. The overproof rates of COD_{Mn} at BSZS, XGSG and XTBM are high, with values bigger than 20%. The overproof rates of NH_3-N at XTBM, BCSD, BSZS, SGSX and XTLT are high, with values bigger than 25% (Table 4).

Table 2. Trend analysis of regional annual mean COD_{Mn}, NH_3-N, and CWQI.

Indicator	Z_s Value	Trend Slope	Tendency	Significance
COD_{Mn}	−0.693	−0.015	Decrease	Not significant
NH_3-N	−3.068	−0.01	Decrease	Significant
CWQI	−1.98	−0.013	Decrease	Significant

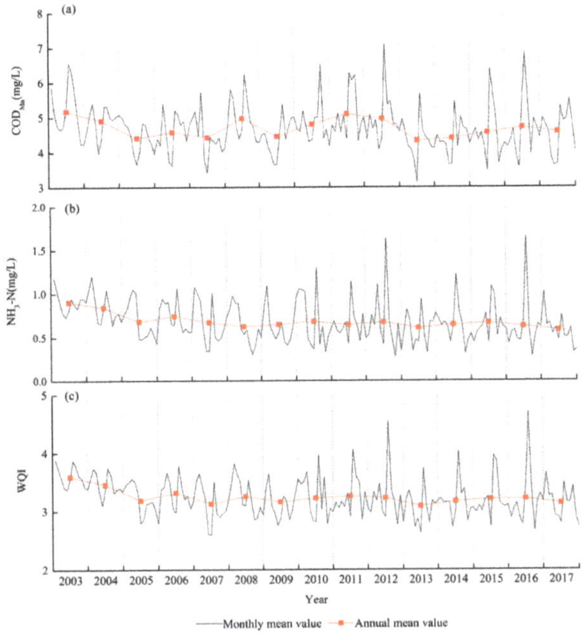

Figure 2. Temporal variations of regional mean COD_{Mn} (**a**), NH_3-N (**b**), and comprehensive water quality index (CWQI) (**c**).

Table 3. Intra-annual variation of water quality.

Month	January	February	March	April	May	June	July	August	September	October	November	December	Mean	Standard Deviation
COD_{Mn} (mg/L)	4.9	4.8	4.8	4.5	4.0	4.1	5.6	5.2	5.0	4.6	4.5	4.4	4.7	0.68
NH_3-N (mg/L)	0.80	0.85	0.77	0.69	0.59	0.54	1.04	0.64	0.52	0.55	0.62	0.61	0.68	0.24
CWQI	3.35	3.42	3.36	3.19	3.00	2.96	3.80	3.34	3.15	3.09	3.16	3.10	3.24	0.35

Table 4. Characteristic values of annual mean water quality of each station.

Region	Monitoring Stations	COD_{Mn} (mg/L)				NH_3-N (mg/L)				CWQI		
		Max.	Min.	Mean	op Rate * (%)	Max.	Min.	Mean	op Rate * (%)	Max.	Min.	Mean
Taizhou	CLDY	6.0	2.5	4.9	16	1.0	0.3	0.5	9	3.7	2.7	3.2
	HGAF	6.1	1.8	5.1	18	1.0	0.2	0.5	7	3.8	1.8	3.2
	LTZZ	4.5	2.0	3.6	6	0.8	0.2	0.5	10	3.3	2.3	2.8
	SGSX	5.8	2.8	4.6	8	1.3	0.5	0.9	32	3.9	3.0	3.5
	TDQT	4.8	1.8	3.4	3	1.1	0.3	0.6	16	3.6	2.1	2.9
	XGSG	8.1	4.8	5.7	28	1.0	0.3	0.6	10	4.1	3.2	3.4
	XTBM	6.7	2.0	5.0	24	2.4	0.3	1.6	73	4.9	2.2	4.2
	XTLT	5.0	3.0	3.5	6	1.1	0.3	0.7	25	3.7	2.5	3.0
Yancheng	GGDG	5.6	4.8	5.3	13	0.7	0.2	0.4	4	3.4	2.9	3.1
	SYYX	5.2	4.4	4.7	4	0.4	0.1	0.3	2	3.0	2.5	2.8
	XTHT	9.3	4.5	5.2	17	0.7	0.3	0.4	2	3.9	2.8	3.1
	YJMD	5.4	3.6	4.4	5	1.4	0.1	0.3	7	3.9	2.1	2.7
Yangzhou	BSZS	8.0	4.4	6.1	42	2.2	0.9	1.1	49	5.0	3.5	4.0
	BCSD	7.4	4.2	5.0	18	1.9	1.0	1.4	62	4.6	3.6	4.0
	SYDG	6.5	2.9	3.9	8	0.7	0.2	0.5	7	3.6	2.3	2.8

* overproof rate: portion of observed water quality data inferior to Grade III.

3.2. Periodic Variation of Water Quality

CWT was employed to the CWQI time series of each station to analyze its periodic characteristics. The CWT results of some stations are presented in Figure 3, and the characteristic period of each station is summarized in Table 5. At a 95% confidence level, the CWQI of each station has a significant period in the time-frequency domain.

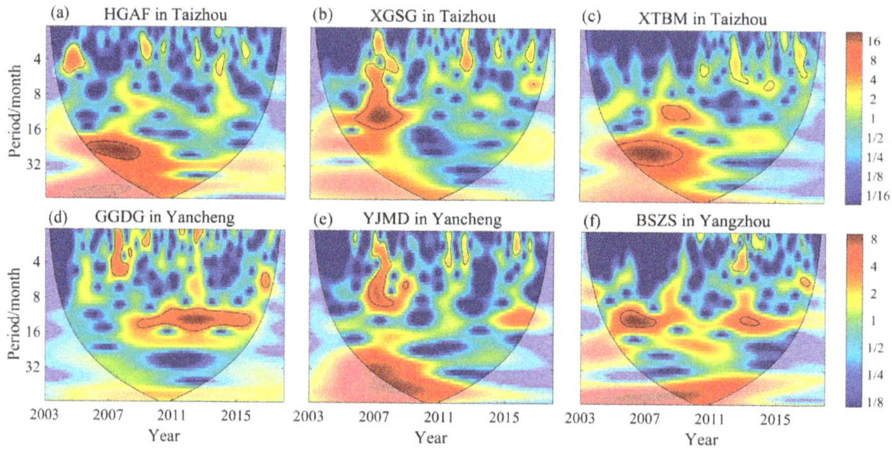

Figure 3. Results of continuous wavelet transform (CWT) of CWQI at six representative stations, (**a**) HGAF, (**b**) XGSG, (**c**) XTBM, (**d**) GGDG, (**e**) YJMD, (**f**) BSZS. Red and blue represent the peak and valley values of energy density respectively, and the color shade represents the relative strength of energy density. The closed area of the black thick solid line has passed the red noise test of the 95% confidence level. The cone area under the black thin solid line is the wavelet influence cone (COI), which indicates the area with great influence of data edge.

In northern Taizhou, CLDY, SGSX, and XGSG have a period of 10–12 months, and the periodic occurrence of each station is different; HGAF has a period of 21–30 months (2005–2008). In southern Taizhou, LTZZ, TDQT, and XTBM have a period of 21–25 months (2005–2008), and XTLT has a period of 2–8 months (2005–2007). In the Yancheng area, GGDG, SYYX, and YJMD have a period of 10 months, and the periods of occurrence of each station are different. XTHT has a period of 30–36 months (2006–2008). In the Yangzhou area, BSZS and SYDG have a 13-month cycle (2005–2007); SYDG has a 4–7-month cycle (2005–2006). In general, the CWQI of each station contains multi-scale significant periodic fluctuations of 3–59 months, and the seasonal variation of 12 months is significant at most stations.

Table 5. The characteristic periods of CWQI.

Region	Monitoring Stations		Characteristic Periods				
Taizhou	CLDY	Time	2010–2012	2012–2013	2015	-	-
		Period (month)	10–13	2–5	2–4	-	-
	HGAF	Time	2004–2005	2005–2008	2008–2009	2013–2014	2016
		Period (month)	2–6	21–30	2–4	4–6	2–4
	LTZZ	Time	2005–2007	2010–2012	2012	2012–2013	2016–2017
		Period (month)	18–25	10–15	2–5	6–7	2–6
	SGSX	Time	2007–2009	2008–2010	2012–2013	2016–2017	-
		Period (month)	5–7	9–14	6	2–8	-
	TDQT	Time	2005–2008	2007–2008	2010–2011	2015	2016
		Period (month)	21–30	5–7	6–7	4–5	2–7
	XGSG	Time	2005–2008	2006–2008	2012	2014	2016
		Period (month)	4–16	2–5	2–5	2–4	2–4
	XTBM	Time	2005–2009	2007–2009	2012–2013	2014	2015–2016
		Period (month)	18–31	9–13	2–6	5–7	2–5
	XTLT	Time	2005–2007	2013	2016	-	-
		Period (month)	2–8	2–3	2–4	-	-
Yancheng	GGDG	Time	2007–2008	2008–2015	2016–2017	-	-
		Period (month)	2–6	10–16	6–7	-	-
	SYYX	Time	2007	2008–2009	2015–2016	2016	-
		Period (month)	2–5	6–7	9–14	4–6	-
	XTHT	Time	2006–2008	2012	2015–2016	-	-
		Period (month)	30–36	2–4	2–5	-	-
	YJMD	Time	2006–2007	2008	2011	-	-
		Period (month)	2–10	6–7	2–4	-	-
Yangzhou	BSZS	Time	2005–2007	2012–2013	2012–2013	2016	-
		Period (month)	10–15	2–5	11–15	5–6	-
	BCSD	Time	2003–2005	2007	2009–2010	2010–2012	2012–2013
		Period (month)	4–9	13	6–8	46–59	2–5
	SYDG	Time	2003–2004	2005–2006	2007–2008	2011	2016
		Period (month)	6–7	4–7	6–7	7	2–6

3.3. Long-Term Trends of Water Quality

Long-term trends of water quality were investigated by applying the Mann–Kendall trend test and Sen's slope estimator on the CWQI. The CWQI of 7 out of 15 monitoring stations has a significant decreasing trend (Table 6). In the Taizhou area, the trend of the CWQI of XTLT is the most significant, showing a rapid decreasing trend, with a trend slope to be −0.056/yr, and the water quality is greatly improved; at SGSX, TDQT, LTZZ, and XGSG, the CWQI variation shows a very significant decreasing trend, with the trend slope to be between −0.046/yr and −0.023/yr, and the water quality is significantly improved. The CWQI variation of CLDY, HGAF, and XTBM is not significant. In the Yancheng area, the trend of CWQI variation is not significant. In the Yangzhou area, the trend of CWQI of BSZS is the most significant, showing a rapid decreasing trend, with the trend slope to be −0.071/yr, which indicates the water quality improvement by one grade in 15 years, and the water quality is greatly improved. The CWQI of SYDG shows a very significant decreasing trend, with the trend slope to be −0.037/yr, and the water quality is obviously improved. The CWQI variation of BCSD is not significant.

The trend slope of CWQI ranges from −0.071/yr to 0.007/yr, and two-thirds of the monitoring stations have a negative trend slope, indicating that the water quality of rivers in the Lixiahe abdominal area is gradually improving, but the spatial difference is large (Figure 4). The water quality of XTLT, TDQT, SGSX, LTZZ, and XGSG in the Taizhou area and SYDG and BSZS in the Yangzhou area, which are located in the main channels of water diversion, has been significantly improved, while in the Yancheng area, which is located downstream of the water diversion, the improvement on water quality is not significant. The water quality of the XTBM in the southeastern area, HGAF and GGDG in the eastern area, and BCSD in the western area, which cannot be reached by the diversion water, even shows a trend of deterioration.

Table 6. The trend and trend slope of the CWQI.

Region	Monitoring Stations	Zs Value	Annual Trend Slope	Tendency	Significance
Taizhou	CLDY	−0.897	−0.007	Decrease	Not significant
	HGAF	0.739	0.006	Increase	Not significant
	LTZZ	−2.734	−0.026	Decrease	Significant
	SGSX	−4.267	−0.042	Decrease	Significant
	TDQT	−2.985	−0.035	Decrease	Significant
	XGSG	−3.144	−0.023	Decrease	Significant
	XTBM	0.170	0.002	Increase	Not significant
	XTLT	−4.015	−0.056	Decrease	Significant
Yancheng	GGDG	0.964	0.007	Increase	Not significant
	SYYX	−0.685	−0.005	Decrease	Not significant
	STHT	−1.851	−0.014	Decrease	Not significant
	YJMD	−1.549	−0.013	Decrease	Not significant
Yangzhou	BSZS	−6.245	−0.071	Decrease	Significant
	BCSD	0.510	0.007	Increase	Not significant
	SYDG	−2.854	−0.037	Decrease	Significant

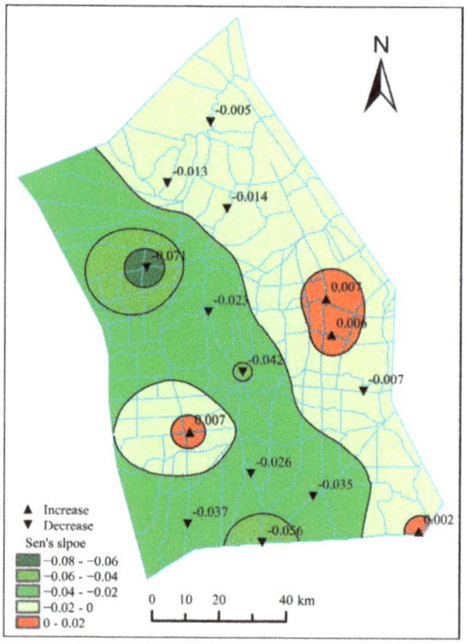

Figure 4. Spatial variation of trend slope of the CWQI.

4. Possible Causes of Water Quality Variation

The temporal and spatial variation of water quality is closely related to the discharge and accumulation of pollutants, the local water resources from precipitation, and the amount of water from external sources. Therefore, the following discussion focuses on the impact of water pollutant input, water level, and the amount of diversion water on water quality.

4.1. Input of Regional Water Pollutants

It is not easy to obtain the regional short-term pollutant input; thus, the annual load of four main indicators (COD_{Mn}, NH_3-N, TN, and TP) of the Lixiahe abdominal area was calculated according to the Jiangsu Statistical Yearbook (Figure 5). During 2003–2017, the discharge of COD_{Mn} and TP increased by 54% and 39% respectively, while the increase in NH_3-N and TN was slightly smaller, 27% and 24%. The growth rate of point source pollutants from industrial and municipal wastewater emissions is about 35%. The main point source pollutant inputs are COD_{Mn} and TP, and the amount of NH_3-N and TN is relatively small. Among the non-point source pollutants, the growth rate of COD_{Mn} and TP is relatively large, 70% and 47% respectively, and the growth rate of NH_3-N and TN is relatively small, about 24%. In general, with the booming of the social economy, the main pollutants discharged to river water have increased significantly, which may induce water environment deterioration.

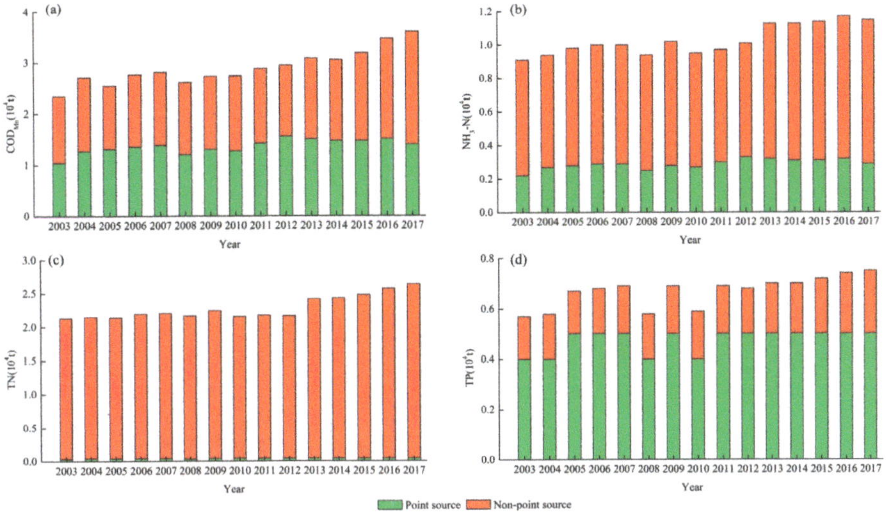

Figure 5. Temporal variation of annual load of four main pollutants, (**a**) COD_{Mn}, (**b**) NH_3-N, (**c**) TN, and (**d**) TP, discharged to the river water of the Lixiahe abdominal area.

4.2. Correlation Analysis of Water Quality and Water Level

The water level reflects the amount of water in the area, which is influenced by multi-processes of precipitation, water diversion, and drainage, and it has an important impact on the water quality. The correlation of the CWQI and water level of each station are analyzed by XWT and WTC analysis by identifying the resonance period. The XWT and WTC reflect the resonance signal characteristics and the correlation coefficient of the CWQI and water level in the high-energy area and low-energy area, respectively. The results of XWT and WTC analysis at some stations are shown in Figure 6, and the correlation between the CWQI and water level at each station is summarized in Table 7.

In the Taizhou area, except for HGAF, the CWQI and water level of the other seven stations have a resonance period of 12–14 months, CLDY, LTZZ, and XTLT have positive correlations, XGSG and XTBM have negative correlations, SGSX has negative and positive correlations in the high-energy area and low-energy area respectively, TDQT has a transition from negative correlation to positive correlation in the high-energy area and positive correlation in the low-energy area. In the Yancheng area, there is a 12–14-month resonance period between the CWQI and water level at four monitoring stations, GGDG and YJMD have positive correlations, XTHT has a negative correlation, SYYX has positive and negative correlations in the high-energy area and low-energy area, respectively. In the Yangzhou area, there are 12–14-month resonance periods between the CWQI and water level of three monitoring stations, BSZS and BCSD have a negative correlation, and SYDG has a positive correlation.

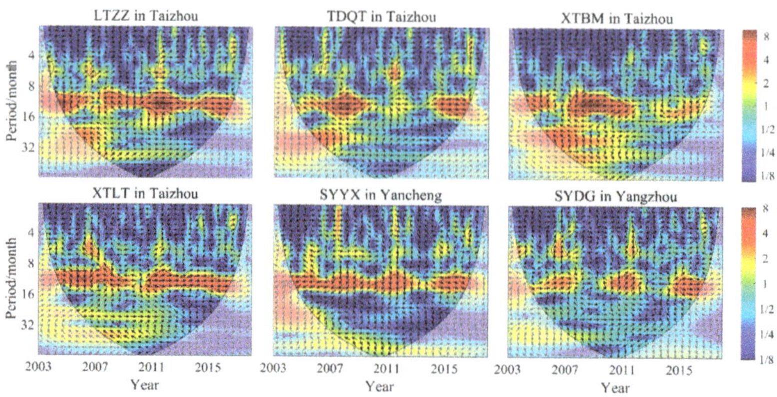

(**a**) Cross wavelet transform (XWT) of CWQI and water level.

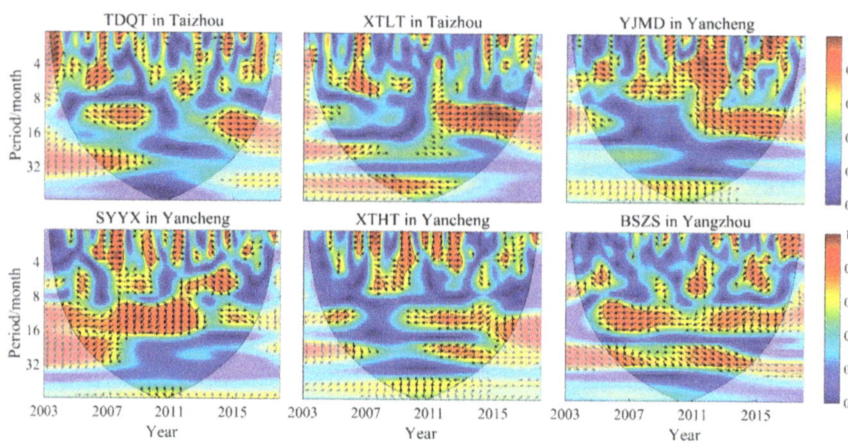

(**b**) WTC of CWQI and water level.

Figure 6. Results of XWT (**a**) and WTC (**b**) of CWQI and water level at six representative stations. The color scale on the right side of the figure represents the density of the cross wavelet power spectrum, and the arrow direction reflects the phase relationship between the CWQI and water level: '→ ' represents the same phase, '← ' represents the opposite phase, '↓ ' represents that the CWQI lags behind the water level by one-quarter of a cycle, and '↑ ' represents that the CWQI advances the water level by one-quarter of a cycle.

Table 7. Correlation between the CWQI and water level.

Region	Monitoring Stations	High-Energy Area			Low-Energy Area			
		Resonance Period (Month)	Time	Phase Relationship	Resonance Period (Month)	Time	Phase Relationship	Correlation Coefficient
Taizhou	CLDY	10–14	2009–2013	→	4–7	2005–2007	↑	0.8
		10–14	2013–2016	→	8–16	2010–2015	→	0.8
	HGAF	11–13	2008	←	1–4	2016–2017	↑	0.9
	LTZZ	10–14	2004–2006	↑	4–7	2006–2007	→	0.8
		9–15	2007–2013	→	39–50	2007–2012	→	0.7
		11–14	2014–2016	→	8–20	2011–2016	→	0.9
	SGSX	10–14	2007–2011	←	2–6	2011	→	0.8
		-	-	-	10–14	2012–2015	→	0.7
	TDQT	10–15	2006–2009	←	10–13	2007–2009	←	0.7
		11–15	2014–2016	→	27–32	2005–2007	↑	0.8
		-	-	-	12–26	2013–2016	→	0.9
	XGSG	9–16	2004–2009	←	10–16	2005–2009	←	0.8
		-	-	-	43–57	2007–2009	←	0.7
	XTBM	10–13	2004–2005	→	8–16	2008–2010	←	0.8
		9–15	2007–2011	→	-	-	-	-
	XTLT	9–14	2004–2008	↓	40–52	2007–2010	←	0.7
		10–15	2010–2016	→	8–16	2011–2016	→	0.9
Yancheng	GGDG	10–15	2009–2013	→	2–8	2004–2006	→	0.9
		12–14	2013–2016	→	8–14	2008–2013	→	0.8
	SYYX	11–15	2004–2009	←	20–30	2005–2006	←	0.8
		10–15	2009–2012	↓	9–18	2004–2012	←	0.9
		11–14	2014–2016	→	-	-	-	-
	XTHT	11–14	2004–2006	←	12–13	2004–2005	←	0.7
		12–14	2008–2010	←	12–14	2011–2014	←	0.7
		-	-	-	12–20	2015–2016	↓	0.8
	YJMD	12–14	2014–2016	→	1–9	2009–2012	→	0.9
		-	-	-	10–16	2011–2016	→	0.8
Yangzhou	BSZS	9–15	2004–2009	←	8–15	2005–2009	←	0.9
		11–14	2011–2015	←	21–31	2011–2014	→	0.8
		-	-	-	11–14	2012–2016	←	0.8
	BCSD	9–15	2004–2010	←	9–16	2004–2012	←	0.9
	SYDG	11–13	2010–2011	→	1–6	2005–2007	↑	0.9
		12–15	2014–2016	→	12–15	2015–2016	→	0.8

In general, the CWQI has a resonance period of about 12 months with water level, with positive correlation accounting for 62% and negative correlation accounting for 38%. The change of correlation in different periods is mainly related to the water level being affected by multiple factors, such as local water resources, water diversion, and drainage. The increase in local water recourse due to precipitation induces the water level rise and water quality deterioration with an increase in the CWQI, due to large amount of non-point source pollutants entering the river with rainfall runoff. The increase in diversion water leads to water level rise and water quality improvement with an increase in the CWQI. Hence, the CWQI and water level is mainly negatively correlated in the wet season, when the water resource is mainly from rainfall runoff, while the CWQI and water level is mainly positively correlated in the dry season, when the water resource is mainly from water diversion.

4.3. Correlation Analysis of Water Quality and Amount of Diversion Water

The Lixiahe abdominal area is in the upstream section of south-to-north water diversion from the Yangtze River in Jiangsu Province. Thus, the water quality of the area is improved by introducing the good-quality Yangtze River water. The correlation of the CWQI and water diversion volume of each station is analyzed by XWT and WTC analysis, by identifying the resonance period. The results of XWT and WTC analysis of the CWQI and water diversion volume at some stations are shown in Figure 7. The correlation between the CWQI and diversion water at each station is summarized in Table 8.

In the Taizhou area, except for the high-energy area of HGAF, there is a resonance period of 11–12 months for the CWQI and water diversion volume of each station; CLDY, HGAF, LTZZ, SGSX and TDQT have inverse correlation, the CWQI of XGSG lags behind the water diversion volume by one-quarter of a cycle, while XTBM and XTLT have different correlation in the high-energy area and low-energy area. In the Yancheng area, there is a 12–13-month resonance period between the CWQI and water diversion volume at four monitoring stations. Here, XTHT has positive correlation, GGDG and YJMD have negative correlation, SYYX has different correlations in the high-energy area and low-energy area. In the Yangzhou area, the resonance period of the CWQI and water diversion volume of three monitoring stations is 12–14 months, BSZS has a positive correlation, SYDG has a negative correlation, and the CWQI of BCSD lags behind the water diversion volume by one-quarter of a cycle.

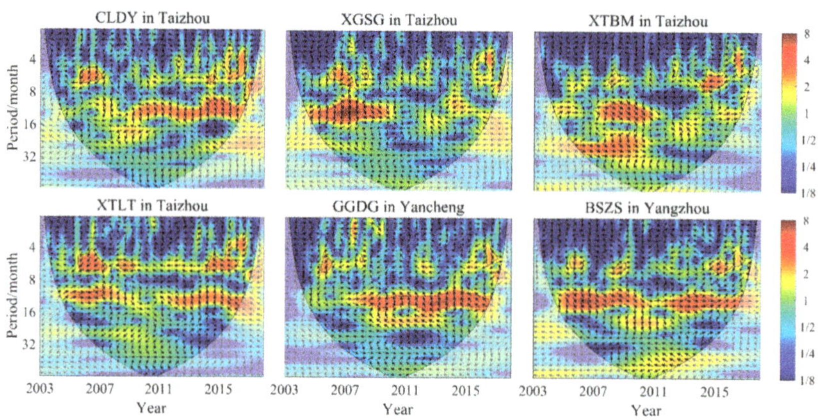

(**a**) XWT of the CWQI and water level.

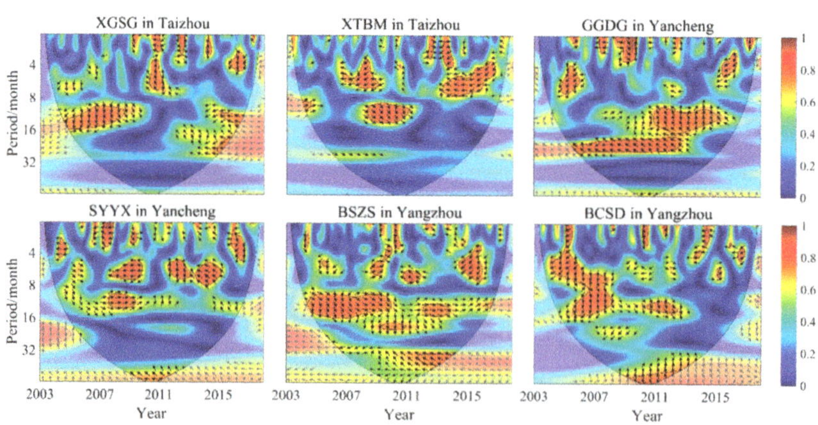

(**b**) WTC of the CWQI and water level.

Figure 7. Results of XWT (**a**) and WTC (**b**) of the CWQI and the amount of diversion water at six representative stations. The color scale on the right side of the figure represents the density of the cross wavelet power spectrum, and the arrow direction reflects the phase relationship between the CWQI and water level: '→ ' represents the same phase, '← ' represents the opposite phase, '↓ ' represents that the CWQI lags behind the water level by one-quarter of a cycle, and '↑ ' represents that the CWQI advances the water level by one-quarter of a cycle.

Table 8. Correlation between the CWQI and the amount of diversion water.

Region	Monitoring Stations	High-Energy Region			Low-Energy Region			
		Resonance Period (Month)	Time	Phase Relationship	Resonance Period (Month)	Time	Phase Relationship	Correlation Coefficient
Taizhou	CLDY	10–13	2013–2015	←	1–7	2004–2007	←	0.9
		-	-	-	8–16	2011–2015	←	0.8
	HGAF	5–7	2013–2014	↑	8–16	2012–2015	←	0.8
	LTZZ	12–14	2011–2012	←	3–7	2005–2008	←	0.9
		11–15	2015–2016	←	5–8	2009–2013	←	0.9
		-	-	-	11–32	2014–2016	←	0.8
		-	-	-	2–7	2015–2016	←	0.8
	SGSX	11–14	2008–2010	↓	4–7	2007–2011	←	0.9
		-	-	-	11–15	2008–2010	↓	0.7
		-	-	-	11–25	2012–2016	←	0.9
	TDQT	12–13	2007–2008	↓	4-8	2005–2007	←	0.9
		2–7	2014–2016	←	4–31	2007–2016	←	0.9
		12–14	2015–2016	←	-	-	-	-
	XGSG	10–15	2005–2008	↓	10–16	2004–2008	↓	0.8
	XTBM	10–13	2007–2009	→	9–14	2008–2011	→	0.9
		-	-	-	4–8	2013–2016	←	0.8
	XTLT	4–7	2005–2007	←	1–8	2004–2013	←	0.9
		11–12	2005–2007	→	10–18	2012–2016	←	0.9
		12–15	2015–2016	←	-	-	-	-
Yancheng	GGDG	11–16	2008–2016	←	10–16	2010–2014	↑	0.8
		-	-	-	18–27	2005–2012	←	0.9
	SYYX	11-14	2014–2016	←	5–8	2007–2009	←	0.9
		-	-	-	10–14	2007–2009	→	0.8
		-	-	-	5–8	2013–2014	←	0.9
	XTHT	12–14	2009–2010	→	11–14	2004–2005	→	0.8
		-	-	-	11–16	2009–2011	→	0.7
		-	-	-	10–13	2013–2014	→	0.7
	YJMD	12–14	2014–2016	←	4–7	2008–2010	←	0.9
				-	9–15	2012–2016	←	0.8
Yangzhou	BSZS	11–14	2005–2008	→	10–16	2004–2009	→	0.9
		12–14	2012–2014	→	12–16	2012–2013	→	0.7
	BCSD	4–8	2004–2006	←	1–15	2003–2007	←	0.9
		12–14	2005–2008	↓	12–14	2010–2012	→	0.7
	SYDG	4–7	2005–2007	←	1–8	2004–2009	←	0.9
		2–7	2014–2016	←	2–8	2010–2014	←	0.9
		12–15	2015–2016	←	2–15	2014–2016	←	0.8

In general, there is a resonance period of about 12 months between the CWQI and water diversion volume at each monitoring station. Two-thirds of the 15 monitoring stations are located in the main channels of water diversion and nearby rivers, and one-third is far away from the main channels of water diversion. The CWQI of the monitoring stations at the water diversion route and its nearby river is inversely related to the water diversion volume. The larger the water diversion volume, the better the water quality. The water diversion from the Yangtze River plays an important role in improving the regional water environment.

Due to the complexity of the water quality variation and the limitation on the methodology, this study was qualitative when assessing the possible causes of water quality variation. Nevertheless, water diversion is proved to be an important contributor to improving the regional water environment. Further work is required to investigate the quantitative relationship between the water quality and intensity and duration of water diversion to provide scientific guidance on the optimization of water diversion operation.

5. Conclusions

From 2003 to 2017, the water quality variation in the Lixiahe abdominal area contains multi-scale periodic fluctuations of 3–59 months, and the seasonal variation of 12 months is significant at most stations. The CWQI of 7 out of 15 monitoring stations has a significant decreasing trend, and the trend slope ranges from −0.071/yr to 0.007/yr. The water quality of the main routes of the water diversion and the nearby rivers has significantly improved, while the water quality of rivers far away from the main routes, which is less affected by the water diversion, has no obvious improvement, or even becomes worse.

The CWQI and water level is mainly positively correlated in the wet season, when the water resource is mainly from rainfall runoff that brings many non-point pollutants, while the CWQI and water level are mainly inversely correlated in the dry season, when the water resource is mainly from water diversion from the Yangtze River. The CWQI and the water diversion volume are inversely related at monitoring stations in the main routes of water diversion and its nearby river. Hence, water diversion plays an important role in improving the regional water environment.

With the booming of the social economy in the Lixiahe abdominal area, the main pollutants discharged to river water have increased, but the water quality has generally improved, especially in the main routes of water diversion and its nearby rivers, due to the water diversion from the Yangtze River. Hence, the key to regional water environment improvement lies in the systematic control of point and non-point source pollutants and the optimization of water diversion operation.

Author Contributions: Conceptualization, C.J., G.F. and J.Z. (Jiren Zhou); methodology, C.J., J.Z. (Jia'nan Zhou) and J.Z. (Jiren Zhou); software, J.Z. (Jia'nan Zhou); validation, J.Z. (Jia'nan Zhou); formal analysis, C.J.; investigation, C.J.; data curation, G.F.; writing—original draft preparation, C.J.; writing—review and editing, J.W. and J.Z. (Jiren Zhou); supervision, C.J.; project administration, G.F. All authors have read and agreed to the published version of the manuscript.

Funding: This research was funded by Priority Academic Program Development of Jiangsu Higher Education Institutions (PAPD) and Jiangsu water conservancy science and technology project (2015024).

Acknowledgments: The authors wish to thank the Taizhou Branch, Yancheng branch, and Yangzhou Branch of The Jiangsu Bureau of hydrological and water resources survey for permission to use the hydrological data and water quality data.

Conflicts of Interest: The authors declare no conflict of interest. The funders had no role in the design of the study; in the collection, analyses, or interpretation of data; in the writing of the manuscript, or in the decision to publish the results.

References

1. Zhai, X.Y.; Xia, J.; Zhang, Y.Y. Integrated approach of hydrological and water quality dynamic simulation for anthropogenic disturbance assessment in the Huai River Basin, China. *Sci. Total Environ.* **2017**, *598*, 749–764. [CrossRef] [PubMed]
2. Hussein, H. The Guarani Aquifer System, highly present but not high profile: A hydropolitical analysis of transboundary groundwater governance. *Environ. Sci. Policy.* **2018**, *83*, 54–62. [CrossRef]
3. Odeh, T.; Mohammad, A.H.; Hussein, H.; Ismail, M.; Almomani, T. Over-pumping of groundwater in Irbid governorate, northern Jordan: A conceptual model to analyze the effects of urbanization and agricultural activities on groundwater levels and salinity. *Environ. Earth Sci.* **2019**, *78*, 40. [CrossRef]
4. Mohammad, A.H.; Jung, H.C.; Odeh, T.; Bhuiyan, C.; Hussein, H. Understanding the impact of droughts in the Yarmouk Basin, Jordan: Monitoring droughts through meteorological and hydrological drought indices. *Arab. J. Geosci.* **2018**, *11*, 103. [CrossRef]
5. Cullaj, A.; Hasko, A.; Miho, A.; Schanz, F.; Brandl, H.; Bachofen, R. The quality of Albanian natural waters and the human impact. *Environ. Int.* **2005**, *31*, 133–146. [CrossRef]
6. Bordalo, A.A.; Teixeira, R.; Wiebe, W.J. A water quality index applied to an international shared river basin: The case of the Douro River. *Environ. Manag.* **2006**, *38*, 910–920. [CrossRef]
7. Xu, Z.-X. Single factor water quality indentification index for environmental quality assessment of surface water. *J. Tongji Univ. (Nat. Sci.)* **2005**, *33*, 321–325. (In Chinese with English abstract)
8. Lermontov, A.; Yokoyama, L.; Lermontov, M.; Machado, M.A.S. River quality analysis using fuzzy water quality index: Ribeira do Iguape river watershed, Brazil. *Ecol. Indic.* **2009**, *9*, 1188–1197. [CrossRef]
9. Zhang, Y.; Guo, F.; Meng, W.; Wang, X.-Q. Water quality assessment and source identification of Daliao river basin using multivariate statistical methods. *Environ. Monit. Assess.* **2009**, *152*, 105–121. [CrossRef]

10. Hurley, T.; Sadiq, R.; Mazumder, A. Adaptation and evaluation of the Canadian Council of Ministers of the Environment Water Quality Index (CCME WQI) for use as an effective tool to characterize drinking source water quality. *Water Res.* **2012**, *46*, 3544–3552. [CrossRef]
11. Ahmed, S.; Khurshid, S.; Madan, R.; Amarah, B.A.A.; Naushad, M. Water quality assessment of shallow aquifer based on Canadian Council of Ministers of the environment index and its impact on irrigation of Mathura District, Uttar Pradesh. *J. King Saud Univ. Sci.* **2020**, *32*, 1218–1225. [CrossRef]
12. Ajorlo, M.; Abdullah, R.B.; Yusoff, M.K.; Halim, R.A.; Hanif, A.H.M.; Willms, W.D.; Ebrahimian, M. Multivariate statistical techniques for the assessment of seasonal variations in surface water quality of pasture ecosystems. *Environ. Monit. Assess.* **2013**, *185*, 8649–8658. [CrossRef] [PubMed]
13. Sojka, M.; Siepak, M.; Ziola, A.; Frankowski, M.; Murat-Blazejewska, S.; Siepak, J. Application of multivariate statistical techniques to evaluation of water quality in the Mala Welna River (Western Poland). *Environ. Monit. Assess.* **2008**, *147*, 159–170. [CrossRef] [PubMed]
14. Zhou, L.; Xu, S. Application of grey clustering method in eutrophication assessment of wetland. *J. Am. Sci.* **2006**, *2*, 53–58.
15. Wątor, K.; Kmiecik, E.; Lipiec, I. The use of principal component analysis for the assessment of the spatial variability of curative waters from the Busko-Zdrój and Solec-Zdrój region (Poland)–preliminary results. *Water Supply* **2018**, *19*, 1137–1143. [CrossRef]
16. Xia, R.; Chen, Z. Integrated water-quality assessment of the Huai River Basin in China. *J. Hydrol. Eng.* **2014**, *20*, 05014018. [CrossRef]
17. Sarkar, A.; Pandey, P. River water quality modelling using artificial neural network technique. *Aquat. Procedia* **2015**, *4*, 1070–1077. [CrossRef]
18. Hind, J.A.; Marwan, A. Assessing groundwater vulnerability in Azraq Basin area by a modified drastic Index. *J. Water Resour. Prot.* **2010**, *2*, 944–951.
19. Merchant, J.M. GIS-Based groundwater pollution hazard assessment: A critical review of the DRASTIC Model. *Photograomm. Eng. Rem. Sens.* **1994**, *60*, 1117–1127.
20. Torrence, C.; Comopo, G.P. A practical guide to wavelet analysis. *Bull. Am. Meteorol. Soc.* **1998**, *79*, 61–78. [CrossRef]
21. Grinsted, A.; Moore, J.C.; Jevrejeva, S. Application of the cross wavelet transform and wavelet coherence to geophysical time serial. *Nonlinear Proc. Geoph.* **2004**, *11*, 505–514. [CrossRef]
22. Smith, L.C.; Turcotte, D.L.; Isacke, B. Stream flow characterization and feature detection using a discrete wavelet transform. *Hydrol. Processes.* **1998**, *12*, 233–249. [CrossRef]
23. Labat, D. Recent advances in wavelet analyses: Part 1. A review of concepts. *J. Hydrol.* **2005**, *314*, 275–288. [CrossRef]
24. Kang, S.-J.; Lin, H. Wavelet analysis of hydrological and water quality signals in an agricultural watershed. *J. Hydrol.* **2007**, *338*, 1–14. [CrossRef]
25. Hatvani, I.G.; Clement, A.; Korponai, J.; Kern, Z.; Kovács, Z. Periodic signals of climatic variables and water quality in a river-eutrophic pond-wetland cascade ecosystem tracked by wavelet coherence analysis. *Ecol. Indic.* **2017**, *83*, 21–31. [CrossRef]
26. Barzegar, R.; Adamowski, J.; Moghaddam, A.A. Application of wavelet-artificial intelligence hybrid models for water quality prediction a case study in Aji-Chay River. *Stoch. Environ. Res. Risk Assess.* **2016**, *30*, 1797–1819. [CrossRef]
27. Barzegar, R.; Moghaddam, A.A.; Adamowski, J.; Ozga-Zielinski, B. Multi-step water quality forecasting using a boosting ensemble multi-wavelet extreme learning machine model. *Stoch. Environ. Res. Risk Assess.* **2018**, *32*, 799–813. [CrossRef]
28. Parmar, K.S.; Makkhan, S.J.S.; Kaushal, S. Neuro-fuzzy-wavelet hybrid approach to estimate the future trends of river water quality. *Neural Comput. Appl.* **2019**, *31*, 8463–8473. [CrossRef]
29. Najah, A.A.; EI-Shafie, A.; Karim, O.A. Water quality prediction model utilizing integrated wavelet-ANFIS model with cross-validation. *Neural Comput. Appl.* **2012**, *21*, 833–841. [CrossRef]
30. Ghassemi, F.; White, I. *Inter-basin Water Transfer: Case Studies from Australia, United States, Canada, China, and India*; Cambridge University Press: New York, NY, USA, 2007.
31. Zeng, Q.H.; Qin, L.H.; Li, X.Y. The potential impact of an inter-basin water transfer project on nutrients (nitrogen and phosphorous) and chlorophyll a of the receiving water system. *Sci. Total Environ.* **2015**, *536*, 675–686. [CrossRef]

32. Hoekstra, A.Y.; Mekonnen, M.M. Imported water risk: The case of the UK. *Environ. Res. Lett.* **2016**, *11*. [CrossRef]
33. Xu, Z.-X. Comprehensive water quality identification index for environmental quality assessment of surface water. *J. Tongji Univ. (Nat. Sci.)* **2005**, *33*, 482–488. (In Chinese with English abstract)
34. GB3838-2002. *The State Standards of the People's Republic of China: Standard for Surface Water Environmental Quality Assessment*; China Environmental Science Press: Beijing, China, 2002. (In Chinese)
35. Liu, F.; Chen, H.; Cai, H.-Y.; Luo, X.-X.; Ou, S.-Y.; Yang, Q.-S. Impacts of ENSO on multi-scale variations in sediment discharge from the Pearl River to the south China sea. *Geomorphology* **2017**, *293*, 24–36. [CrossRef]
36. Gocic, M.; Trajkovic, S. Analysis of changes in meteorological variables using Mann-Kendall and Sen's slope estimator statistical tests in Serbia. *Glob. Planet. Chang.* **2013**, *100*, 172–182. [CrossRef]
37. Douglas, E.M.; Vogel, R.M.; Kroll, C.N. Trends in floods and low flows in the United States: Impact of spatial correlation. *J. Hydrol.* **2000**, *240*, 90–105. [CrossRef]
38. Tabari, H.; Talaee, P.H. Analysis of trends in temperature data in arid and semi-arid regions of Iran. *Glob. Planet. Chang.* **2011**, *79*, 1–10. [CrossRef]
39. Rahman, A.; Dawood, M. Spatio-statistical analysis of temperature fluctuation using Mann-Kendall and Sen's slope approach. *Clim. Dyn.* **2017**, *48*, 783–797. [CrossRef]
40. Zhou, J.-N.; Fu, G.S.; An, H.; Jiang, C.J. Analysis of Temporal and spatial variation characteristics of water quality in Lixiahe Abdominal Area. *China Rural Water Hydropower* **2020**, *4*, 22–29. (In Chinese with English Abstract)

water

MDPI

Article

Assessment of Water Quality Using Chemometrics and Multivariate Statistics: A Case Study in Chaobai River Replenished by Reclaimed Water, North China

Yilei Yu [1,2], Xianfang Song [3,*], Yinghua Zhang [3] and Fandong Zheng [4]

1 Institute of Wetland Research, Chinese Academy of Forestry, Beijing 100091, China; yuyilei1222@126.com
2 Beijing Key Laboratory of Wetland Services and Restoration, Beijing 100091, China
3 Key Laboratory of Water Cycle and Related Land Surface Processes, Institute of Geographic Sciences and Natural Resources Research, Chinese Academy of Sciences, Beijing 100101, China; zhangyinghua@igsnrr.ac.cn
4 Beijing Water Science and Technology Institute, Beijing 10048, China; zfd@bwsti.com
* Correspondence: songxf@igsnrr.ac.cn; Tel.: +86-010-6488-9849

Received: 3 August 2020; Accepted: 10 September 2020; Published: 12 September 2020

Abstract: Dry rivers could be effectively recovered by reclaimed water in North China, while river water quality would be an important issue. Therefore, it is important to understand the spatiotemporal variation and controlling factors of river water. Water samples were collected during March, May, July, September, and November in the year 2010, then 20 parameters were analyzed. The water environment was oxidizing and alkaline, which was beneficial for nitrification. Nitrate was the main nitrogen form. Depleted and enriched isotopes were found in reclaimed water and river water, respectively. Total nitrogen (TN) and total phosphorus (TP) of reclaimed water exceed the threshold of reclaimed water reuse standard and Class V in the surface water quality criteria. Most river water was at the severe eutrophication level. The sodium adsorption ratio indicated a medium harmful level for irrigation purpose. Significant spatial and temporal variation was explored by cluster analysis. Five months and nine stations were both classified into two distinct clusters. It was found that 6 parameters (chloride: Cl^-, sulphate: SO_4^{2-}, potassium: K^+, sodium: Na^+, magnesium: Mg^{2+}, and total dissolved solids: TDS) had significant upward temporal variation, and 12 parameters (dissolved oxygen: DO, electric conductivity: EC, bicarbonate: HCO_3^-, K^+, Na^+, Ca^{2+}, TDS, nitrite-nitrogen: NO_2-N, nitrate nitrogen: NO_3-N, TN, TP, and chlorophyll a: *Chl.a*) and 4 parameters (Mg^{2+}, ammonia nitrogen: NH_3-N, and the oxygen-18 and hydron-2 stable isotope: $\delta^{18}O$ and δ^2H) had a significant downward and upward spatial trend, respectively. The Gibbs plot showed that river water chemistry was mainly controlled by a water–rock interaction. The ionic relationship and principal component analysis showed that river water had undergone the dissolution of carbonate, calcite, and silicate minerals, cation exchange, a process of nitrification, photosynthesis of phytoplankton, and stable isotope enrichment. In addition, gypsum and salt rock have a potential dissolution process.

Keywords: water chemistry; river water; reclaimed water; multivariate statistics; Chaobai River

1. Introduction

North China has been facing serious water resources shortage in recent decades, as a result of continual drought, large consumption of water resources, water pollution, and economic development [1,2]. The groundwater table continued to decline and many rivers have been cut off or dried up for years [3]. Beijing as the capital of China, i.e., a big city located in North China, has also been facing massive water shortage. Multi-year average precipitation and evaporation are about 590 mm and 1800 mm, respectively [4,5]. Surface water resources of Beijing was 7.22×10^8 m^3, and 14.32 × 10^8 m^3 in the year 2010 and 2018, respectively [4]. In 2018, the inflow to Beijing from the middle route of the South-to-North Water Diversion Project was 11.92 × 10^8 m^3. Consumption of Beijing water resources was 35.2 × 10^8 m^3 and 39.3 × 10^8 m^3 in 2010 and 2018, respectively. Then, the shortage gap was satisfied by excessive use of groundwater. At the same time, utilization of reclaimed water rapidly increased from 2010 (6.8 × 10^8 m^3) to 2018 (10.8 × 10^8 m^3) [4,5]. Now, reclaimed water has become the "stable second water source of Beijing", which was used for industry reuse, agricultural irrigation, river and lake landscape, and municipal utility [6]. As reclaimed water originated from treated wastewater, water quality was significantly different from natural surface water. High and complex content of salinity, nutrients (nitrogen and phosphorus), metals, and organic matter were remarkable features of reclaimed water. This could lead to soil salination [7], accumulation of heavy metals [8], groundwater pollution risk [9,10], antibiotics risk [11], and nutrient load in surface water [12].

Therefore, an understanding of water quality is very important for better use of reclaimed water. At present, chemometrics and multivariate statistics could provide powerful exploration for revealing water chemistry/quality characteristics. The Gibbs plot was depicted by drawing the relationship of the major ion ratio vs. total dissolved solids (TDS) of water, which included major rivers, rainfall, and seawater in the world [13]. It was powerful to determine the controlling mechanisms, which contained natural processes (atmospheric precipitation, rock weathering, and evaporation–crystallization) and anthropogenic activities [14]. A complex interaction among lithosphere, atmosphere, hydrosphere, and biosphere usually caused lithological weathering, which was the source of river water chemistry [14–16]. Stoichiometric analysis, e.g., the relationship of different combinations of dissolved cations and anions, could provide qualitative sources of ions in river water, such as evaporites, carbonates, and silicates [17–19]. Multivariate statistical analyses are particularly useful to explore the water chemistry/quality data set. Correlation analysis is powerful to interpret the relationship of water quality data and to infer specific water chemical processes [20]. The inner characteristics and distribution rule of water quality could be explored using a cluster analysis, and the similarity and dissimilarity also can be clarified. It usually contains a cluster of water chemical variables and samples. Consequently, ionic transformation and spatiotemporal variation could be clearly delineated [21,22]. Analysis of variance can be used to identify whether the difference of water parameters is significant, furtherly the spatiotemporal variation could be ascertained quantificationally [23]. Water quality has an extent of random and uncertainty, which will increase the difficulty of understanding the data. By dimensionality reduction of a large amount of data, the principal component analysis could determine the key water quality parameters, which are used to identify pollutant sources and the transforming mechanism of water chemistry [24,25]. In the past, research of river water quality mainly focused on natural rivers or rivers polluted by different sources of pollutants. Less studies were performed on rivers mainly replenished by reclaimed water. Generally, the single research method was usually applied to explore the river water quality. In this study, we try to combine the application of chemometrics and multivariate statistics methods for river water quality. So, the problem of river water quality/chemistry, e.g., the relationship of different water quality parameters, spatiotemporal variation, and evolution of water chemistry, could be quantitatively solved. Therefore, the combination of chemometrics and multivariate statistics would be more effectively to clarify the water chemical composition and governing factors. This study would provide sufficient suggestions for reclaimed water reuse and management of river water quality, even for water quality control issues of discharge from wastewater treatment plants.

For recovery of the dry Chaobai River, reclaimed water with a flow of 1.0 × 10^5 m^3/d, treated from Wenyu River water using membrane bioreactor technology, was moved to Chaobai River for ecological implementation. So, we researched the water quality of reclaimed water and river water for better use of reclaimed water. The main objectives of our study are (1) to understand the physical and chemical composition of reclaimed water and river water; (2) to clarify the spatial and temporal variation of water parameters; and (3) to ascertain the governing factors of water chemical evolution in the Chaobai River replenished by reclaimed water.

2. Materials and Methods

2.1. Study Site

The study site, located on the northeast of Beijing city, is the water course in Shunyi County, belonging to Chaobai River (Figure 1). The climate here has a seasonal temperature with a semi-humid monsoon climate with four distinct seasons. The multi-annual average temperature, annual rainfall, and evaporation are 11.8 °C, 614.9 mm, and 1175 mm, respectively. Meanwhile, the precipitation was concentrated mostly from June to September [26].

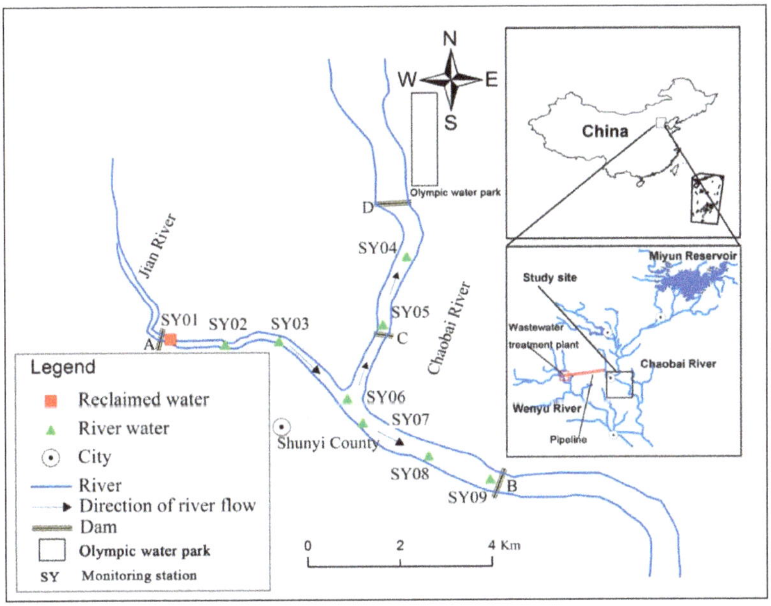

Figure 1. Location of the study site and distribution of monitoring stations.

Chaobai River has been dry as a result of continual dry weather and the impoundment of the Miyun Reservoir since 1999, which belongs to the Quaternary Holocene alluvial–diluvial strata. The riverbed is mainly composed of fine sand, silt, and gravel layers. The thickness of the sand layer is generally from ten to a few dozen meters. Floodplain and terraces on both banks of the river channel are loam and sand [26]. While, Wenyu River (Figure 1) near the Chaobai River always has water flow throughout the year, which is mainly composed of domestic wastewater from surrounding communities. Wastewater in Wenyu River was treated using membrane bioreactor technology (MBR). Consequently, treated wastewater (reclaimed water) was transported to the Jian River (length: 4 km and width: 50–90 m; tributary of Chaobai River) by a water pipeline since the year 2007 [27]. Then, the Chaobai River channel from dam B to dam D was replenished by reclaimed water with a flow of 2.5 m^3/s. Meanwhile, river flows freely through dam C. Most reclaimed water was stored in this section of the river as a result of the dams, except for evaporation and infiltration. The main channel of the river has a length of 7.3 km, width of 200–400 m, and average water depth of 2.5 m. River water flow was also changed by the dams after the replenishment of reclaimed water. One direction was the original river flow, i.e., southeast (to SY09), the other was northeast (to SY04; Figure 1).

2.2. Methods

2.2.1. Water Sampling

For investigating the water quality and controlling factors of river water replenished by reclaimed water in the Chaobai River, samples of reclaimed water and river water along the river channel were collected during March, May, July, September, and November in 2010. Nine monitoring stations (SY01–SY09) are shown in Figure 1, which included one reclaimed water and eight river water stations. In addition, one sample of reclaimed water in March was missed. Sample bottles with different volumes (100 mL and 500 mL) were made of polyethylene. The bottles were cleaned once with detergent, then cleaned once with tap water, and finally cleaned with deionized water 3 times. Water samples were collected at a depth of 50 cm below the river water surface using a plexiglass water collector. Three bottles of the water sample were collected for each monitoring station every time. The polyethylene bottles were prerinsed with water samples three times, before the final water sample was collected. The sampling frequency was five times for one year, and the sampling time was from 9:00 am to 5:00 pm. Three bottles of samples were collected at each station. The 100 mL sample was used for the determination of water stable isotopes, and the cap of the bottle was sealed with tape to prevent evaporation. A sample of 500 mL for the determination of anions, cations, nitrogen, and phosphorus was used and another sample of 500 mL was used for chlorophyll a (*Chl. a*) determination. All samples were stored in a portable cooler containing ice packs under 4 °C.

The precipitation data was measured using the tipping bucket automatic rain sensor (CG-04-D1, Hebei Yiqing Electronic Technology Company, Handan, China) on the roof of the Geographical Science Museum of the Institute of Geographical Sciences and Natural Resources Research, Chinese Academy of Sciences (40°00′11″ N, 116°23′07″ E, 45 m above sea level, about 10 m above the ground) from January to December 2010. This sampling point was 28 km from the study area in a straight line. A total of 17 precipitation events were collected, with a total precipitation of 412.7 mm. Then, monthly distribution data of precipitation in 2010 was obtained by accumulating precipitation events into the month.

2.2.2. Analytical Techniques

pH, water temperature (T, °C), dissolved oxygen (DO), and electric conductivity (EC) were measured by the portable multi-parameter water quality analyzer (American Hach HQ-40d), which was produced by HACH Company (Loveland, CO, USA). Water samples were taken back to the laboratory under 4 °C cold storage and analyzed within 24 h. The bicarbonate (HCO_3^-) was determined by titration under the addition of sulfuric acid (0.02 mol/L), which the endpoint of titration had methyl orange as an indicator. Before further ionic analysis, water samples were filtered through a 0.45 μm Millipore membrane. Major cations including potassium (K^+), sodium (Na^+), calcium (Ca^{2+}), and magnesium (Mg^{2+}) were measured by inductively coupled plasma spectroscope (ICP-OES Optima 5300DV), produced by Perkinelmer Instruments Co., LTD (Norwalk, Connecticut, USA), with a detection limit of 0.01 mg/L. Major anions including chloride (Cl^-), sulphate (SO_4^{2-}), and nitrate (NO_3^-) were measured by ion chromatograph (Thermo Fisher ICS2100) produced by the DIONEX company (Sunnyvale, CA, USA), and the detection limit was 0.01 mg/L. Ammonia nitrogen (NH_3-N), nitrite nitrogen (NO_2-N), total nitrogen (TN), and total phosphorus (TP) were measured using AMS's Smartchem 200 batch analyzer produced by Alliance company (Paris, France), with a detection limit of 0.01 mg/L. Stable isotopes (δ^2H and $\delta^{18}O$) were measured by a laser spectroscopic instrument (LGR DLT-100, Los Gatos Research, Mountain View, CA, USA), with the standard of Vienna standard mean ocean water (VSMOW). The precisions of $\delta^{18}O$ and δ^2H were 0.2‰ and 0.6‰, respectively. TDS was measured by the gravimetric method. The filtered water sample (200 mL) was placed in an evaporating dish weighed to a constant weight, and then baked to a constant weight at 103 ~ 105 °C. The TDS value was calculated by the increased weight [28]. The collected chlorophyll a (*Chl. a*) sample samples were stored at 4 °C, and 1 mL of 1% magnesium carbonate suspension was added to each liter of water samples to prevent pigmentation caused by acidification. In the laboratory, samples were filtered and concentrated, and the filter membrane was fully grinded and extracted, then dissolved in acetone to a constant volume, and finally the supernatant was measured by spectrophotometry. These experimental details were referred from the analytical book [28].

2.3. Data Processing

Data of the major ions need to be balanced with an error of less than 5% before further analysis. The summary statistics (e.g., mean, max, min, and coefficient of variation) of water chemistry were performed by the descriptive statistics package in SPSS 16.0 software (International Business Machines Corporation, Armonk, NY, USA) [29]. Analysis of variance (ANOVA), hierarchical cluster analysis (HCA), and principal component analysis (PCA) were all performed by SPSS [29]. Analysis of variance needs two assumptions about the data, which includes normal distribution and homogeneity of variance, respectively. These two parameters could be judged by the coefficients of skewness and Kurtosis, and Levene. The data could be considered to obey the hypothesis of normal distribution and homogeneity of variance, if the p value was more than 0.05 ($p > 0.05$). PCA is a way of selecting factors belonging to the factor analysis. Whether the data are suitable for factor analysis, it could be identified by KMO (Kaiser–Meyer–Olkin) and its significance (sig.) level test. If the KMO test value is greater than 100 and $p < 0.05$, then PCA could be performed. Respect to hierarchical clustering, especially for R-type, the data need to be standardized for avoiding the difference of dimension and orders of magnitude for variables [29]. The normality of water quality data was checked firstly by SPSS 16.0 software [29]. Then, all the data (except pH) were log-transformed and standardized before further analysis [30,31].

In addition, index, i.e., $\overline{R^2}$ was used for judging the selection of key variable in HCA, that was calculated as follows:

$$\overline{R^2} = \frac{\sum r^2}{m \quad 1} \tag{1}$$

where, r^2 represents the correlation coefficient between different variables in clusters, and m represents the number of variables in one cluster.

Meanwhile, PHREEQC (Version 3, United States Geological Survey) is a hydrogeochemical simulation software based on C language [32]. It is mainly used to solve the analysis of chemical components, solute transport, and dynamic chemical reactions in the interaction system of water, gas, and rock–soil. The saturation index (SI) of major minerals was calculated by PHREEQC software.

Excessive sodium and salinity in irrigation water would result in sodium hazard. Calcium and magnesium in soil could be replaced by sodium, which leads to the reduction of permeability and soil harden [33]. Sodium adsorption ratio (SAR) calculated based on chemical variables was used to assess irrigation water quality, which was an effective evaluation index [34].

$$SAR = Na^+ / \sqrt{(Ca^{2+} + Mg^{2+})/2} \tag{2}$$

where, ionic concentrations are expressed in milliequivalent per liter (meq/L).

3. Results and Discussion

3.1. Water Chemical Composition

Physical and chemical compositions of water samples in the Chaobai River are given in Table 1. pH ranged from 7.65 to 9.45, with an average value of 8.37, which showed reclaimed water and river water were all alkaline. The average value in river water was higher than that in reclaimed water. Water temperature was mainly controlled by the operation of the wastewater treatment plant (WWTP) and air temperature of seasonal variation. Electric conductivity (EC) and TDS were the similar parameters indicating the total dissolved solids in aqueous solution, which in reclaimed water were both slightly higher than in river water. The order of nitrogen forms in reclaimed water and river water was: NO_3-N > NO_2-N > NH_3-N, and NO_3-N > NH_3-N > NO_2-N, respectively. The coefficient of variation (C.V.) of NH_3-N and NO_2-N was much higher than others. It indicated that they were prone to chemical transformation in the water environment. Considering the average, higher percentage of NO_3-N /TN in reclaimed water (85%) and river water (78%) indicated nitrate was the major nitrogen form. The order of anions in reclaimed water was: HCO_3^- > SO_4^{2-} > Cl^-; while, the order in river water was: HCO_3^- > Cl^- > SO_4^{2-}. Whereas, the order of cations (Na^+ >Ca^{2+} >Mg^{2+} > K^+) was the same in both waters. In terms of the average, seven indicators of river water were higher than in reclaimed water, which contained pH, Cl^-, Mg^{2+}, NH_3-N, *Chl.a*, $\delta^{18}O$, and δ^2H. While, the remaining indexes were the opposite. A reduction of nutrients (NO_3-N, NO_2-N, TN, and TP) may be caused by the consumption of phytoplankton or dilution [35,36]. External input from surface runoff in the rainy season could lead to the increased content of Cl^- and Mg^{2+} [37,38]. Increased ammonia content may be caused by the mineralization of organic nitrogen [39,40]. High value of *Chl.a* indicated the reproduction of phytoplankton, e.g., alga in river water [36]. Stable isotopes ($\delta^{18}O$, δ^2H) were depleted in reclaimed water, while enriched in river water (Table 1), and the similar enrichment phenomenon was also found in Huai River [10].

Table 1. Water chemical composition of reclaimed water and river water in the Chaobai River (March–November 2010).

	Reclaimed Water					River Water				
	Min.	Max.	Mean	SD	C.V. (%)	Min.	Max.	Mean	SD	C.V. (%)
pH	7.65	8.21	7.93	0.25	3.16	8.08	9.45	8.81	0.41	4.68
T (°C)	6.90	28.60	21.03	9.65	45.92	1.90	29.70	17.42	10.30	59.12
DO (mg/L)	4.73	8.50	7.33	1.75	23.84	1.25	10.30	6.20	2.50	40.24
EC (μS/cm)	937	1025	982	36.88	3.75	935	1047	859	90.72	10.56
Cl^-(mg/L)	72.90	114.00	87.65	18.65	21.27	65.2	148.0	98.15	19.96	20.34
HCO_3^-(mg/L)	219.00	299.00	261.75	33.24	12.70	230	386	227.78	49.01	21.52
SO_4^{2-}(mg/L)	81.90	99.50	89.33	7.97	8.92	64.30	120.00	87.04	13.66	15.69
K^+(mg/L)	12.90	18.60	16.25	2.83	17.44	6.32	25.80	15.25	4.82	31.60
Na^+(mg/L)	80.20	108.00	94.45	11.62	12.30	57.3	135.0	88.46	17.51	19.79
Ca^{2+}(mg/L)	60.10	66.40	63.47	3.17	5.00	23.7	75.2	48.92	12.80	26.17
Mg^{2+}(mg/L)	22.60	28.60	26.00	2.71	10.42	23.3	39.9	29.10	3.43	11.80
NH_3-N(mg/L)	0.03	0.11	0.08	0.03	43.92	0.02	1.81	0.38	0.42	109.57
NO_2-N(mg/L)	0.22	0.76	0.38	0.26	67.66	0.00	0.74	0.16	0.15	91.60
NO_3-N(mg/L)	10.60	19.40	14.75	3.83	25.99	0.03	19.90	7.03	5.95	84.66
TN (mg/L)	11.50	22.10	17.18	4.47	26.01	2.30	20.00	9.62	5.29	54.94
TP (mg/L)	0.44	1.52	1.05	0.46	43.8	0.09	2.76	0.60	0.55	92.00
Chl.a (μg/L)	0.50	3.58	2.03	1.77	87.03	1.96	175.00	58.87	45.18	76.74
TDS (mg/L)	568	655	601	39.35	6.55	419	794	540	80.07	14.82
$\delta^{18}O$ (‰)	−8.55	−7.25	−7.99	0.63	−7.91	−8.64	−5.14	−6.85	0.98	−14.34
δ^2H (‰)	−61.34	−55.36	−58.39	2.73	−4.68	−63.64	−43.70	−53.48	5.28	−9.86

A Pearson correlation analysis was applied to explore the relationship of water chemical parameters by SPSS, and the results are given in Table 2. Carbonate was observably positively correlated with K^+, Na^+, and Ca^{2+}, which indicated the dissolution of minerals [41]. Chloride was prominently and positively related with SO_4^{2-}, K^+, Na^+, Mg^{2+}, and TDS. TDS was significantly and positively correlated with EC, Cl^-, HCO_3^-, SO_4^{2-}, K^+, Na^+, Ca^{2+}, NO_3-N, TN, and TP, respectively. Meanwhile, the same correlation was also found between EC with these ions. It indicated these ions were the major composition of TDS and EC. While, TDS were significantly and negatively correlated with stable isotopes and *Chl.a*. This may be due to the consumption of nitrogen and phosphorus by phytoplankton [35,36]. TN was significantly and positively related with TP, indicating their common origin. At the same time, they were both significantly and positively related with EC, HCO_3^-, K^+, Na^+, Ca^{2+}, TDS, and NO_3-N. In addition, TN was also positively and negatively significantly correlated with NH_3-N and NO_2-N, respectively. TN, TP, and other cations were mainly from reclaimed water, which could explain their significant relationship. The major nitrogen forms (NO_3-N) were furtherly confirmed by its significant and positive relationship with TN (Tables 1 and 2). Part of NH_3-N will be released into the atmosphere [42–44], which could be inferred by its significant and negative relationship with TN.

Table 2. Correlation of water chemical parameters in water samples in the Chaobai River (March–November 2010).

	pH	T	DO	EC	Cl^-	HCO_3^-	SO_4^{2-}	K^+	Na^+	Ca^{2+}	Mg^{2+}	TDS	NH_3-N	NO_2-N	NO_3-N	TN	TP	*Chl.a*	$\delta^{18}O$	δ^2H
pH	1.00																			
T	0.10	1.00																		
DO	0.16	0.37 *	1.00																	
EC	−0.18	0.09	0.16	1.00																
Cl^-	0.37 *	0.00	0.11	0.10	1.00															
HCO_3^-	−0.25	0.14	0.40 **	0.66 **	0.21	1.00														
SO_4^{2-}	0.29	−0.15	−0.10	0.24	0.71 **	−0.04	1.00													
K^+	0.15	−0.08	0.38 *	0.59 **	0.59 **	0.66 **	0.51 **	1.00												
Na^+	0.11	0.06	0.27	0.70 **	0.60 **	0.59 **	0.61 **	0.93 **	1.00											
Ca^{2+}	−0.37 *	−0.04	0.42 **	0.77 **	−0.20	0.82 **	−0.16	0.51 **	0.46 **	1.00										
Mg^{2+}	0.08	0.03	0.21	−0.21	0.58 **	0.27	0.01	0.21	0.11	0.11	1.00									
TDS	−0.05	−0.06	0.31 *	0.77 **	0.49 **	0.73 **	0.43 **	0.89 **	0.89 **	0.72 **	0.20	1.00								
NH_3-N	−0.26	0.07	−0.13	−0.35	−0.20	−0.06	−0.20	−0.23	−0.24	−0.15	0.07	−0.21	1.00							
NO_2-N	−0.07	0.19	−0.02	0.45 **	−0.17	0.34 *	−0.04	0.22	0.28	0.25	−0.36 *	0.23	−0.07	1.00						
NO_3-N	−0.15	−0.20	0.24	0.85 **	0.10	0.50 **	0.20	0.64 **	0.64 **	0.75 **	−0.20 **	0.77 **	−0.41 **	0.28	1.00					
TN	−0.02	0.08	0.30	0.83 **	0.01	0.50 **	0.15	0.54 **	0.57 **	0.64 **	−0.37	0.63 **	−0.48 **	0.41 **	0.88 **	1.00				
TP	−0.14	0.30	0.33	0.77 **	0.26	0.67 **	0.22	0.56 **	0.62 **	0.62 **	−0.02	0.70 **	−0.30	0.26	0.62 **	0.67 **	1.00			
Chl.a	0.60 **	0.14	0.01	−0.44 **	0.25	−0.34 **	0.12	−0.03	−0.08	−0.53 **	0.15	−0.27	−0.09	−0.16	−0.38	−0.26	−0.29	1.00		
$\delta^{18}O$	0.33	0.31	0.01	−0.70 **	0.29	−0.28	0.09	−0.19	−0.26	−0.67 **	0.36	−0.44 **	0.26	−0.21	−0.79 **	−0.63 **	−0.41 **	0.45 **	1.00	
δ^2H	0.32	0.23	−0.03	−0.77 **	0.31	−0.39	0.13	−0.27	−0.33	−0.71 **	0.34	−0.48 **	0.30	−0.28	−0.80 **	−0.66 **	−0.46 **	0.44 **	0.94 **	1.00

*: Correlation is significant at $p < 0.05$; **: Correlation is significant at $p < 0.01$. The observation includes samples of nine stations in five months except SY01 (four months), and the numbers are 44 ($n = 44$).

Chl.a was remarkably and negatively related with EC, HCO_3^-, Ca^{2+}, $\delta^{18}O$, and δ^2H. While, it was significantly and positively related with pH. CO_2 in river water could be converted into organic matter by the photosynthesis of alga, and HCO_3^- was the carbon form in the carbonate system [45,46]. While, the consumption of nitrogen, phosphorus, and carbon would lead to the decrease of EC. In addition, the amount of O_2 produced by the photosynthesis of alga was much greater than the one required for respiration, which increased DO content in the water [47,48]. Consequently, the reduction of HCO_3^- and the increase of DO would together raise the pH value (Table 1) [49,50]. Stable isotopes were both dramatically and negatively correlated with EC, Ca^{2+}, TDS, NO_3-N, TN, and TP. Significant and positive correlation of $\delta^{18}O$ and δ^2H was determined by a stable isotope fractionation mechanism [51–53].

3.2. *Spatial and Temporal Variation*

3.2.1. pH, T, DO, EC, and TDS

Spatial and temporal variation of pH, T, DO, EC, and TDS is given in Figure 2. A first gradual upward and then downward trend were the clear spatial variation of pH, and the rising stage mainly occurred during stations from SY04 to SY06. The lowest pH was found in reclaimed water i.e., water sources. Moreover, pH in September was lower than in other months. Meanwhile, the pH value was very close in other months. River water received more replenishment of precipitation and surface runoff with low pH with range of 4.35–5.70 [54]. Obvious temporal variation of water temperature was found, that was mainly affected by air temperature. Its order was: July > May > September > March > November. Microbial activity was strongly influenced by water temperature. Consequently, the element cycle driven by a microbe, e.g., nitrogen would be affected by temperature variation [55]. Lower DO content was found in November, and a higher value was found in May and July (Figure 2c), while DO in other months was close. DO content varied drastically among monitoring stations. EC and TDS had the similar downward trend of spatial variation (Figure 2d,e). They both were higher in reclaimed water, and gradually decreased along the river. Furthermore, the lowest EC value (718 μS/cm) in March was found at one end of the river flow (SY04), and EC was close in different months. However, TDS in March, May, and July were slightly higher than that in September and November. Similar temporal variation of water temperature was also found in the Yongding River [56], while the spatial variation of these five parameters was different, due to the diverse water quality of reclaimed water [56] or treated wastewater [57], hydraulic conditions, and the geomorphic feature [58].

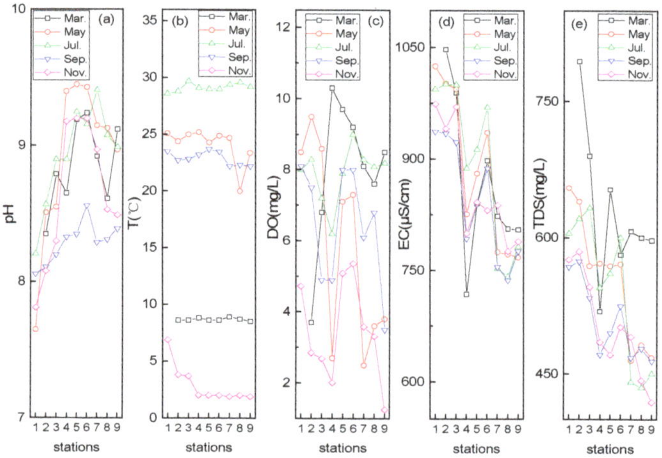

Figure 2. Spatial and temporal variation of pH (**a**), water temperature (T; **b**), dissolved oxygen (DO; **c**), electric conductivity (EC; **d**), and total dissolved solids (TDS; **e**) in the Chaobai River.

3.2.2. Major Cations and Anions

Figure 3 shows the spatial and temporal variation of major cations and anions. Cl^- has apparent temporal variation, and its order among months was: May > March > July > November > September (Figure 3a). Cl^- content increased gradually along the river water flow direction, except reclaimed water in March. Elevated chloride indicated the external input, e.g., dissolution of soil and/or sediment, or surface runoff, as chloride was the conservative ion [59]. The average value in this study was much less than the mean value in Kakoba sewage effluents (833.33 mg/L) [58]. While, the temporal variation was not significant [58]. The content of SO_4^{2-} was high in May, and low in July and September (Figure 3b). Meanwhile, it varied greatly among monitoring stations. Similar spatiotemporal variation of K^+ and Na^+ was found in Figure 3c,d. Their high values were found in March, while low values in September and November, and the medium in May and July. Their downward trend of spatial variation may be caused by ion exchange and/or dilution [19]. Variation of Ca^{2+} showed a remarkable downward spatial trend (Figure 3e). High Ca^{2+} content was firstly found in reclaimed water. However, its concentration gradually decreased along river flow, especially in SY04 and SY07 in May. The observed decrease may be the result of calcium precipitation, and/or calcium ions exchange with soils/minerals [17,60]. The content of Ca^{2+} was high in March and July, and low in May, while medium in other months. Gradual rising was the primary feature of spatial variation of Mg^{2+}, except a high value of reclaimed water in March (Figure 3f). External input and/or dissolution of mineral with magnesium may contribute to an increase [61]. High and low value of Mg^{2+} was found in March and November, respectively. While, the medium one was in other months. Major cations and anions in the river impacted by sewage effluents in the Mediterranean had significant spatiotemporal variation [57]. Complexity and high variability occurred in different rivers with diverse conditions.

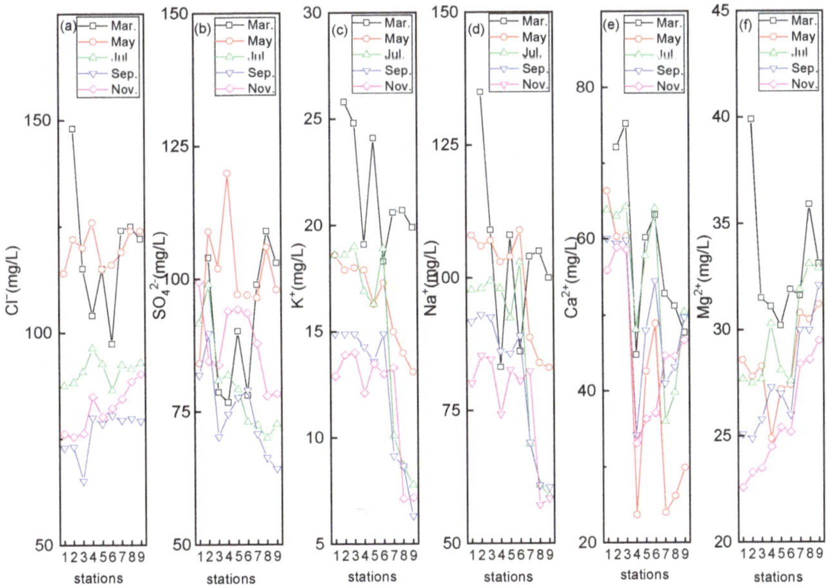

Figure 3. Spatial and temporal variation of Cl^- (**a**), SO_4^{2-} (**b**), K^+ (**c**), Na^+ (**d**), Ca^{2+} (**e**), and Mg^{2+} (**f**) in the Chaobai River.

3.2.3. Nitrogen, Phosphorus, and *Chl.a*

Spatial and temporal variations of nitrogen, phosphorus, and *Chl.a* are given in Figure 4. Spatiotemporal variation of NH_3-N was obvious (Figure 4a), with the highest and the second highest value in SY01 (September) and SY04 (May), respectively. NH_3-N content in May and September was higher than in other months, while most values of NH_3-N were lower than 0.5 mg/L. Most NO_2-N content was lower than 0.4 mg/L (Figure 4b). Generally, spatial variation of NO_2-N was decreasing except for four peak values (SY01 and SY03 in November and SY04 and SY05 in July). These nitrogen forms also exhibited very high variability Rwizi River (Uganda) [58]. Gradual downward spatial variation of NO_3-N, TN, and TP was apparent. The evident features of NO_3-N and TN were the slight increase from SY05 to SY06 and sharp decrease in SY04. A gradual decrease was also found in the Sand River (Limpopo, South Africa) impacted by sewage effluents, and this may indicate the self-purification capacity of the river [62]. The peak value of TP was found in SY02 (May), which was higher than in reclaimed water. It may be due to the release of sediments and/or sudden external input [63,64]. While, TP in the Sand River fluctuated across the different sites [62]. First an increase and then a decrease were the feature of *Chl.a* spatial variation (Figure 4f). Meanwhile, it was higher downstream than upstream, with a peak value of SY08 in May (167 μg/L) and July (175 μg/L). High *Chl.a* was found in May and July, while a low value in November. There was a slightly low concentration of *Chl.a* in September corresponding to July, which may be caused by the dilution. Stations with a high peak value of *Chl.a* were the corresponding stations with low nitrogen and phosphorus. It further confirmed that the absorption and utilization of phytoplankton was the reason for nutrients reduction.

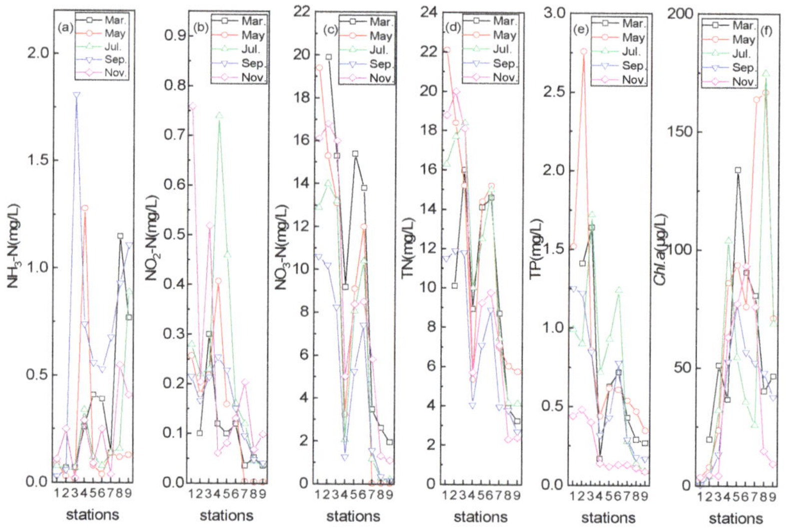

Figure 4. Spatial and temporal variation of NH_3-N (**a**), NO_2-N (**b**), NO_3-N (**c**), TN (**d**), TP (**e**), and *Chl.a* (**f**) in the Chaobai River.

3.2.4. δ^2H and $\delta^{18}O$

Figure 5 shows the spatiotemporal variation and the relationship of δ^2H and $\delta^{18}O$ in reclaimed water and river water. δ^2H and $\delta^{18}O$ had the similar variation (Figure 5a,b). The clear spatial trend was increasing variation of δ^2H and $\delta^{18}O$ from reclaimed water to the river channel end. It also indicated the enrichment process of stable isotopes, and the depleted and enriched isotopes in reclaimed water and the two ends (SY04 and SY09), respectively. Stable isotopes in March and May were higher, and gradually increased from July to November. In the dry season, e.g., March and May, high air temperature and less rainfall contributed to isotope enrichment fractionation. However, stable isotopes of river water would be prone to being depleted as a result of rainfall and surface runoff [65]. According to our observation, the precipitation in July was the largest (124.6 mm), followed by August (101.8 mm), which accounted for 30.19% and 24.67% of the annual precipitation respectively. During the rainy season (July–September), the cumulative precipitation was 297.9 mm, accounting for 72.18%. Therefore, isotopes in November also would be affected, as it was just after the rainy season. Almost all samples were located below the global meteoric water line (GMWL) and local meteoric water line (LMWL), except several samples on and near the LMWL line (Figure 5c), which indicated that most reclaimed water and river water were influenced by strong evaporation [52]. Additionally, it was consistent with the results of the isotope enrichment feature (March and May) in Figure 5a,b. Samples located on LMWL showed their source of atmospheric precipitation and less evaporation [53,66]. The evaporation line of stable isotopes among different months were different, and the order of the slope was: July (6.13) > March (5.72) > November (5.21) > May (4.85) > September (4.02).

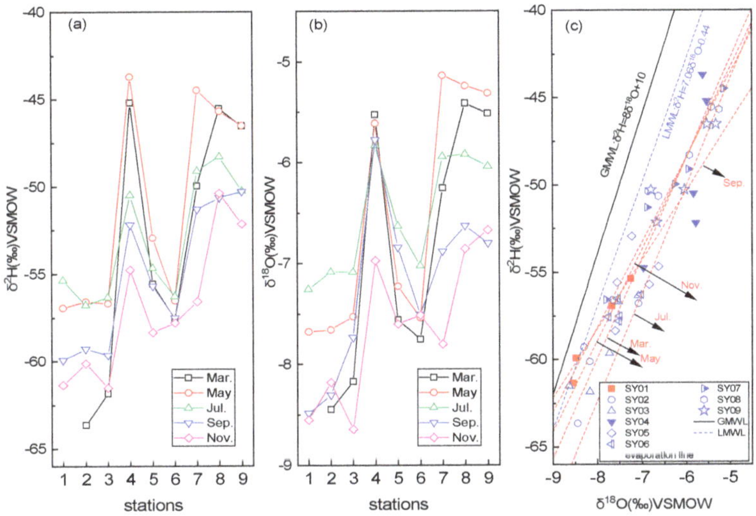

Figure 5. Spatiotemporal variation of δ^2H (**a**) and δ^{18}O (**b**), and their relationship (**c**) in the Chaobai River (GMWL (global meteoric water line) [53] and LMWL (local meteoric water line) [67] in Figure 5c were cited from published articles, respectively. Red dashed lines represent evaporation line of river water in different months).

3.3. Processes Controlling Water Chemistry

3.3.1. Gibbs Plot and Bivariate Plot

Mechanisms controlling river water chemistry included atmospheric precipitation, rock weathering, evaporation–crystallization processes [13], and anthropogenic activities [14]. The high ratio of $Na^+/(Na^++Ca^{2+})$, low ratio of $Cl^-/(Cl^-+HCO_3^-)$, and medium TDS concentration were found in the Gibbs plot (Figure 6). Meanwhile, distributions of reclaimed water and river water samples in cation and anion plots were both located in the rock dominance area. It indicated that river water chemistry was mainly controlled by reclaimed water or the interaction of river water with soil/rock. While, the effect of precipitation and evaporation was weak. The discharge of reclaimed water from WWTP was about 1.0×10^5 m^3/day. As a result, the TDS load was about 60.1 Ton/day, according to the average value of the TDS concentration (601 mg/L). Similar Gibbs plot results were found in the Yongding River, which was also replenished by reclaimed water [56]. Water chemistry in headwaters of the Yangtze River [15] and Yellow River [68] in China was governed by rock weathering, which was less impacted by human activities. While, evaporation–crystallization plays an important role in water chemistry in the river of an arid watershed, e.g., Northern Xinjiang, China [16].

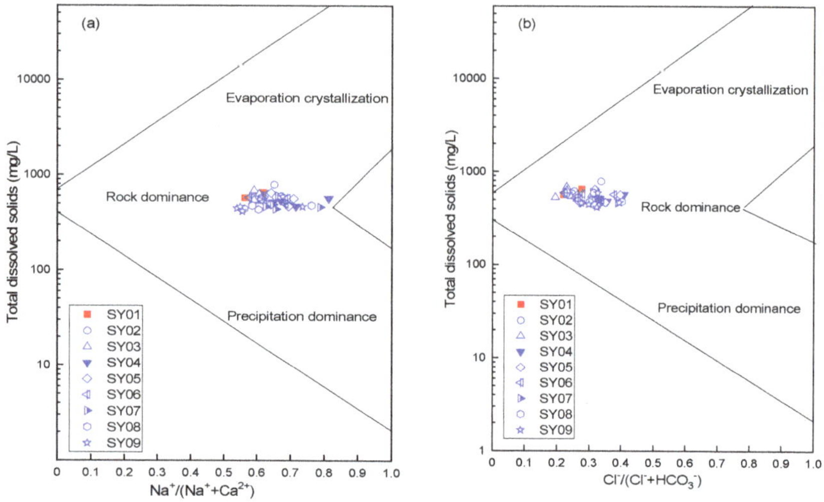

Figure 6. Gibbs graph of major cations (**a**) and anions (**b**) in the Chaobai River channel replenished by reclaimed water.

Mineral dissolution, precipitation, and redox reaction in the water environment could be inferred by the relationship of different dissolved ions in waters [13,69]. The ratio of Ca^{2+}/Mg^{2+} could indicate the dissolution of carbonate minerals [69]. The ratio value was about equal to 1, indicating the dissolution of dolomite, that included SY04 (March and July), four samples in November (SY05, SY06, SY07, and SY08), SY09 (July, September, and November; Figure 7a). The value was between 1 and 2, indicating the dissolution of calcite, which included reclaimed water (SY01), river water (SY02 and SY03) river samples of SY05 (March and May) and SY06 (Figure 7a). The ratio value of the remaining samples was less than 1, showing the decrease of Ca^{2+}, which may be caused by ion exchange. Values of Ca^{2+}+Mg^{2+} vs. the cation of all reclaimed water and river water samples were located above the equilibrium line (1:1; Figure 7b). It indicated that Ca^{2+} and Mg^{2+} were mainly from the dissolution of carbonate rock and calcite [18,41]. All samples of Na^++K^+ vs. Cl^- (Figure 7c) were located below the equilibrium line, indicating that sodium and potassium ions were also affected by the dissolution of silicate minerals except the dissolution of salt rock [68]. Samples of Na^++Ca^{2+} vs. HCO_3^- (Figure 7d) were all located below the equilibrium line, showing that sodium and calcium were more than bicarbonate, which indicated the dissolution of calcium-bearing minerals. Most of SO_4^{2-}+Cl^- vs. HCO_3^- were located below the equilibrium line (Figure 7e), indicating the influence of strong evaporation. Some of SO_4^{2-}+HCO_3^- vs. Ca^{2+}+Mg^{2+} were located near the line (Figure 7e), indicating the dissolution of carbonate minerals [70]. While, most were below the line (Figure 7e), indicating the dissolution of the silicate mineral. The same silicate weathering location was found between the plot of Ca^{2+}+/Na^+ vs. Mg^{2+}/Na^+ (Figure 7g), and plot of Ca^{2+}/Na^+ vs. HCO_3^-/Na^+ (Figure 7h), indicating the weathering and dissolution of the silicate mineral.

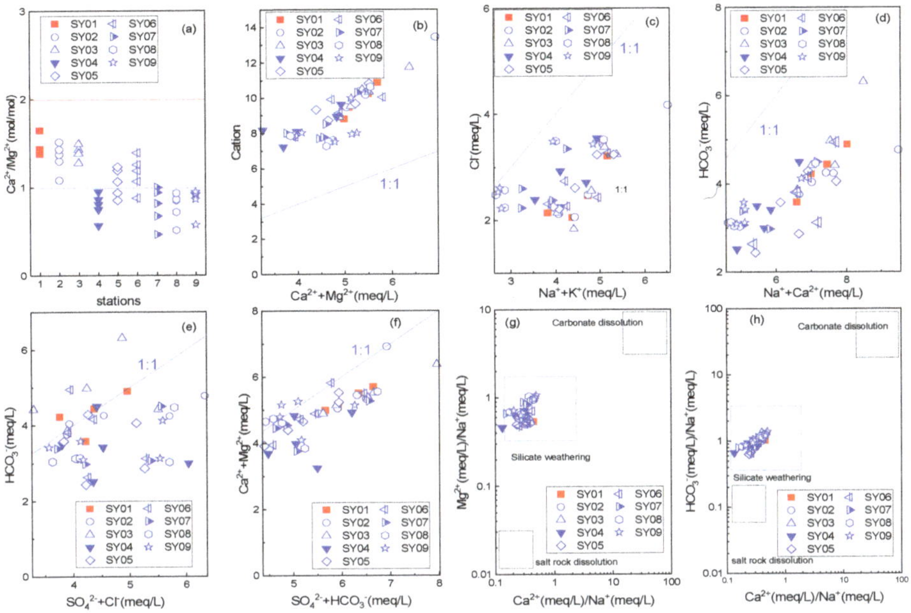

Figure 7. Plot of the ratio of Ca^{2+}/Mg^{2+} (**a**), bivariate plot of $Ca^{2+}+Mg^{2+}$ vs. the cation (**b**), $Na^{+}+K^{+}$ vs. Cl^{-} (**c**), $Na^{+}+Ca^{2+}$ vs. HCO_3^{-} (**d**), $SO_4^{2-}+Cl^{-}$ vs. HCO_3^{-} (**e**), $SO_4^{2-}+HCO_3^{-}$ vs. $Ca^{2+}+Mg^{2+}$ (**f**), $Ca^{2+}+/Na^{+}$ vs. Mg^{2+}/Na^{+} (**g**), and Ca^{2+}/Na^{+} vs. HCO_3^{-}/Na^{+} (**h**) in all samples.

Samples of $1/2HCO_3^{-} + SO_4^{2-}$ vs. $Ca^{2+} + Mg^{2+}$ were located near the equilibrium line (Figure 8a), indicating the dissolution of calcite, dolomite, and gypsum [71,72]. All samples were below the line except one sample, showing the excess $Ca^{2+}+Mg^{2+}$. These cations would be balanced by other anions, which may be silicate. Variation of the ratio of $Na^{+}/Ca^{2+}+Mg^{2+}$ (Figure 8b) was first to increase and then to decrease, which indicated the occurrence of cation exchange with clay mineral or soils [70]. Lower values were found in SY08 and SY09 stations. The process of weathering and hydrolysis of carbonate rock or silicate minerals produced equal amounts of divalent cations, HCO_3^{-} and SO_4^{2-}. As a result, $Ca^{2+}+Mg^{2+}$-HCO_3^{-}-SO_4^{2-}, and Na^{+}-Cl^{-} were used to indicate the participation of cation exchange, respectively [71]. Some samples of $Ca^{2+} + Mg^{2+}$- HCO_3^{-}-SO_4^{2-} vs. Na^{+}-Cl^{-} were located around the line (y = −x; Figure 8c), which indicated the clear cation exchange. Others showed the excess sodium ions from reclaimed water. In Figure 8d, the ratio of Na^{+}/Cl^{-} samples was all more than 1, except three samples (SY08 and SY09 in November and SY09 in July). The average value of Na^{+}/Cl^{-} in reclaimed water was 1.69, indicating the excess sodium. Values of Na^{+}/Cl^{-} in most river water samples were close to or less than the reclaimed water, which indicated the external chloride ion. The ratio of Na^{+}/Cl^{-} was greater than 1, indicating the possible cation exchange [73]. The downward trend of the Na^{+}/Cl^{-} value may be caused by the following reason. A higher Na^{+} content exceeded the equilibrium concentration of exchangeable cations in the medium. Then, the exchange of Ca^{2+} and adsorbed Na^{+} would be suppressed, and even reverse cations exchange would occur. Na^{+} and Ca^{2+} in the water body exchange and Mg^{2+} may also participate in the exchange [17,19].

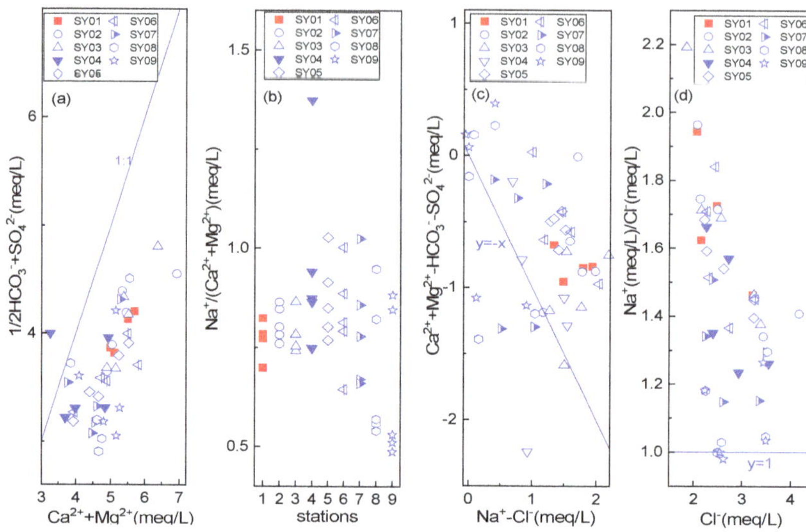

Figure 8. Plot of the ratio of Ca^{2+} + Mg^{2+} vs. 1/2HCO_3^- + SO_4^{2-} (**a**), Na^+/Ca^{2+} + Mg^{2+} (**b**), Na^+-Cl^- vs. Ca^{2+} + Mg^{2+}- HCO_3^--SO_4^{2-} (**c**), and Cl^- vs. Na^+/Cl^- (**d**) in all samples.

Saturation index (SI) of reclaimed water and river water samples is given in Table 3. Potential dissolution and precipitation processes in the aqueous solution could be inferred by SI values. A zero SI value indicated the equilibrium state of the mineral to the aqueous, and a positive value showed the supersaturated state, while a negative value indicated the unsaturated state [32]. The SI value of gypsum ($CaSO_4$ and $CaSO_4{\cdot}2H_2O$) and halite (NaCl) was negative, while the SI value of calcite ($CaCO_3$) and dolomite (Ca Mg $(CO_3)_2$) was positive. This indicated the potential dissolution process of gypsum and salt rock, and precipitation of calcite and dolomite in reclaimed water and river water. The excessive Na^+ of water samples may require silicate dissolution to balance it.

Table 3. Saturation index (SI) of samples in monitoring stations in the Chaobai River (March–November 2010).

	SY01	SY02	SY03	SY04	SY05	SY06	SY07	SY08	SY09
$CaSO_4$	−1.99	−1.97	−2.04	−2.23	−2.13	−2.12	−2.23	−2.20	−2.17
$CaCO_3$	0.63	0.94	1.19	1.09	1.36	1.46	1.16	1.00	1.12
Ca Mg $(CO_3)_2$	1.16	1.76	2.26	2.32	2.73	2.89	2.47	2.12	2.32
$CaSO_4{\cdot}H_2O$	−1.76	−1.73	−1.81	−2.00	−1.90	−1.89	−1.19	−1.96	−1.94
NaCl	−6.68	−6.58	−6.62	−4.39	−6.62	−4.26	−6.67	−6.71	−6.72

The numbers of measurements of all monitoring stations were five, except SY01 station (n = 4).

3.3.2. Redox Condition and Nitrogen Forms

Transformation and species of nitrogen in aqueous solution were strongly impacted by the redox environment [74,75], which could be characterized by DO and/or ORP measured in the field. The corresponding relationship of DO and three nitrogen forms (NO_3-N, NH_3-N and NO_2-N) is given in Figure 9. Nitrite was usually the intermediate of the nitrification reaction, which was not stable and easy to be oxidized [76]. In Figure 9, average DO content of river water in March, May, July, September, and November was 7.99 ± 2.07 mg/L, 5.64 ± 2.79 mg/L,7.90 ± 0.85 mg/L, 6.21 ± 1.66 mg/L, and 3.27 ± 1.41 mg/L, respectively. The saturated oxygen content of water was about 10 mg/L at normal temperature and atmospheric pressure [48,77]. DO content was more than 5 mg/L except the one in November. Additionally, the oxidizing environment contributes to nitrification [78]. The decrease in November probably was due to low temperature and less aquatic plants [49,79].

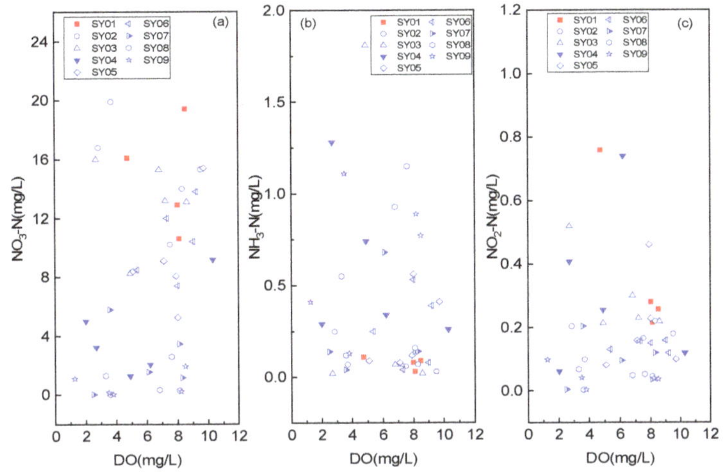

Figure 9. Relationship of DO with NO_3-N (**a**), NH_3-N (**b**), and NO_2-N (**c**) in reclaimed water and river water.

pH also played an important role in nitrogen transformation, and the double influences of pH and the redox environment could be evaluated by the pH–pE plot [78]. Therefore, samples of reclaimed water and river water were projected into Figure 10. All samples were located around the NO_3^- line, where the stable nitrogen form was nitrate (NO_3^-, +5). High DO value of reclaimed water (7.33 ± 1.75) and river water (6.20 ± 2.47) showed the oxidizing water environment, which was beneficial for nitrate stable and nitrification [78,80].

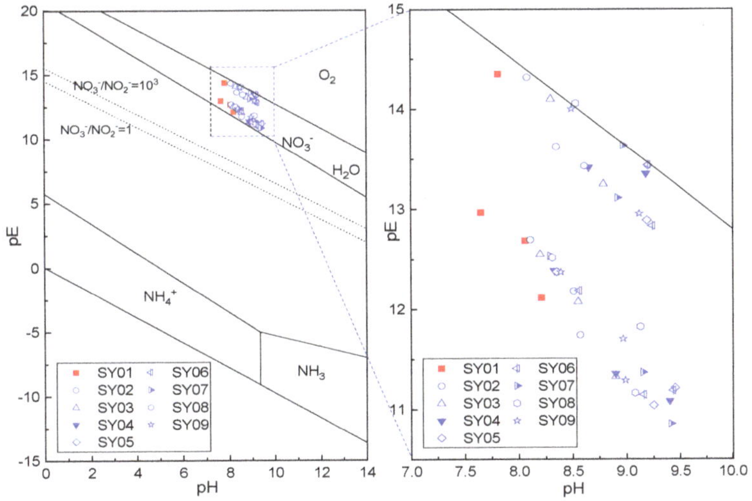

Figure 10. pH–pE diagram of reclaimed water and river water samples in the Chaobai River.

3.3.3. Cluster Analysis and Spatiotemporal Similarity

For identifying the spatiotemporal variation, the key variables should be firstly confirmed. Hence, three cluster analysis results are shown in Figure 11. In Figure 11c, twenty water chemical parameters were divided into seven clusters by the R-type cluster method for variables. Clusters were as follows.

I: $\delta^{18}O$, $\delta^{2}H$, pH, and *Chl.a*; II: Mg^{2+}; III: Cl^{-}, $SO_4{}^{2-}$; IV: NH_3-N; V: T, DO; VI: K^{+}, Na^{+}, TDS, $HCO_3{}^{-}$, Ca^{2+}, NO_3-N, TN, EC, and TP; VII: NO_2-N. The R^2 value is calculated according to equation 1 if the quantity of parameters in one cluster was more than 2. The parameter with the highest R^2 value was retained for the key parameter. The R^2 value of $\delta^{18}O$ and TDS was 0.45 and 4.80, respectively. As a result, nine parameters ($\delta^{18}O$, Mg^{2+}, Cl^{-}, $SO_4{}^{2-}$, NH_3-N, T, DO, TDS, and NO_2-N) were selected as the key ones for further spatiotemporal cluster analysis. Five months were classified into two groups, Cluster I includes March and May, and Cluster II includes July, September, and November. These two distinct groups show the significant temporal variation. Spatial variation and similarity were analyzed by an HCA analysis (Figure 11c). All monitoring stations were classified into two groups, which include Cluster I (SY04, SY07, SY08, andSY09) and Cluster II (SY01, SY02, SY03, SY05, and SY06). Stations in Cluster II and Cluster I represent the upstream and downstream (Figure 1), respectively. The flow direction of reclaimed water along the river channel would eventually go in two directions, one was SY04 station and the other was SY09 station, because of the blocking effect of the rubber dam (Figure 1).

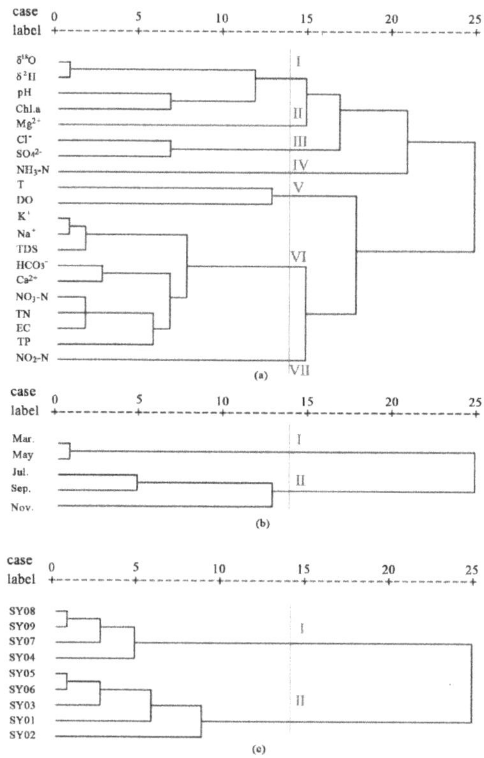

Figure 11. Cluster analysis of water quality variables (**a**), months (**b**), and monitoring stations (**c**) in the Chaobai River.

Parameters with a significant difference in different clusters were identified using an ANOVA analysis by SPSS software [29], and the corresponding results are given in Table 4. Whereas, parameters with no significant difference were not listed. In temporal clusters, six parameters in Cluster I were significantly higher than the ones in Cluster II, which included Cl^{-}, $SO_4{}^{2-}$, K^{+}, Na^{+}, Mg^{2+}, and TDS. In the spatial clusters, 12 parameters in Cluster I were significantly less than in Cluster II, which included DO, EC, $HCO_3{}^{-}$, K^{+}, Na^{+}, Ca^{2+}, TDS, NO_2-N, NO_3-N, TN, TP, and *Chl.a*. While, four parameters in Cluster I were significantly higher than in Cluster II, which included Mg^{2+}, NH_3-N, $\delta^{18}O$, and δ^2H. As a result, the parameters with a significant difference had remarkable spatiotemporal variation. $\delta^{18}O$ and δ^2H of reclaimed water (−7.99‰ ± 0.6‰ and −58.39‰ ± 2.73‰) were more depleted than the ones in downstream stations (SY04: −5.94‰ ± 0.58‰ SY07: −49.26‰ ± 4.68‰; SY09: −6.07‰ ± 0.66‰, and SY08: −49.15‰ ± 2.51‰). The evaporation process along the river may lead to more isotope enrichment. Reclaimed water was the main source of river water, besides the precipitation and surface runoff from precipitation. Stable isotopes of precipitation were more depleted compared with reclaimed water [10]. Therefore, strong evaporation was the controlling reason for stable isotope enrichment downstream [53,66].

Table 4. Parameters in spatiotemporal clusters with a significant difference in the Chaobai River.

		Cl^-	SO_4^{2-}	K^+	Na^+	Mg^{2+}	TDS		
Temporal clusters	I	119.44 [a]	96.97 [a]	18.91 [a]	101.37 [a]	30.70 [a]	590.12 [a]		
	II	83.19 [b]	81.12 [b]	13.10 [b]	81.21 [b]	27.63 [b]	518.00 [b]		
		DO	EC	HCO_3^-	K^+	Na^+	Ca^{2+}	Mg^{2+}	TDS
Spatial Cluster	I	5.47 [b]	786 [b]	208 [b]	13.10 [b]	79.36 [b]	40.59 [b]	30.36 [a]	495 [b]
	II	7.00 [a]	940 [a]	250 [a]	17.21 [a]	97.04 [a]	58.22 [a]	27.51 [b]	588 [a]
		NH_3-N	NO_2-N	NO_3-N	TN	TP	*Chl.a*	$\delta^{18}O$	δ^2H
Spatial Cluster	I	0.51 [a]	0.124 [b]	2.04 [b]	5.33 [b]	0.281 [b]	71.00 [b]	−6.11 [a]	−49.20 [a]
	II	0.22 [b]	0.234 [a]	12.47 [a]	14.46 [a]	0.944 [a]	39.29 [a]	−7.66 [b]	−57.87 [b]

Note: Average values of the same parameters in corresponding groups with different letters are significantly different ($p < 0.05$). Temporal clusters/ Spatial Cluster

3.3.4. Principal Component Analysis and Controlling Factors

For inferring the controlling factors, two clusters were further analyzed by PCA (Table S1), and the relationship plots are given in Figure 12. The principal components would be retained according to the corresponding eigenvalue >1 [24], and the critical related parameters could be kept in terms of an explaining proportion >0.60 [25]. In Cluster I, six principal components with a total explaining 90.46% were retained. PC1 had significant and positive loading with Cl^-, SO_4^{2-}, K^+, Na^+, TDS, $\delta^{18}O$, and δ^2H. PC2 was highly and positively related with HCO_3^-, Ca^{2+}, and Mg^{2+}. PC3 had a high and positive relation with EC, NO_2-N, and TP. High and positive loading with these three components showed main water chemical composition of TDS and EC, and mineral dissolution. PC4 had high and positive loading with $\delta^{18}O$ and δ^2H. PC5 was positively related with Na^+, TN, and *Chl.a*, while negatively related with NH_3-N. PC6 was highly and positively related with NO_3-N and TN. NO_3-N was the main nitrogen formation formed by nitrification of NH_3-N [78]. As a result, the negative loading with NH_3-N could be found. Meanwhile, nitrate was also the nitrogen consumed by phytoplankton [35]. Therefore, PC5 and PC6 mainly indicated the nitrogen transformation and the photosynthesis by phytoplankton. In Cluster II, five principal components explaining 87% were retained. PC1 had high and positive loading with EC, Cl^-, HCO_3^-, K^+, Na^+, Ca^{2+}, Mg^{2+}, and TDS, which indicated reclaimed water was the main source of major ions. Additionally, these stations were strongly impacted by reclaimed water. PC2 was highly related with *Chl.a*, $\delta^{18}O$, and δ^2H. PC4 had a negative loading with SO_4^{2-}. There were no highly related parameters to PC3 and PC5. High and positive loading of $\delta^{18}O$ and δ^2H, e.g., PC1, PC4 of Cluster I, and PC2 of Cluster II together showed that stable isotope compositions were controlled by reclaimed water (main water source) and strong evaporation.

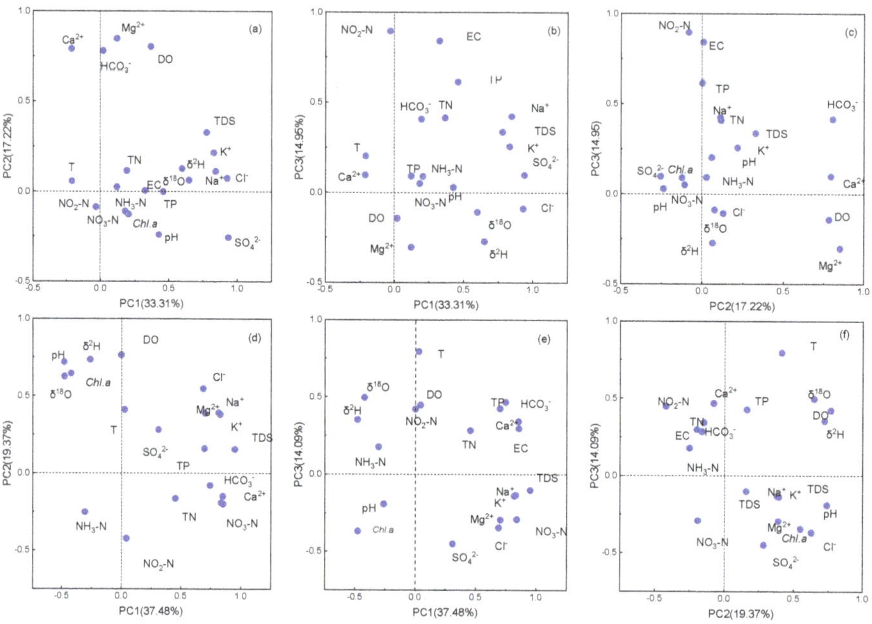

Figure 12. Relationship of principal components in spatial cluster I (**a**–**c**) and II (**d**–**f**).

Nitrogen and phosphorus are the nutrient elements for phytoplankton growth. Consequently, N and P contents would be affected by the biomass of phytoplankton [49]. Therefore, the relationship of nutrients with *Chl.a* were complicated. Sometimes, a significant and positive correlation in a certain period and a disproportionate relationship both could be found in the water environment [35,81]. There must be a relatively excessive amount of N (for P-restricted water) or P (for N-restricted water). Nutrients would be generally consumed by phytoplankton according to the Redfield ratio [82,83]. This case could be identified by the ratio of N/P. As a result, the surplus nutrient would not contribute to eutrophication. In fresh water, the N/P ratio was less than 7, indicating that N was the possible restrictive nutrient, and if the N/P ratio was greater than 7, then P was the possible restrictive one. Except SY02 in May, the ratio values of all monitoring stations were much higher than 7. Therefore, P was the restricted nutrient. Besides, hydrodynamic conditions such as temperature, light, water volume, and flow rate were also important influencing factors [49,81].

3.4. Assessment of Water Quality

According to the water quality standard for reclaimed water used for a scenic environment (GB/T18921-2019) in China [84], the average pH of reclaimed water (Table 1) was 7.93 ± 0.25, which was in the range of 6–9; the means of NH_3-N, TN, and TP were 0.08 ± 0.03 mg/L, 17.18 ± 4.47 mg/L, and 1.05 ± 0.46 mg/L, respectively. Correspondingly, their threshold values were 5 mg/L, 15 mg/L, and 0.5 mg/L. Therefore, TN and TP of reclaimed water should be reduced to meet the current standard requirement [84]. In terms of the surface water environment quality standard (GB3838-2002) in China [85], Class I is for the source of drinking water, the National Nature Reserve; Class II is for the centralized drinking water surface water source with the first level protected areas, rare aquatic habitat; and Class V is for the agricultural water area and the general landscape requirements of the waters. Average value of TN (9.62 ± 5.29 mg/L) and TP (0.60 ± 0.55 mg/L) in river water exceeded their limited value of the V class (2.0 mg/L, 0.4 mg/L). While, NH_3-N (0.38 ± 0.42) in river water was in the range of the I (0.15 mg/L) and II (0.50 mg/L) class, and DO (6.20 ± 2.50 mg/L) was higher than the threshold value in the II (6.00 mg/L) class.

On the basis of the sodium adsorption ratio (SAR), proposed by Richards [86], levels were divided into four classes (S1: <40; S2: 40–90; S3: 90–150; and S4: >150), and the corresponding harmful extents were low, medium, high, and very high, respectively. SAR of reclaimed water and river water were 57.80 ± 5.43 and 56.84 ± 10.27, respectively. Both were in the range of 46.46 ± 12.66–62.02 ± 6.92. As a result, the level was S2, indicating a medium harmful level. TP = 0.02 mg/L and TN = 0.2 mg/L, which were the internationally recognized threshold for eutrophication [87]. TP and TP in reclaimed water and river water were significantly higher than these threshold values. *Chl.a* in almost all samples were higher than 40 μg/L, except for three monitoring stations (SY01, SY02, and SY03) and some individual samples (SY04 in March; SY06 and SY07 in July; SY09 in September and November; and SY08 in November), which indicated severe eutrophication [87,88]. Except for SY01 and SY02, the remaining samples were at the eutrophication level, with *Chl.a* > 7 μg/L.

4. Conclusions

Chemometrics and multivariate statistics were used to study the characteristics and controlling factors in Chaobai River water, replenished by reclaimed water. The main conclusions were as follows. All water was oxidized and alkaline, which was beneficial for nitrification. Nitrate was the main nitrogen form in reclaimed and river water. Depleted and enriched stable isotopes were in reclaimed water and river water, respectively. TN and TP of reclaimed water exceeded the threshold of the reclaimed water reuse standard and Class V in the surface water quality criteria. Most river water was at the severe eutrophication level. The sodium adsorption ratio indicated a medium harmful level for irrigation purposes. Significant spatial and temporal variation was explored by a cluster analysis. Five months are classified into two distinct groups (I: March, May; II: July, September, and November). Nine stations were classified into two clusters (upstream: SY01, SY02, SY03, SY05, and SY06 and downstream: SY04, SY07, SY08, andSY09). Six parameters (Cl^-, SO_4^{2-}, K^+, Na^+, Mg^{2+}, and TDS) had significant upward temporal variation. Twelve parameters (DO, EC, HCO_3^-, K^+, Na^+, Ca^{2+}, TDS, NO_2-N, NO_3-N, TN, TP, and *Chl.a*.) had a significant downward spatial trend. While, four parameters (Mg^{2+}, NH_3-N, $\delta^{18}O$, and δ^2H) were the opposite. The Gibbs plot showed that river water chemistry was mainly controlled by reclaimed water or the interaction of river water with soil/rock. The ionic relationship and principal component analysis showed that river water had undergone the dissolution of carbonate, calcite, and silicate minerals, cation exchange, a process of nitrification, photosynthesis of phytoplankton, and stable isotope enrichment by strong evaporation, and gypsum and salt rock have a potential dissolution process, after reclaimed water was replenished to the river. We think that water quality of reclaimed water should be furtherly improved to avoid river water eutrophication, especially for nitrogen and phosphorus.

Supplementary Materials: The following are available online at http://www.mdpi.com/2073-4441/12/9/2551/s1, Table S1: Principal component analysis of spatial clusters including Group I and Group II.

Author Contributions: Y.Y. and X.S. had the original idea for the study, and carried out the design with all authors. F.Z. and Y.Z. had collected water samples in field and participate laboratory analysis. Y.Y. drafted the manuscript, and other authors gave advices. All authors have read and agreed to the published version of the manuscript.

Funding: This research was funded by the Government Funded Abroad Program (CAFYBB2019GC001-22) and the Basic Research Project (CAFYBB2017ZA007) of Chinese Academy of Forestry and the National Natural Science Foundation of China (41601037).

Acknowledgments: We wish to thank the two anonymous reviewers for their invaluable comments and constructive suggestions used to improve the quality of the manuscript.

Conflicts of Interest: The authors declare no conflict of interest.

References

1. Xia, J. A perspective on hydrological base of water security problem and its application study in North China. *Process. Geogr.* **2002**, *21*, 517–526. (In Chinese)
2. Xia, J.; Liu, M.; Jia, S.; Song, X.; Luo, Y.; Zhang, S. Water security problem and research perspective in North China. *J. Nat. Resour.* **2004**, *19*, 550–560. (In Chinese)
3. Wenquan, G.; Dongguo, S. Risk Evaluation of Water Shortage in Source Area of Middle Route Project for South-North Water Transfer in China. Available online: https://ideas.repec.org/a/spr/waterr/v26y2012i12p3479-3493.html (accessed on 12 September 2020).
4. Fan, Q.; Dai, L.; Liu, W.; Jiao, Z.; Jiang, T.; Bai, G.; Liu, B. *Beijing Water Resources Bulletin in 2010*; Beijing Municipal Bureau of Water Affairs; China Water Conservancy and Hydropower Press: Beijing, China, 2011. (In Chinese)
5. Zhou, Z.; Liu, W.; Tang, N.; Jiao, Z.; Li, M.; Zhao, H. *Beijing Water Resources Bulletin in 2018*; Beijing Municipal Bureau of Water Affairs; China Water Conservancy and Hydropower Press: Beijing, China, 2019; pp. 1–20. (In Chinese)
6. Yang, H.; Abbaspour, K.C. Analysis of wastewater reuse potential in Beijing. *Desalination* **2007**, *212*, 238–250. [CrossRef]
7. Mohammad, M.J.; Mazahreh, N. changes in soil fertility parameters in response to irrigation of forage crops with secondary treated wastewater. *Commun. Soil Sci. Plant. Anal.* **2011**, *34*, 1281–1294. [CrossRef]
8. Asgari, K.; Cornelis, W.M. Heavy metal accumulation in soils and grains, and health risks associated with use of treated municipal wastewater in subsurface drip irrigation. *Environ. Monit. Assess.* **2015**, *187*, 4565. [CrossRef] [PubMed]
9. Yu, Y.; Song, X.; Zhang, Y.; Zheng, F.; Liang, J.; Han, D.; Ma, Y.; Bu, H. Identification of key factors governing chemistry in groundwater near the water course recharged by reclaimed water at Miyun County, Northern China. *J. Environ. Sci.* **2013**, *25*, 1754–1763. [CrossRef]
10. Yu, Y.; Song, X.; Zhang, Y.; Zheng, F.; Liu, L. Impact of reclaimed water in the watercourse of Huai River on groundwater from Chaobai River basin, Northern China. *Front. Earth Sci.* **2017**, *11*, 643–659. [CrossRef]
11. Iwane, T.; Urase, T.; Yamamoto, K. Possible impact of treated wastewater discharge on incidence of antibiotic resistant bacteria in river water. *Water Sci. Technol. J. Int. Assoc. Water Pollut. Res.* **2001**, *43*, 91–99. [CrossRef]
12. Qin, C.; Liu, H.; Liu, L.; Smith, S.; Sedlak, D.L.; Gu, A.Z. Bioavailability and characterization of dissolved organic nitrogen and dissolved organic phosphorus in wastewater effluents. *Sci. Total Environ.* **2015**, *511*, 47–53. [CrossRef]
13. Gibbs, R.J. Mechanisms controlling world water chemistry. *Science* **1970**, *170*, 1088–1090. [CrossRef]
14. Meybeck, M. Global occurrence of major elements in rivers. *Treatise Geochem.* **2003**, *5*, 207–223.
15. Qu, B.; Sillanpää, M.; Zhang, Y.; Guo, J.; Wahed, M.S.M.A.; Kang, S. Water chemistry of the headwaters of the Yangtze River. *Environ. Earth Sci.* **2015**, *74*, 1–16. [CrossRef]
16. Zhu, B.; Yu, J.; Qin, X.; Rioual, P.; Xiong, H. Climatic and geological factors contributing to the natural water chemistry in an arid environment from watersheds in northern Xinjiang, China. *Geomorphology* **2012**, *153–154*, 102–114. [CrossRef]
17. Cerling, T.E.; Pederson, B.L.; Von Damm, K.L. Sodium-calcium ion exchange in the weathering of shales: Implications for global weathering budgets. *Geology* **1989**, *17*, 552. [CrossRef]
18. Liu, B.; Liu, C.Q.; Zhang, G.; Zhao, Z.Q.; Li, S.L.; Hu, J.; Ding, H.; Lang, Y.C.; Li, X.D. Chemical weathering under mid-to cool temperate and monsoon-controlled climate: A study on water geochemistry of the Songhuajiang River system, northeast China. *Appl. Geochem.* **2013**, *31*, 265–278. [CrossRef]
19. Grasby, S.E.; Hutcheon, I.; Mcfarland, L. Surface-water-groundwater interaction and the influence of ion exchange reactions on river chemistry. *Geology* **1999**, *27*, 223. [CrossRef]
20. Azhar, S.C.; Aris, A.Z.; Yusoff, M.K.; Ramli, M.F.; Juahir, H. Classification of river water quality using multivariate analysis. *Procedia Environ. Ences* **2015**, *30*, 79–84. [CrossRef]
21. Costa, M.; Gonçalves, A.M. Clustering and forecasting of dissolved oxygen concentration on a river basin. *Stoch. Environ. Res. Risk Assess.* **2010**, *25*, 151–163. [CrossRef]
22. Alberto, W.D.; del Pilar, D.M.; Valeria, A.M.; Fabiana, P.S.; Cecilia, H.A.; de los Ángeles, B.M. Pattern recognition techniques for the evaluation of spatial and temporal variations in water quality. a case study Suquía River Basin (Córdoba–Argentina). *Water Res.* **2001**, *35*, 2881–2894. [CrossRef]

23. Rong, Y. Statistical methods and pitfalls in environmental data analysis. *Environ. Forensics* **2000**, *1*, 213–220. [CrossRef]
24. Kaiser, H.F. The varimax criteria for analytical rotation in factor analysis. *Psychometrika* **1958**, *23*, 187–200. [CrossRef]
25. Mazlum, N.; Özer, A.; Mazlum, S. Interpretation of water quality data by principal components analysis. *J. Eng. Environ. Sci.* **1999**, *23*, 19–26.
26. Zhang, A.; Ye, C.; Li, Y.; Xie, Z. *Groundwater in Beijing*; China Land Press: Beijing, China, 2008. (In Chinese)
27. Zhang, T.; Li, M.; Yang, Q.; Yu, F.; Qian, T. Scheme of water delivery in the water transfer project from Wenyu River to Chaobai River. *South. North Water Divers. Water Sci. Technol.* **2012**, *10*, 118–120. (In Chinese)
28. SEPA (State Environmental Protection Administration). *Water and Wastewater Monitoring and Analysis Methods (Version 4)*; China Environmental Science Press: Beijing, China, 2002. (In Chinese)
29. Norusis, M. *SPSS 16.0 Guide to Data Analysis*; Prentice Hall Press: Upper Saddle River, NJ, USA, 2008.
30. Mujica, L.E.; Vehí, J.; Ruiz, M.; Verleysen, M.; Staszewski, W.; Worden, K. Multivariate statistics process control for dimensionality reduction in structural assessment. *Mech. Syst. Signal. Process.* **2008**, *22*, 155–171. [CrossRef]
31. Reimann, C.; Filzmoser, P. Normal and lognormal data distribution in geochemistry death of a myth. Consequences for the statistical treatment of geochemical and environmental data. *Environ. Geol.* **2000**, *39*, 1001–1014. [CrossRef]
32. Parkhurst, D.L. User's guide to PHREEQC a computer program for speciation, batch-reaction, one-dimensional transport, and inverse geochemical calculations SuDoc I 19.42/4. *U.S. Geol. Surv. Water Resour. Investig. Rep.* **1999**, *312*. [CrossRef]
33. Shaki, A.A.; Adeloye, A.J. Evaluation of quantity and quality of irrigation water at Gadowa irrigation project in Murzuq basin, southwest Libya. *Agric. Water Manag.* **2006**, *84*, 193–201. [CrossRef]
34. Ayers, R.S.; Westcot, D.W. *Water Quality for Agriculture*; FAO: Rome, Italy, 1985.
35. Kuenzler, E.J.; Stone, K.L.; Albert, D.B. *Phytoplankton Uptake and Sediment Release of Nitrogen and Phosphorus in the Chowan River, North Carolina*; Water Resources Research Institute of the University of North Carolina: Raleigh, NC, USA, 1982.
36. Quiblier, C.; Leboulanger, C.; Sane, S.; Dufour, P. Phytoplankton growth control and risk of cyanobacterial blooms in the lower Senegal River delta region. *Water Res.* **2008**, *42*, 1023–1034. [CrossRef]
37. Deletic, A. The first flush load of urban surface runoff. *Water Res.* **1998**, *32*, 2462–2470. [CrossRef]
38. Vrebos, D.; Beauchard, O.; Meire, P. The impact of land use and spatial mediated processes on the water quality in a river system. *Sci. Total Environ.* **2017**, *601*, 365–373. [CrossRef]
39. Strauss, E.A.; Richardson, W.B.; Bartsch, L.A.; Cavanaugh, J.C.; Bruesewitz, D.A.; Imker, H.; Heinz, J.A.; Soballe, D.M. Nitrification in the Upper Mississippi River: Patterns, controls, and contribution to the NO_3−budget. *J. North Am. Benthol. Soc.* **2004**, *61*, 1102–1112. [CrossRef]
40. Brion, N.; Billen, G. Wastewater as a source of nitrifying bacteria in river systems: The case of the River Seine downstream from Paris. *Water Res.* **2000**, *34*, 3213–3221. [CrossRef]
41. Zhu, B.; Yu, J.; Qin, X.; Rioual, P.; Zhang, Y.; Liu, Z.; Yan, M.; Li, H.; Ren, X.; Xiong, H. Identification of rock weathering and environmental control in arid catchments (northern Xinjiang) of Central Asia. *J. Asian Earth Sci.* **2013**, *66*, 277–294. [CrossRef]
42. Garcia-Ruiz, R.; Pattinson, S.N.; Whitton, B.A. Denitrification in river sediments: Relationship between process rate and properties of water and sediment. *Freshw. Biol.* **1998**, *39*, 467–476. [CrossRef]
43. Grischek, T.; Hiscock, K.M.; Metschies, T.; Dennis, P.F.; Nestler, W. Factors affecting denitrification during infiltration of river water into a sand and gravel aquifer in Saxony, Germany. *Water Res.* **1998**, *32*, 450–460. [CrossRef]
44. Jia, Z.; Liu, T.; Xia, X.; Xia, N. Effect of particle size and composition of suspended sediment on denitrification in river water. *Sci. Total Environ.* **2016**, *541*, 934–940. [CrossRef]
45. Cohen, R.R.H. Biochemical oxygen demand and algae: Fractionation of phytoplankton and nonphytoplankton respiration in a large river. *Water Resour. Res.* **1990**, *26*, 671–678. [CrossRef]
46. Gelda, R.K.; Effler, S.W. A river water quality model for chlorophyll and dissolved oxygen that accommodates Zebra Mussel metabolism. *Water Qual. Ecosyst. Model.* **2000**, *1*, 271–309. [CrossRef]

47. Zhou, W.; Yuan, X.; Long, A.; Huang, H.; Yue, W. Different hydrodynamic processes regulated on water quality (nutrients, dissolved oxygen, and phytoplankton biomass) in three contrasting waters of Hong Kong. *Environ. Monit. Assess.* **2014**, *186*, 1705–1718. [CrossRef]
48. Huang, J.; Yin, H.; Chapra, S.C.; Zhou, Q. Modelling dissolved oxygen depression in an urban river in China. *Water* **2017**, *9*, 520. [CrossRef]
49. Kotovshchikov, A.V.; Dolmatova, L.A. Dynamics of Chlorophyll a content in the Ob River and its relationship with abiotic factors. *Inland Water Biol.* **2018**, *11*, 21–28. [CrossRef]
50. Abonyi, A.; Ács, É.; Hidas, A.; Grigorszky, I.; Várbíró, G.; Borics, G.; Kiss, K.T. Functional diversity of phytoplankton highlights long-term gradual regime shift in the middle section of the Danube River due to global warming, human impacts and oligotrophication. *Freshw. Biol.* **2018**. [CrossRef]
51. Craig, H.; Gordon, L.I. Deuterium and Oxygen 18 Variations in the Ocean and Marine Atmosphere. Available online: https://www.worldcat.org/title/deuterium-and-oxygen-18-variations-in-the-ocean-and-the-marine-atmosphere/oclc/8019537 (accessed on 12 September 2020).
52. Craig, H.; Gordon, L.I.; Horibe, Y. Isotopic exchange effects in the evaporation of water: 1. Low-temperature experimental results. *J. Geophys. Res.* **1963**, *68*, 5079–5087. [CrossRef]
53. Craig, H. Isotopic variations in meteoric waters. *Science* **1961**, *133*, 1702–1703. [CrossRef]
54. Wang, Y.; Li, X.; Yao, L.; Zhao, Y.; Pan, Y. Variation of pH and chemical composition of precipitation by multi2step sampling in summer of Beijing 2007. *Environ. Sci.* **2009**, *30*, 2715–2721. (In Chinese)
55. Kuypers, M.M.; Marchant, H.K.; Kartal, B. The microbial nitrogen-cycling network. *Nat. Rev. Microbiol.* **2018**, *16*, 263. [CrossRef]
56. Yu, Y.; Ma, M.; Zheng, F.; Liu, L.; Zhao, N.; Li, X.; Yang, Y.; Guo, J. Spatio-temporal variation and controlling factors of water quality in Yongding river replenished by reclaimed water in Beijing, North China. *Water* **2017**, *9*, 453. [CrossRef]
57. David, A.; Tournoud, M.-G.; Perrin, J.-L.; Rosain, D.; Rodier, C.; Salles, C.; Bancon-Montigny, C.; Picot, B. Spatial and temporal trends in water quality in a Mediterranean temporary river impacted by sewage effluents. *Environ. Monit. Assess.* **2013**, 2517–2534. [CrossRef]
58. Atwebembeire, J.; Andama, M.; Yatuha, J.; Lejju, J.B.; Rugunda, G.K.; Bazira, J. The physico-chemical quality of effluents of selected sewage treatment plants draining into River Rwizi, Mbarara Municipality, Uganda. *J. Water Resour. Prot.* **2019**, *11*, 23–36. [CrossRef]
59. Marei, A.; Khayat, S.; Weise, S.; Ghannam, S.; Sbaih, M.; Geyer, S. Estimating groundwater recharge using the chloride mass-balance method in the West Bank, Palestine. *Hydrol. Sci. J.* **2010**, *55*, 780–791. [CrossRef]
60. Selim, H.M.; Schulin, R.; Flühler, H. Transport and ion exchange of calcium and magnesium in an aggregated soil. *Soil Sci. Soc. Am. J.* **1987**, *51*, 876. [CrossRef]
61. Reid, J.; MacLeod, D.; Cresser, M.S. Factors affecting the chemistry of precipitation and river water in an upland catchment. *J. Hydrol.* **1981**, *50*, 129–145. [CrossRef]
62. Seanego, K.G.; Moyo, N.A.G. The effect of sewage effluent on the physico-chemical and biological characteristics of the Sand River, Limpopo, South Africa. *Phys. Chem. Earth Parts A B C* **2013**, *66*, 75–82. [CrossRef]
63. Perkins, R.G.; Underwood, G.J.C. The potential for phosphorus release across the sediment–water interface in an eutrophic reservoir dosed with ferric sulphate. *Water Res.* **2001**, *35*, 1399–1406. [CrossRef]
64. Kim, L.H.; Choi, E.; Stenstrom, M.K. Sediment characteristics, phosphorus types and phosphorus release rates between river and lake sediments. *Chemosphere* **2003**, *50*, 53–61. [CrossRef]
65. Liu, J.; Song, X.; Fu, G.; Liu, X.; Zhang, Y.; Han, D. Precipitation isotope characteristics and climatic controls at a continental and an island site in Northeast Asia. *Clim. Res.* **2011**, *49*, 29–44. [CrossRef]
66. Wang, L.; Caylor, K.K.; Villegas, J.C.; Barron-Gafford, G.A.; Breshears, D.D.; Huxman, T.E. Partitioning evapotranspiration across gradients of woody plant cover: Assessment of a stable isotope technique. *Geophys. Res. Lett.* **2010**, 37. [CrossRef]
67. Song, X.; Tang, Y.; Zhang, Y.; Ma, Y.; Han, D.; Bu, H.; Yang, L.; Liu, F. Using stable isotopes to study vapor transport of continuous precipitation in Beijing. *Adv. Water Sci.* **2017**, *28*, 387–495. (In Chinese)
68. Fan, B.L.; Zhao, Z.Q.; Tao, F.X.; Liu, B.J.; Tao, Z.H.; Gao, S.; Zhang, L.H. Characteristics of carbonate, evaporite and silicate weathering in Huanghe River basin: A comparison among the upstream, midstream and downstream. *J. Asian Earth Sci.* **2014**, *96*, 17–26. [CrossRef]

69. Smolders, A.J.P.; Hudson-Edwards, K.A.; Velde, G.V.d.; Roelofs, J.G.M. Controls on water chemistry of the Pilcomayo river (Bolivia, South-America). *Appl. Geochem.* **2004**, *19*, 1745–1758. [CrossRef]
70. Li, J.; Yuan, G.L.; Deng, X.R.; Jing, X.M.; Sun, T.H.; Lang, X.X.; Wang, G.H. Major ion geochemistry of the Nansihu Lake basin rivers, North China: Chemical weathering and anthropogenic load under intensive industrialization. *Environ. Earth Sci.* **2016**, *75*, 453. [CrossRef]
71. Reddy, A.G.S.; Saibaba, B.; Sudarshan, G. Hydrogeochemical characterization of contaminated groundwater in Patancheru industrial area, southern India. *Environ. Monit. Assess.* **2012**, *184*, 3557. [CrossRef] [PubMed]
72. Singh, C.K.; Shashtri, S.; Mukherjee, S. Integrating multivariate statistical analysis with GIS for geochemical assessment of groundwater quality in Shiwaliks of Punjab, India. *Environ. Earth Sci.* **2010**, *62*, 1387–1405. [CrossRef]
73. Moon, S.; Huh, Y.; Qin, J.; Pho, N.V. Chemical weathering in the Hong (Red) River basin: Rates of silicate weathering and their controlling factors. *Geochim. Et Cosmochim. Acta* **2007**, *71*, 1411–1430. [CrossRef]
74. Jahangir, M.M.; Khalil, M.I.; Johnston, P.; Cardenas, L.M.; Hatch, D.J.; Butler, M.; Barrett, M.; O'flaherty, V.; Richards, K.G. Denitrification potential in subsoils: A mechanism to reduce nitrate leaching to groundwater. *Agric. Ecosyst. Environ.* **2012**, *147*, 13–23. [CrossRef]
75. Gomez-Velez, J.D.; Harvey, J.W.; Cardenas, M.B.; Kiel, B. Denitrification in the Mississippi River network controlled by flow through river bedforms. *Nat. Geosci.* **2015**, *8*, 941–945. [CrossRef]
76. Zhao, L.; Delatolla, R.; Mohammadian, A. Nitrification Kinetics & Modified Model for the Rideau River, Canada. *Water Qual. Res. J. Can.* **2013**, *48*, 192–201.
77. Radwan, D.M.; Willems, D.P.; El Sadek, D.A.; Berlamont, P.J. Modelling of dissolved oxygen and biochemical oxygen demand in river water using a detailed and a simplified model. *Int. J. River Basin Manag.* **2003**, *1*, 97–103. [CrossRef]
78. Yang, J.; Trela, J.; Plaza, E.; Wahlberg, O.; Levlin, E. Oxidation-reduction potential (ORP) as a control parameter in a single-stage partial nitritation/anammox process treating reject water. *J. Chem. Technol. Biotechnol.* **2016**, *91*, 2582–2589. [CrossRef]
79. Tromans, D. Temperature and pressure dependent solubility of oxygen in water: A thermodynamic analysis. *Hydrometallurgy* **1998**, *48*, 327–342. [CrossRef]
80. Weissenbacher, N.; Loderer, C.; Lenz, K.; Mahnik, S.N.; Wett, B.; Fuerhacker, M. NOx monitoring of a simultaneous nitrifying–denitrifying (SND) activated sludge plant at different oxidation reduction potentials. *Water Res.* **2007**, *41*, 397–405. [CrossRef] [PubMed]
81. Mortazavi, B.; Iverson, R.L.; Landing, W.M.; Lewis, F.G.; Huang, W.R. Control of phytoplankton production and biomass in a river-dominated estuary: Apalachicola Bay, Florida, USA. *Mar. Ecol. Prog.* **2000**, *198*, 19–31. [CrossRef]
82. Redfield, A.C.; Ketchum, B.H.; Richards, F.A. The influence of organisms on the composition of sea-water. In *The Sea Water*; Interscience Publishers: New York, NY, USA, 1963.
83. Bulgakov, N.G.; Levich, A.P. The nitrogen: Phosphorus ratio as a factor regulating phytoplankton community structure. *Arch. Hydrobiol.* **1999**, *146*, 3–22. [CrossRef]
84. MHURD, C. *The Reuse of Urban Recycling Water—Water Quality Standard For Scenic Environment Use (Gb/T 18921–2019)*; Chinese Specification Press: Beijing, China, 2020; p. 5. (In Chinese)
85. EPA, C. *Surface Water Environmental Quality Standard (GB3838-2002)*; China Environmental Science Press: Beijing, China, 2002. (In Chinese)
86. Richards, L.A. *Diagnosis And Improvements of Saline And Alkali Soils*; National Agricultural Library: Washington, DC, USA, 1954.
87. Brown, C.D.; Canfield, D.E.; Bachmann, R.W.; Hoyer, M.V. Seasonal patterns of chlorophyll, nutrient concentrations and Secchi disk transparency in Florida lakes. *Lake Reserv. Manag.* **1998**, *14*, 60–76. [CrossRef]
88. Wang, X.J.; Liu, R.M. Spatial analysis and eutrophication assessment for chlorophyll a in Taihu Lake. *Environ. Monit. Assess.* **2005**, *101*, 167.

water

MDPI

Article

Assessing Land-Cover Effects on Stream Water Quality in Metropolitan Areas Using the Water Quality Index

TaeHo Kim, YoungWoo Kim, Jihoon Shin, ByeongGeon Go and YoonKyung Cha *

School of Environment Engineering, University of Seoul, 163, Seoulsiripdae-ro, Dongdaemun-gu, Seoul 02504, Korea; willy1995@uos.ac.kr (T.K.); youngwoo0508@uos.ac.kr (Y.K.); sjh3473@uos.ac.kr (J.S.); rhqudrjs7@uos.ac.kr (B.G.)

* Correspondence: ykcha@uos.ac.kr

Received: 7 October 2020; Accepted: 21 November 2020; Published: 23 November 2020

Abstract: This study evaluated the influence of different land-cover types on the overall water quality of streams in urban areas. To ensure national applicability of the results, this study encompassed ten major metropolitan areas in South Korea. Using cluster analysis, watersheds were classified into three land-cover types: Urban-dominated (URB), agriculture-dominated (AGR), and forest-dominated (FOR). For each land-cover type, factor analysis (FA) was used to ensure simple and feasible parameter selection for developing the minimum water quality index (WQI_{min}). The chemical oxygen demand, fecal coliform (total coliform for FOR), and total nitrogen (nitrate-nitrogen for URB) were selected as key parameters for all land-cover types. Our results suggest that WQI_{min} can minimize bias in water quality assessment by reducing redundancy among correlated parameters, resulting in better differentiation of pollution levels. Furthermore, the dominant land-cover type of watersheds, not only affects the level and causes of pollution, but also influences temporal patterns, including the long-term trends and seasonality, of stream water quality in urban areas in South Korea.

Keywords: urban stream; factor analysis; land-cover type; metropolitan area; minimum water quality index; pollution

---

1. Introduction

Global urbanization is an ongoing trend, with 55% to 68% of the world's population projected to reside in urban areas by 2050 [1]. Urbanization induces multiple stressors, especially land-use/land-cover changes such as deforestation and the growth of industrial and residential areas, resulting in increased impervious surfaces [2–5]. Consequently, urbanization leads to a deterioration of water quality in streams through an increase in pollution sources and various hydromorphological changes [6–8]. Despite their at-risk status, streams in urban areas are crucial water resources with a number of designated uses, such as drinking water supply, recreation, and wildlife conservation [9–12].

Therefore, it is vital to establish management strategies for preventing or alleviating water quality problems; this requires efforts to monitor and assess stream water quality in urban areas. The water quality index (WQI), an approach that quantitatively integrates a number of chemical, physical, and biological water quality parameters, has been widely used to assess the water quality status of both surface and groundwater systems [13–17]. In recent years the advent of big data and the accumulation of monitored multivariate data has prompted a substantial increase in the application of WQI to environmental and ecological studies [18–20]. In many of these studies, the developed WQI has been used to capture long-term trends [21,22], seasonal fluctuations [23,24], or spatial variations [25,26] in the overall stream water quality in urban areas. As well as determining the spatiotemporal patterns of stream water quality in urban areas, previous WQI-based research has also determined pollution sources and anthropogenic effects [27–29] and selected the key parameters that represent variations in water quality [30–33].

Recent assessments of urban stream water quality have increasingly employed parameter selection using a number of statistical methods, highlighting the advantages of this process for cost and time saving during assessment. For example, Wu et al. [33] used stepwise multiple regression, which assumes linearity between the WQI and each parameter, to select five parameters representing the water quality of streams in the highly developed area of Lake Taihu Basin, China. Tripathi and Singal [31] used both principal component analysis (PCA) and correlation analysis to select nine parameters to develop a WQI for the Ganga River, which flows through some highly polluted cities of India. Moreover, linear discriminant analysis was applied by Han et al. [34] to select parameters that most effectively differentiate temporal groups (wet versus dry period) and spatial groups (east vs. west parts of the lake) in the Fu River and Baiyangdian Lake, both of which are located in a highly populated region of northern China.

However, the spatial scales of previous parameter selection studies have been limited to single water bodies or single basins; thus, the parameters selected in these studies have limited applicability to other urban stream ecosystems. Furthermore, the effects of different types of anthropogenic activities (e.g., industry, cultivation, or forestation), on stream water quality in urban areas has rarely been considered [26,35]. To overcome these limitations, this study presents the first attempt, to our knowledge, to explicitly account for the effects of different land-cover types on the water quality response and key water quality parameters of urban streams. This study was conducted on a national scale, encompassing a wide range of hydromorphological and geographical characteristics and socioeconomic backgrounds, which are also key factors influencing water quality [36–40]. Therefore, this study aimed to provide parameter selection results that are both informative and applicable to other unexplored streams in urban areas of South Korea.

Streams across ten major metropolitan areas of South Korea were investigated. Cluster analysis was performed to classify stream watersheds based on their land-cover characteristics. Then, the objective WQI (WQI_{obj}) was calculated for each land-cover type using all available water quality parameters. The long-term trends of WQI_{obj} were evaluated using the seasonal Mann-Kendall (SMK) test, and only periods exhibiting temporal stability were used in further analyses. For each land-cover type, key parameters were selected using factor analysis (FA) to develop the minimum WQI (WQI_{min}). The objectives of this study were: (1) To assess the long-term trends and seasonality of the overall stream water quality in metropolitan areas in South Korea; (2) to analyze how different land-cover types affect stream water quality in urban areas and key water quality parameters; and (3) to evaluate the correlation between WQI_{obj} and WQI_{min} and relationships between WQI_{min} and land-covers.

2. Materials and Methods

2.1. Study Area and Data Description

Ten major metropolitan areas across South Korea, with populations of greater than one million, were included in this study (Seoul, Busan, Incheon, Daegu, Daejeon, Gwangju, Suwon, Ulsan, Changwon, and Goyang (Figure 1)) [41]. Within the study area, 81 water quality monitoring sites were selected at tributaries that directly or indirectly flow into either the Han, Geum, Nakdong, or Yeongsan Rivers, the four major rivers of South Korea. The selected monitoring sites covered 35 standard watersheds with the range of watershed area from 39 to 294.9 km^2, and a mean area of 103.29 km^2, the smallest unit of the drainage area division system in South Korea (http://wamis.go.kr). Water quality data were provided by the National Institute of Environmental Research of the Ministry of Environment (http://water.nier.go.kr). The data spanned the time period from 2007 to 2018, and the monitoring frequency varied by site from weekly to monthly. Among the 54 water quality parameters initially included in the data, heavy metals and other toxic chemicals, such as mercury, cadmium, arsenic, and cyanide, were not included because at least 99.5% of the values for these parameters were either missing or below the detection limit. Furthermore, parameters without available reference values were not included in the analyses. The reference values (i.e., normalization factors and weights) required to develop the Bascarón WQI were provided by previous studies [27,42–44].

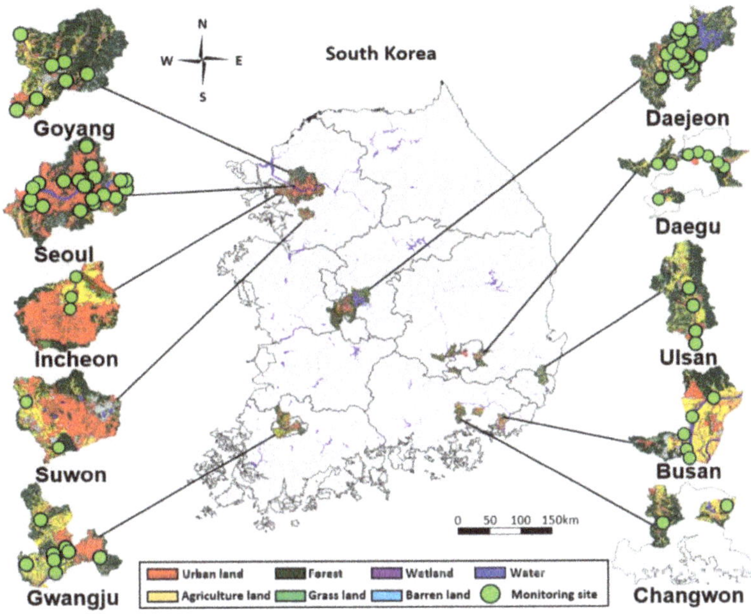

Figure 1. Location of monitoring sites in ten major metropolitan areas of South Korea.

Fourteen water quality parameters were included in the analyses: Water surface temperature (Temp), electrical conductivity (EC), pH, dissolved oxygen (DO), five-day biochemical oxygen demand (BOD_5), chemical oxygen demand (COD), suspended solids (SS), total nitrogen (TN), ammonium nitrogen (NH_4^+-N), nitrate nitrogen (NO_3^--N), total phosphorus (TP), orthophosphate phosphorus (PO_4^{3-}-P), total coliform (TC), and fecal coliform (FC) (Table 1). Among the 81 monitoring sites initially selected for our study, 58 were included for the water quality assessment as they had measurements for all 14 water quality parameters. Land-cover data were provided by the Environmental Geographic Information System; the year of data collection varied from 2010 to 2018 depending on the region (https://egis.me.go.kr). The land-cover data involved seven categories: urban (or built-up) land, agricultural land, forested land, grassland, wetland, barren land, and water. For each of the 35 watersheds, the relative proportions of the seven land-cover categories were calculated using QGIS 2.18.16 [45] and ArcGIS 10.3 software [46].

Table 1. Reference values (i.e., normalization factors, C_i and weights, P_i) required to develop the Bascarón WQI and provided by previous studies [27,42–44].

Parameter	Unit	Relative Weight (P_i)	Normalization Factor (C_i)										
			100	90	80	70	60	50	40	30	20	10	0
Temp	°C	1	21/16	22/15	24/14	26/12	28/10	30/5	32/0	36/−2	40/−4	45/−6	>45/<−6
pH	-	1	7	7–8	7–8.5	7–9	6.5–7	6–9.5	5–10	4–11	3–12	2–13	1–14
EC	µS/cm	1	<750	<1000	<1250	<1500	<2000	<2500	<3000	<5000	<8000	≤12,000	>12,000
DO	mg/L	4	≥7.5	>7	>6.5	>6	>5	>4	>3.5	>3	>2	≤1	<1
BOD_5	mg/L	3	<0.5	<2	<3	<4	<5	<6	<8	<10	<12	≤15	>15
COD	mg/L	3	<5	<10	<20	<30	<40	<50	<60	<80	<100	≤150	>150
SS	mg/L	4	<20	<40	<60	<80	<100	<120	<160	<240	<320	≤400	>400
TN	mg/L	2	<0.8	<3.8	<7.5	<13	<18	<27	<48	<85	<149	≤265	>265
NH_4^+-N	mg/L	3	<0.01	<0.05	<0.1	<0.2	<0.3	<0.4	<0.5	<0.75	<1	≤1.25	>1.25
NO_3^--N	mg/L	2	<0.5	<2	<4	<6	<8	<10	<15	<20	<50	≤100	>100
TP	mg/L	1	<0.2	<1.6	<3.2	<6.4	<9.6	<16	<32	<64	<96	≤160	>160
PO_4^{3-}-P	mg/L	1	<0.025	<0.05	<0.1	<0.2	<0.3	<0.5	<0.75	<1	<1.5	≤2	>2
TC	CFU/100 mL	3	<50	<500	<1000	<2000	<3000	<4000	<5000	<7000	<10,000	≤14,000	>14,000
FC	CFU/100 mL	3	<5	<50	<100	<200	<300	<400	<500	<700	<1000	≤1400	>1400

2.2. Statistical Analyses

2.2.1. Cluster Analysis (CA)

CA is an unsupervised pattern recognition technique, whereby individual objects are grouped into a number of clusters whose objects are more similar than those in other clusters. Among the available CA methods, hierarchical agglomerative CA (HACA) was used in this study. HACA is a successive process, in which two objects in the closest proximity form a cluster at the lowest hierarchy. In the next step, the newly generated two clusters in the closest proximity form a combined cluster. Merging continues until all objects are linked to form a single cluster at the highest hierarchy. The squared Euclidean distance was used as a measure for calculating the proximity between objects/clusters.

Furthermore, we employed the Ward's minimum variance linkage function, which uses distance information to merge objects into a hierarchical cluster tree and is visually represented by a dendrogram [47]. As HACA results in a single cluster, the dendrogram needs to be divided at a specific height to generate multiple clusters. The height in the dendrogram can be defined as $(D_{link}/D_{max})\cdot 100$, where D_{link} is the linkage distance for a pair of objects/clusters and D_{max} is the maximum linkage distance. According to previous studies, the height for dendrogram partitioning was set to 60; that is, $(D_{link}/D_{max})\cdot 100 > 60$ [48,49]. The CA was performed using 'dendrogram' function from the 'SciPy' library [50] in Python 3.6 [51]. To generate clusters based on land-cover type, HACA was performed using the relative proportions of the six land-cover types for each standard watershed (excluding water) as variables.

The differences in water quality parameters and WQI among different clusters were assessed using summary statistics and non-parametric tests, i.e., Kruskal Wallis *H* and Mann-Whitney *U* tests. Non-parametric tests were selected due to the non-normality of water quality parameters. The Kruskal Wallis *H* test examined the differences in distributions for the three clusters. When the significant differences occurred, as a post-hoc analysis, the Mann-Whitney *U* test was used to identify which cluster(s) revealed the significant difference in distribution from the other cluster(s). The Kruskal Wallis *H* and Mann-Whitney *U* tests were performed using 'kruskal' and 'mannwhitneyu' from the 'SciPy' library [50] Python 3.6 [51]. Statistical significance was indicated by p-value < 0.05.

2.2.2. Water Quality Index (WQI) Development

The method for WQI development used in this study builds on the WQI_{obj} [43], a modification of the *Bascarón* WQI, also known as subjective WQI [13], which excludes the constant term multiplied to WQI_{obj}, which reflects the subjective judgment of overall water quality. The WQI_{obj} is calculated as follows,

$$WQI_{obj} = \frac{\sum_{i=1}^{n} C_i P_i}{\sum_{i=1}^{n} P_i} \tag{1}$$

where n is the number of available water quality parameters, C_i is a normalization factor that converts the value of a parameter into a common scale ranging from 0 to 100 with an interval of 10 (Table 1), P_i is the weight indicating the relative importance of parameters, which ranges from 1 to 4 (Table 1), and WQI_{min} is a simplification of WQI_{obj} indicating the minimum WQI [42,43] and is calculated as,

$$WQI_{min} = \frac{\sum_{i=1}^{n_{min}} C_i}{n_{min}} \tag{2}$$

Note that Equation (2) for WQI_{min} does not include the weight term, indicating that the parameters included in WQI_{min} assessment are considered equally important. Here, n_{min} is the number of key parameters, which is a subset of all n available parameters. The WQI_{obj} and WQI_{min} scores were graded into five classes to indicate the overall water quality status: excellent (91–100), good (71–90), medium (51–70), bad (26–50), and very bad (0–25) [42,43,52]. Also, when comparing WQI_{min} with

WQI_{obj} for evaluating whether they are well-correlated, linear regression ($WQI_{obj} = a \cdot WQI_{min} + b$) was performed using 'linregress' function from the 'SciPy' library [50] in Python 3.6 [51].

2.2.3. Seasonal Mann-Kendall (SMK) Test

The Mann-Kendall (MK) test is a non-parametric test that assesses if the temporal trend of a variable exhibits a monotonic increase or decrease [53,54]. The SMK test is an extension of the MK test that accounts for the effect of seasonality by performing the test separately for each pre-defined season [55]. In this study, the SMK was employed to identify the point in time at which WQI no longer shows a significant increasing or decreasing trend; this stabilized time period was divided into training and test sets, and further analyses were performed. CA results were used to assess the trend of monthly averaged WQI_{obj} for each land-cover cluster, initially using the data for the entire time period (2007–2018). When the trend showed a significant increase or decrease (p-values < 0.05), the SMK was performed excluding a 1-year period of data from the starting year. The test was iteratively performed until the trend appeared to be insignificant. The SMK test was performed using 'seasonal_test' function from the 'pyMannKendall' library [56] in Python 3.6 [51].

2.2.4. Factor Analysis (FA)

FA attempts to account for the structure (i.e., correlation and variation) of data, consisting of measured variables with a reduced number of factors, which are also termed latent variables. Exploratory FA (EFA), which does not assume a priori relationships among the factors and measured variables, was used to reveal the underlying factors behind the correlations among measured water quality parameters. Contrary to confirmatory FA, EFA does not posit any relationship between specific factors and measured variables. Therefore, it suits the purpose of this analysis.

Prior to analysis, the values of all water quality parameters except Temp and pH were log-transformed, and the values of all parameters were standardized to have a distribution with a mean of zero and standard deviation of one. To examine whether water quality data were suitable for FA, the Kaiser-Mayer-Olkin (KMO) test [57] and Bartlett's test [58] were performed. The FA was assumed to be valid when the KMO value exceeded 0.5 and the Bartlett's test result was significant (p-value < 0.05).

To determine the number of factors retained in the FA, Horn's parallel analysis (PA) was used [59]. PA compares the eigenvalues (which indicate the relative importance of a factor in explaining the variance of measured variables) from measured data with the eigenvalues from random data, which have the same sample size and number of variables as the measured data and are obtained using a Monte-Carlo simulation. The differences between the eigenvalues from measured data and the mean eigenvalues from random data were calculated. Factors with differences greater than zero were retained in the FA.

As a method for factor extraction, principal component analysis (PCA) was used [60,61]. The maximum likelihood method, another common method for factor extraction, was not selected because of its multivariate normality requirement, which is often not met for water quality parameters even after the transformation (e.g., log-transformation) of values. Squared factor loading, which reflects the proportion of variance in a measured variable explained by each factor, was calculated as a result of PCA implementation. The communality was calculated by summing the squared factor loadings of a given variable across all factors to indicate the proportion of variance in a measured variable explained by all factors. Moreover, the uniqueness was calculated by subtracting the communality from the total variance of a variable.

Factor rotation (i.e., the change in the axes of factors) was implemented to yield interpretable factors by attaining a simple structure for factor loadings. Without rotation, most variables load heavily onto the first and early factors, whereas rotation yields a simple structure in which each variable loads heavily onto only one factor, while loading lightly onto the other factors. Varimax rotation was used as a rotation method, which is a common type of orthogonal rotation. Orthogonal rotation assumes that factors remain uncorrelated with one another. FA was performed using 'principal' function from 'psych' [62] packages in R 3.5.3 [63].

For each land-cover type, the water quality parameter that showed the highest loading factor associated with each retained factor was interpreted as a key water quality parameter. Accordingly, the number of retained factors corresponded to the number of key parameters representing the overall stream water quality in urban areas. The FA procedure was performed for each land-cover type determined by CA, and the key parameters for different land-cover types were used for the WQI_{min} calculation.

3. Results

3.1. Land-Cover Characteristics of Metropolitan Areas in South Korea

Using the HACA, three clusters were generated based on the land-cover characteristics of 35 watersheds in ten major metropolitan areas (Figure 2a). Notably, each of the watersheds included in each of the three clusters had a single dominant land-cover: Urban, agriculture, and forest, respectively (Figure 2b, Table S1). The mean proportion of urban land for the 15 watersheds with urban-dominated land-cover (URB) was 0.50 (± one standard deviation of 0.12), which was higher than that of agricultural (0.06 ± 0.05) and forested (0.30 ± 0.10) land. In contrast, the five watersheds with agriculture-dominated land-cover (AGR) had a mean relative area of 0.44 (± 0.08) for agricultural land-cover, which was more dominant than urban (0.16 ± 0.07) and forested (0.24 ± 0.07) land-cover. The 15 watersheds with forest-dominated land-cover (FOR) were mainly composed of forested land, with a mean proportion of 0.60 (± 0.08), whereas the proportion of urban (0.12 ± 0.06) and agricultural (0.16 ± 0.04) land was relatively minor.

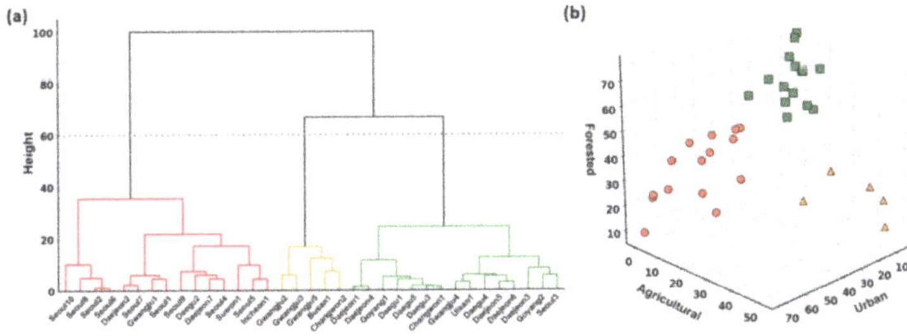

Figure 2. Clustering results of 35 watersheds, named metropolitan area with numbering, based on six land-cover categories. (**a**) Dendrogram exhibiting three clusters generated from hierarchical agglomerative cluster analysis. The horizontal dashed gray line represents the height for dendrogram partitioning, $(D_{link}/D_{max})\cdot 100 > 60$. (**b**) Percentage (%) of the dominant land-cover type for each of the three clusters. The red circle, yellow triangle, and green square denote watersheds that are urban-dominated, agriculture-dominated, and forest-dominated, respectively.

The three land-cover types (URB, AGR, and FOR) were unevenly distributed across the metropolitan areas. Among the URB, 73.3% were concentrated in Seoul (nine watersheds) and its adjacent cities, Suwon (one watershed) and Incheon (one watershed). Three of the five AGR were located in Gwangju, whereas the other two were located in Busan and Changwon. The spatial distribution of FOR was also concentrated, with 33.3% in Daejeon and 26.7% in Daegu.

3.2. Land-Cover Effects on Stream Water Quality in Urban Areas

The long-term trends of overall water quality calculated using all available parameters (WQI_{obj}), based on the results of SMK tests, differed by land-cover type (Figure 3). For URB, WQI_{obj} values gradually improved until becoming stable in 2015 (Figure 3a). In comparison, WQI_{obj} values for AGR showed a greater improvement in early years before becoming stable in 2012 (Figure 3b). For FOR,

WQI_{obj} values did not show any significant trend during the entire period from 2007 to 2018 (Figure 3c). In more recent years (2015–2018), during which all land-cover types exhibited a stable trend, the overall water quality was worst for URB (p-values < 0.05 as a result of Kruskal Wallis H and Mann-Whitney U tests), as indicated by lower WQI_{obj} values (75.04 ± 9.90) than those for AGR (78.91 ± 8.31) and FOR (82.82 ± 7.97). Regardless of the land-cover type and time period, WQI_{obj} values tended to be lower during the wet season (July to September) than during the dry season (Figure 3).

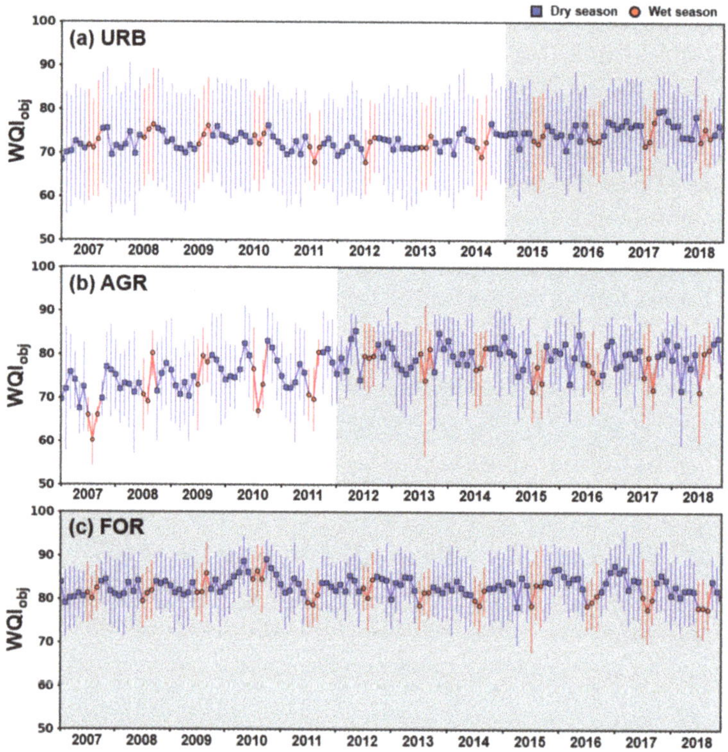

Figure 3. Long-term (2007–2018) trends of objective water quality index (WQI_{obj}) for watersheds with (**a**) urban-dominated, (**b**) agriculture-dominated, and (**c**) forest-dominated land-cover. Blue and red circles denote the mean monthly WQI_{obj} for dry and wet seasons, respectively, and vertical lines denote one standard deviation of monthly WQI_{obj}. The gray area represents a period exhibiting no significant increase or decrease in WQI_{obj} based on the results of seasonal Mann-Kendall tests.

The land-cover types of the watersheds influenced most water quality parameters in urban streams except for pH, EC, DO, and PO_4^{3-}-P, which were similar regardless of the dominant land-cover (Table 2). Compared with URB and AGR, FOR exhibited the lowest level of contamination for the majority of water quality parameters. The level of contamination between URB and AGR differed depending on the water quality parameter. In terms of nitrogen (TN and NO_3^--N) and microbiological indicators (TC and FC), the streams in URB exhibited significantly worse conditions than those in AGR (Table 2). On the other hand, indicators for organic matter (BOD_5 and COD) and turbidity (SS) indicated significantly higher levels of water contamination in AGR than URB (Table 2).

Table 2. Summary statistics (mean ± one standard deviation) of 14 water quality parameters from 2015 to 2018 for watersheds with urban-dominated (URB), agricultural-dominated (AGR), and forest-dominated (FOR) land-cover. Asterisks (*) denote parameters whose mean value for either URB or AGR is significantly higher (or lower in the case of DO) than the other (p-value < 0.05 based on Kruskal Wallis H and Mann-Whitney U tests).

Parameter	Unit	Watershed Type		
		URB	AGR	FOR
Temp	°C	16.33 ± 1.56	16.99 ± 0.69	15.56 ± 1.52
pH	-	7.81 ± 0.32	7.76 ± 0.29	7.79 ± 0.34
EC	µS/cm	455.51 ± 175.19	497.27 ± 257.41	384.76 ± 209.78
DO	mg/L	10.52 ± 1.46	10.49 ± 0.88	11.03 ± 1.04
* BOD_5	mg/L	3.05 ± 2.37	4.00 ± 0.49	1.69 ± 1.01
* COD	mg/L	5.79 ± 3.04	8.17 ± 0.9	4.38 ± 2.16
* SS	mg/L	7.37 ± 5.82	17.29 ± 3.06	6.51 ± 5.27
* TN	mg/L	5.92 ± 3.18	3.49 ± 1.42	3.24 ± 1.43
NH_4^+-N	mg/L	0.87 ± 1.35	0.52 ± 0.44	0.22 ± 0.32
* NO_3^--N	mg/L	3.86 ± 1.83	2.12 ± 0.78	2.26 ± 0.75
TP	mg/L	0.11 ± 0.10	0.10 ± 0.02	0.06 ± 0.03
PO_4^{3-}-P	mg/L	0.05 ± 0.07	0.03 ± 0.01	0.03 ± 0.02
* TC	CFU/100 mL	$49.20 \times 10^3 \pm 69.12 \times 10^3$	$18.09 \times 10^3 \pm 24.97 \times 10^3$	$11.73 \times 10^3 \pm 10.49 \times 10^3$
* FC	CFU/100 mL	$14.01 \times 10^3 \pm 36.73 \times 10^3$	$22.14 \times 10^2 \pm 27.98 \times 10^2$	$16.44 \times 10^2 \pm 22.51 \times 10^2$

3.3. Key Water Quality Parameters for Different Land-Cover Types

The water quality data were suitable for the application of FA, as indicated by the results of the KMO test (0.82 for URB, 0.67 for AGR, and 0.73 for FOR) and Barlett's test (p-value < 0.05 for all land-cover types). To perform the FA, the data measured during the more recent years (2015–2018), when the WQI_{obj} values stabilized for all land-cover types, were divided into training (2015–2016) and testing (2017–2018) data sets. The results of FA using the training data indicated that three factors apiece should be retained for URB, AGR, and FOR (Table S2). For each land-cover type, the water quality parameters with the highest factor loading, associated with each of the retained factors, were selected as the key parameters for the WQI_{min} calculation (Table 3). Frequently, for a given factor, more than one water quality parameter had a factor loading greater than 0.75 [64], which is indicative of a strong correlation between the factor and the parameter (Table 3). In such cases, the parameters were generally highly correlated to each other, with a Pearson's correlation coefficient ranging from 0.49 to 0.88 (Figure S1). Consequently, the three key parameters selected for URB were COD, FC, and NO_3^--N, in order of corresponding factors (Table 3). Three parameters were selected for AGR were FC, COD, and TN (Table 3). The three parameters selected for FOR were COD, TN, and TC (Table 3).

3.4. Comparison between WQI_{obj} and WQI_{min}

Using the test data, the relationships between monthly WQI_{min} and WQI_{obj} values were assessed; WQI_{min} and WQI_{obj} generally exhibited moderate to strong, linear relationships with R^2 values of 0.66 for URB, 0.78 for AGR, and 0.73 for FOR (Figure 4). For both WQI_{obj} and WQI_{min}, URB was generally associated with the poorest overall water quality, with mean WQI values of 75.79 and 67.20, respectively. Further, based on both WQI_{obj} and WQI_{min}, the overall water quality for AGR (mean WQI values of 78.86 and 73.39) was generally poorer than that for FOR (mean WQI values of 82.41 and 77.41). The location of intersection, where the regression line and one-to-one line cross, differed by land-cover type: 87.48 for URB, 81.98 for AGR, and 88.14 for FOR (Figure 4). Below the intersection, WQI_{obj} values tended to be higher than WQI_{min} scores, whereas the opposite was true above the intersection (Figure 4). As the proportion of values below the intersection was greatest for URB, the positive difference between the mean WQI_{obj} and WQI_{min} values for URB (8.59) was greater than that for AGR (5.47) and FOR (5.00). Within each land-cover type, the variation of WQI_{min} values, with one standard deviation of 13.61 for URB, 13.05 for AGR, and 12.09 for FOR, was greater than the variation of WQI_{obj} values, with one standard deviation of 9.27 for URB, 8.62 for AGR, and 7.97 for

FOR. Note that the degree of variation in WQI values, in descending order, was URB, AGR, and FOR for both WQI_{obj} and WQI_{min}.

Table 3. Factor loadings for 14 water quality parameters for watersheds with urban-dominated (URB), agricultural-dominated (AGR), and forest-dominated (FOR) land-cover. Asterisks (*) indicate a factor loading greater than 0.75 or the highest factor loading in the factor. Var (%) represents the explained variance of total variance for each factor.

Parameter	Watershed Type								
	URB			AGR			FOR		
	Factor 1	Factor 2	Factor 3	Factor 1	Factor 2	Factor 3	Factor 1	Factor 2	Factor 3
Temp	0.261	0.241	−0.639	0.243	0.643	−0.578	0.220	0.473	−0.580
pH	0.205	−0.661	−0.213	−0.666	0.264	−0.434	−0.567	0.021	0.040
EC	0.553	0.108	0.568	−0.340	0.295	0.612	0.342	0.021	0.605
DO	−0.164	−0.677	0.379	−0.634	−0.383	−0.071	−0.441	−0.578	0.340
BOD_5	* **0.867**	0.148	0.062	−0.192	* **0.797**	0.031	* **0.751**	0.004	0.079
COD	* **0.905**	0.155	0.015	−0.185	* **0.879**	0.018	* **0.897**	0.081	−0.027
SS	* **0.860**	0.050	−0.129	0.028	* **0.791**	0.073	* **0.825**	0.144	−0.033
TN	0.416	0.330	0.742	0.074	0.015	* **0.930**	0.108	0.127	* **0.930**
NH_4^+-N	0.449	0.503	0.353	0.280	0.279	0.685	0.657	0.144	0.236
NO_3^--N	−0.006	0.145	* **0.806**	0.040	−0.359	* **0.819**	−0.123	0.065	* **0.910**
TP	0.696	0.560	0.102	0.609	0.557	−0.098	0.651	0.411	−0.030
PO_4^{3-}-P	0.340	* **0.751**	0.125	0.704	−0.150	−0.095	0.141	0.667	0.028
TC	0.158	* **0.810**	0.048	* **0.862**	−0.157	−0.033	0.062	* **0.825**	0.077
FC	0.226	* **0.831**	0.025	* **0.865**	−0.024	0.084	0.099	* **0.825**	0.091
Var (%)	27.2	25.9	16.4	25.3	23.7	21.0	26.0	18.7	18.5

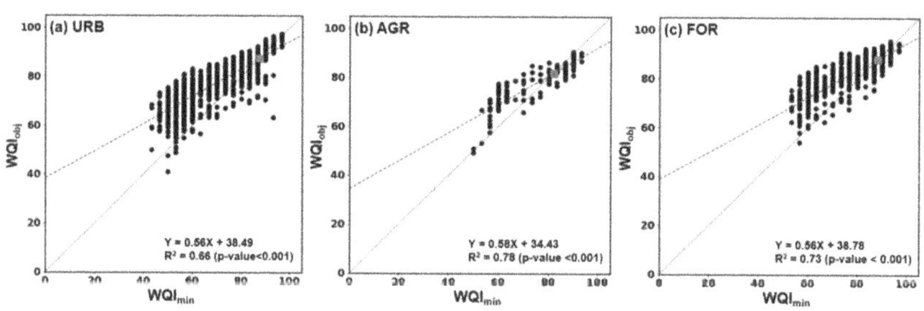

Figure 4. Relationships between objective water quality index (WQI_{obj}) and minimum WQI (WQI_{min}) for watersheds with, (**a**) urban-dominated, (**b**) agriculture-dominated, and (**c**) forest-dominated land use. Black circles denote WQI values calculated using the testing data set (2017–2018). Black dotted and blue dashed lines represent one-to-one, and regression lines, respectively. Red square represents the point of intersection between the one-to-one line and regression line.

3.5. Spatial Distribution of Overall Stream Water Quality in Urban Areas

WQI values by site, calculated for 2015–2018, indicated that WQI_{obj} and WQI_{min} values were highly linearly correlated, with an R^2 value of 0.84 (Figure 5b). However, there was a clear tendency for WQI_{obj} values to be higher than WQI_{min} values (Figure 5a,b). The difference in values between WQI_{obj} and WQI_{min} led to differences in WQI classification in 25.9% of the 58 monitoring sites (Figure 5a). In Seoul, the change in calculation method from WQI_{obj} to WQI_{min} yielded a change of classification from good to medium in 33.3% of 18 monitoring sites. In the other five metropolitan areas (i.e., Daejeon, Gwangju, Daegu, Busan, and Ulsan), a change in classification occurred in one or two sites, accounting for 7.7–40.0% of the sites in each area (Figure 5a). In the remaining five metropolitan areas (i.e., Goyang, Suwon, Incheon, and Changwon), no change in WQI classification occurred in response to application of the WQI_{min} (Figure 5a).

3.6. Seasonality of Overall Stream Water Quality in Urban Areas

From 2015 to 2018, the monthly patterns of overall water quality calculated using WQI_{min} differed by land-cover type (Figure 6). For URB, which exhibited the worst overall water quality, the proportion of WQI_{min} values corresponding to equal to or worse than medium status increased during the wet season (July to September), whereas the proportion of good to excellent status sites increased during the dry season (all other months) (Figure 6a). For FOR, the WQI_{min} status was consistently better than or equal to medium, and the proportion of medium status sites increased during the wet season (Figure 6c). For AGR, the WQI_{min} status tended to worsen during the wet season, with an increase in the proportion of medium status sites; however, this seasonality was less consistent compared with other land-cover types (Figure 6b).

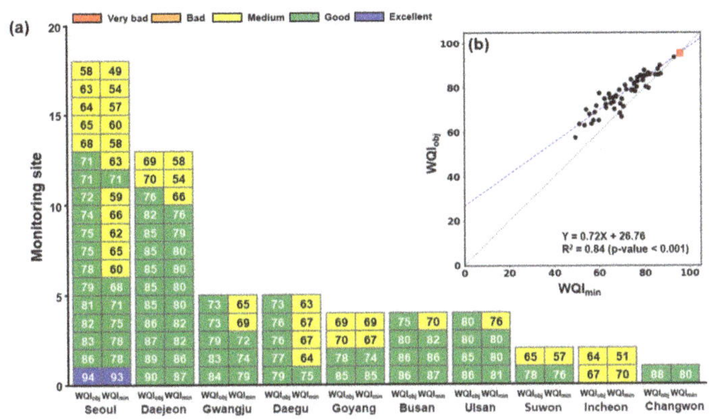

Figure 5. Spatial distribution of the water quality index (WQI) in ten major metropolitan areas of South Korea. (**a**) Mean objective WQI (WQI_{obj}) and minimum WQI (WQI_{min}) values and grades from 2015 to 2018 for each of the 58 monitoring sites. (**b**) Relationship between mean WQI_{obj} and WQI_{min} values.

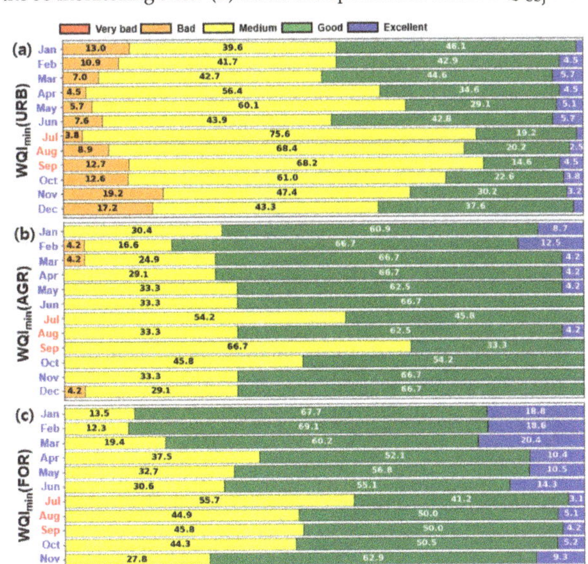

Figure 6. Monthly distribution (%) of minimum water quality index (WQI_{min}) grades from 2015 to 2018 for watersheds with, (**a**) urban-dominated, (**b**) agriculture-dominated, and (**c**) forest-dominated land-cover. Month names for dry and wet seasons are colored blue and red, respectively.

4. Discussion

4.1. Suitability of FA as a Parameter Selection Method

In this study, FA, which involves factor extraction and rotation processes, was used to reduce multiple intercorrelated physical, chemical, and biological water quality parameters into a smaller number of latent factors, and to select key water quality parameters that had the strongest correlation with a given latent factor. In previous studies, along with subjective judgments [27,33,65–69], multivariate statistical techniques were employed to select parameters on an objective basis. For example, stepwise multiple regression has been used [33,69,70] to determine the set of parameters that could best explain the variance of WQI_{obj}. Compared with unsupervised learning (e.g., FA), regression is a supervised method that requires reference values; in this case, WQI_{obj} values for training data. However, because of the multi-collinearity and the resulting bias, WQI_{obj} is not often a suitable reference.

Furthermore, previous studies have used PCA at the first step followed by Pearson's correlation analysis to extract water quality parameters that showed high contributions to selected components and low correlations with other parameters [31,66]. Post-hoc correlation analysis was required, since few first factors derived from PCA are strongly associated with most of the correlated parameters. Therefore, the application of PCA alone is not sufficient to attain key parameters that represent extracted factors. To address this limitation, in this study PCA was conducted in conjunction with factor rotation, which yields a simple structure for the factor loading matrix, in which only a small number of variables have high loadings onto a given factor and do not overlap among the factors. As a result, parameters with high loadings on a given factor appear to be more distinct and homogeneous. Therefore, a set of parameters with high loadings across all factors are expected to represent multifaceted aspects of water quality. Furthermore, the use of varimax rotation as a factor rotation method ensures the extracted factors are uncorrelated with one another, facilitating the selection of key parameters, the relationships among which can be assumed to be independent. Therefore, factor rotation used in conjunction with PCA does not require subsequent correlation analysis, which simplifies parameter selection to a single-step process.

4.2. Key water Quality Parameters

Selected key water quality parameters were similar among different land-cover types (COD, FC, and NO_3^--N for URB; COD, FC, and TN for AGR; COD, TC, and TN for FOR), indicating that the relationships among parameters were consistent regardless of land-cover type. For example, across all land-cover types, COD, BOD_5, and SS were closely correlated (Figure S1) and had high loadings with the same factor (Table 3). The high correlations were shown, since the three parameters commonly account for biodegradable organic matter. In addition, for one being the subset of the other, FC and TC, and NO_3^--N and TN, were closely related to each other (Figure S1) and had the highest loadings onto the same factor for all land-cover types (Table 3). Note that phosphorus parameters showed moderate to strong associations with TC and FC within the same factor for all land-cover types (Table 3). Therefore, rather than phosphorus parameters, either TC or FC, which showed higher loadings with the factor than the phosphorus parameters, was selected as the key parameter. A possible speculation over this co-occurrence tendency is that phosphorus and fecal indicator bacteria may originate from the same pollution source (e.g., domestic sewage and agricultural runoff) or the same mechanism (e.g., sediment release), but future research will be necessary for interpreting the causal relationships.

The presence of multiple parameters with almost equally high loadings onto a given factor necessitated comparisons between WQI_{min} and modified WQI_{min}, in which a key parameter (e.g., COD) is replaced by its surrogate parameter (e.g., BOD_5) that was strongly related to the key parameter within the same factor. The results illustrated that modified WQI_{min} was generally in close agreement with WQI_{min} (Figure S2), suggesting that a set of parameters that shows high loadings within the same factor can be used interchangeably. Note that, compared with other sets of parameters, linear relationships between WQI_{min} and modified WQI_{min} for fecal indicator bacteria were weaker because of the large variability inherent in FC and TC concentrations. Nonetheless, given the marginal differences in factor loading between TC and FC regardless of land-cover type (Table 3), between the two parameters, the key parameter should be selected depending on management focus or data availability.

The results of FA need to be interpreted and applied with care. The factor extraction process of FA determines the factors worth retaining, and the subsequent factor rotation, whereby the factors become least correlated with each other, yields the proportion of variance explained by a given factor to be distributed more evenly among the factors. Therefore, it is not particularly valid to prioritize the factors and the consequent key parameters. Instead, the selected key parameters should be considered independently of each other and as equally important. In this regard, assigning different weights to key water quality parameters with equal importance should not be included as a step for WQI_{min} development. Previous studies reported that using weights improved the linearity between WQI_{min} and WQI_{obj} [33,70]. In contrast to these findings, we found that the use of weights, which were estimated based on two methods, the relative weight [33,70] and the percentage of variance explained by the given factor (Table 3), yielded only slight differences in the WQI_{min}-WQI_{obj} relationships (Figure S3).

It should be acknowledged that the water quality data, used in this study, did not include several widely measured parameters, such as parameters for minerals, salts, metals and flow rate. If such parameters were added to the data, FA may include additional factors and key parameters. Moreover, the results of parameter selection did not contain the basic water quality parameters of Temp and pH in the key parameter list for any land-cover type. In addition, despite being frequently included as a key parameter [42,43,68–70] in previous studies, DO was not selected for any land-cover type in this study (Table 3). Variations in Temp, pH, and DO may be influenced by anthropogenic activities but are also attributable to natural variability. That is, they exhibit diurnal fluctuations and are strongly influenced by meteorological conditions [33,65]. Our results suggest that Temp, pH, and DO, whose patterns are substantially influenced by natural variations, may not successfully capture the total variance of stream water quality in urban areas, and may not be suitable for being included as key parameters.

4.3. Comparison between WQI_{min} and WQI_{obj}

Our results of test data showed that WQI_{min} and WQI_{obj} have close linear relationships across all land-cover types (Figure 4), suggesting that WQI_{min} can be used to predict WQI_{obj} using the established regression model. However, WQI_{min} values tended to be higher than WQI_{obj} above a certain threshold and lower than WQI_{obj} below this threshold. This tendency indicates that the use of WQI_{min} eliminates the "eclipse effect" [71], which arises from the redundancy inherent in WQI_{obj}; accordingly, WQI_{obj} is subject to overestimating bad water quality status and underestimating good water quality status. The removal of redundancy was also evidenced by the larger variance of WQI_{min} compared with that of WQI_{obj} for all land-cover types (Figure 4). Therefore, the development and use of WQI_{min} is expected to improve the identification of the overall water quality status and the level of water pollution in streams across urban areas. Our results demonstrate that the method selection for WQI assessment has important resource and management implications. Changing the method from WQI_{obj} to WQI_{min} altered the spatial distribution of the overall water quality status; this status change occurred in a minor to substantial portion of monitoring sites, depending on the metropolitan area (Figure 5). This change suggests that the use of WQI_{min} instead of WQI_{obj}, which may involve a status change from "good" to "medium" or vice versa, may affect priority setting and resource allocation among individual watersheds or groups of watersheds.

4.4. Land-Cover Effects on Stream Water Quality in Urban Areas

Our results indicate that the dominant land-cover affected the overall stream water quality in urban areas, with mean values of both WQI_{obj} and WQI_{min} decreasing in the order: FOR > AGR > URB (Figure 4). The dominant land-cover type also contributed to the deterioration of differing water quality parameters (i.e., nitrogen and microbiological indicators for URB, but organic matter and turbidity for AGR) (Table 2). The long-term trends of overall water quality differed by land-cover type (Figure 3). Over the last decade, WQI_{obj} trends for URB and AGR exhibited early improvement before becoming stable, whereas the trend for FOR did not change significantly (Figure 3). These patterns support that, across the country, management programs implemented to control point or non-point sources for URB and AGR were effective in improving overall stream water quality [72–75]. Moreover, the implementation of conservation measures against continuing development pressures in metropolitan areas played a role maintaining the water quality in FOR. Furthermore, the land-cover type exerted an influence on the seasonality of overall water quality (Figure 6). In recent years (2015–2018), the seasonal patterns of WQI_{min} have differed for URB and FOR, whereas AGR exhibited less obvious seasonality. The less consistent seasonality for AGR may be partly attributable to the small sample size (n = 287, compared with n for URB = 1881 and n for FOR = 1162) corresponding to AGR. During the wet season, both URB and FOR exhibited a negative change in overall water quality with an increase in the proportion of "medium" and "good" status sites relative to "excellent" status sites (Figure 6). For URB with typically high proportions of impervious surfaces, stormwater runoff may play a significant role in decreasing overall water quality during the wet season [76–78]. Moreover, an increase in sediment discharge as well as sediment perturbation with rainfall events may facilitate the release of pollutants into surface water [79–82], resulting in a decrease in overall water quality during the wet season in both URB and FOR. In contrast, subsequent to the wet season, when dilution effects can occur [83–85], URB alone exhibited an increase in the proportion of "bad" status sites relative to "medium" and "good" status sites (Figure 6). This indicates that, not only non-point sources, but also point sources, such as wastewater treatment plant effluent, are significant forms of pollution for URB.

5. Conclusions

This study provided a statistical framework for implementing parameter selection in order to develop an objective WQI_{min} in a single-step process. Comparisons between WQI_{obj} and WQI_{min} suggested that WQI_{min} calculated with the key parameters yielded comparable results to WQI_{obj}. Furthermore, WQI_{min} reduced the eclipse effects arising from the use of correlated parameters for water quality assessment to result in a better differentiation between good and bad water quality statuses. These results have implications for management authorities, especially those motivated to launch their own monitoring network system but who have limited available resources. In this context, our results can be used to reduce monitoring demands by prioritizing the monitoring importance of a minimal number of water quality parameters. The results of WQI_{min} confirmed that the dominant land-cover type of watersheds influence multidimensional aspects of urban stream water quality; namely, the overall degree and level of pollution as well as long-term and seasonal patterns. To confirm our results, future studies should expand the number of water quality parameters exhibiting various characteristics.

Supplementary Materials: The following are available online at http://www.mdpi.com/2073-4441/12/11/3294/s1. Figure S1: Matrices of the Pearson's correlation coefficient for the period 2015–2016 among 14 water quality parameters for (a) urban-dominated (URB), (b) agricultural-dominated (AGR), and (c) forest-dominated (FOR) land-cover. Water quality parameters with high factor loadings (>0.75) on the same factor are outlined in the same color, Figure S2: Relationships between the minimum water quality index (WQI_{min}) and modified WQI_{min} from 2015 to 2018. To develop the modified WQI_{min}, key parameter values were predicted using the established linear relationship between a key parameter and a surrogate parameter. Then, predicted values were converted into normalization factors for WQI_{min} calculation. In the x-axis label, WQI_{min} (COD → BOD_5) indicates that biochemical oxygen demand (BOD_5) was used as the surrogate for the key parameter of chemical oxygen demand (COD). Black dotted lines indicate 1:1 lines. Figure S3: Relationships between objective and minimum water quality indices (WQI_{obj} and WQI_{min}) from 2017 to 2018. Weights were determined using two methods; for a-c, a relative weight was assigned to each key parameter and for d-f, the percent variance explained by a given extracted factor was assigned to each key parameter. Black dotted lines and blue dashed lines indicate 1:1 lines and regression lines, respectively. Table S1: Proportions of three land-cover categories (urban, agricultural, and forested land) for urban-dominated watersheds (URB), agricultural-dominated watersheds (AGR), and forest-dominated watersheds (FOR). Table S2: Parallel analysis results comparing eigenvalues and simulated mean eigenvalues for urban-dominated (URB), agriculture-dominated (AGR), and forest-dominated (FOR) land-cover. The simulated mean eigenvalue indicates the mean eigenvalue calculated from randomly generated simulation data. Asterisks (*) indicate that the eigenvalue is higher than the corresponding simulated mean eigenvalue.

Author Contributions: Conceptualization, T.K. and Y.C.; data curation, T.K., Y.K., J.S. and B.G.; formal analysis, T.K.; funding acquisition, T.K., Y.K., J.S., B.G. and Y.C.; Investigation, T.K., Y.K., J.S. and B.G.; methodology, T.K. and Y.C.; project administration, T.K. and Y.C.; resources, T.K., Y.K., J.S. and B.G.; software, T.K. and Y.C.; supervision, T.K. and Y.C.; validation, T.K. and Y.C.; visualization, T.K. and Y.C.; writing—original draft, T.K. and Y.C.; writing—review and editing, T.K. and Y.C. All authors have read and agreed to the published version of the manuscript.

Funding: This research was supported by the National Research Foundation of Korea (NRF) grant funded by the Korea government (MSIT) (No. 2020R1A2C1009961).

Acknowledgments: The authors would like to thank the editors and reviewers for useful comments which are helpful in improving the manuscript quality.

Conflicts of Interest: The authors declare no conflict of interest.

References

1. United Nations Department of Economic and Social Affairs. 68% of the World Population Projected to Live in Urban Areas by 2050, Says UN. Available online: https://www.un.org/development/desa/en/news/population/2018-revision-of-world-urbanization-prospects.html (accessed on 20 February 2020).
2. Arnold, C.L., Jr.; Gibbons, C.J. Impervious surface coverage: The emergence of a key environmental indicator. *J. Am. Plan. Assoc.* **1996**, *62*, 243–258. [CrossRef]
3. Defries, R.S.; Rudel, T.; Uriarte, M.; Hansen, M. Deforestation driven by urban population growth and agricultural trade in the twenty-first century. *Nat. Geosci.* **2010**, *3*, 178–181. [CrossRef]
4. Dewan, A.M.; Yamaguchi, Y. Land use and land cover change in Greater Dhaka, Bangladesh: Using remote sensing to promote sustainable urbanization. *Appl. Geogr.* **2009**, *29*, 390–401. [CrossRef]
5. Grimm, N.B.; Faeth, S.H.; Golubiewski, N.E.; Redman, C.L.; Wu, J.; Bai, X.; Briggs, J.M. Global change and the ecology of cities. *Science* **2008**, *319*, 756–760. [CrossRef] [PubMed]
6. Barnett, T.P.; Pierce, D.W.; Hidalgo, H.G.; Bonfils, C.; Santer, B.D.; Das, T.; Bala, G.; Wood, A.W.; Nozawa, T.; Mirin, A.A.; et al. Human-induced changes in the hydrology of the western United States. *Science* **2008**, *319*, 1080–1083. [CrossRef] [PubMed]
7. Carpenter, S.R.; Caraco, N.F.; Correll, D.L.; Howarth, R.W.; Sharpley, A.N.; Smith, V.H. Nonpoint pollution of surface waters with phosphorus and nitrogen. *Ecol. Appl.* **1998**, *8*, 559–568. [CrossRef]
8. Meybeck, M.; Helmer, R. The quality of rivers: From pristine stage to global pollution. *Palaeogeogr. Palaeoclimatol. Palaeoecol.* **1989**, *75*, 283–309. [CrossRef]
9. Everard, M.; Moggridge, H.L. Rediscovering the value of urban rivers. *Urban Ecosyst.* **2012**, *15*, 293–314. [CrossRef]
10. Findlay, S.J.; Taylor, M.P. Why rehabilitate urban river systems? *Area* **2016**, *38*, 312–325. [CrossRef]
11. Francis, R.A. Positioning urban rivers within urban ecology. *Urban Ecosyst.* **2012**, *15*, 285–291. [CrossRef]
12. Paul, M.J.; Meyer, J.L. Streams in the urban landscape. *Annu. Rev. Ecol. Syst.* **2001**, *32*, 333–365. [CrossRef]

13. Bascarón, M. Establishment of a methodology for the determination of water quality. *Bol. Inf. Medio Ambient.* **1979**, *9*, 30–51.
14. Brown, R.M.; McClelland, N.I.; Deininger, R.A.; Tozer, R.G. A water quality index: Do we dare? *Water Sew. Works* **1970**, *117*, 339–343.
15. Canadian Council of Ministers of the Environment. Canadian Water Quality Guidelines for the Protection of Aquatic Life: CCME Water Quality Index 1.0, User's Manual. In *Canadian Environmental Quality Guidelines*; Canadian Council of Ministers of the Environment: Edmonton, AB, Canada, 2001.
16. Cude, C.G. Oregon water quality index: A tool for evaluating water quality management effectiveness. *J. Am. Water. Resour. Assoc.* **2001**, *37*, 125–137. [CrossRef]
17. Ramakrishnaiah, C.R.; Sadashiyaiah, C.; Ranganna, G. Assessment of water quality index for the groundwater in Tumkur Taluk, Karnataka State, India. *J. Chem.* **2009**, *6*, 523–530. [CrossRef]
18. Abbasi, T.; Abbasi, S.A. *Water Quality Indices*; Elsevier: Amsterdam, The Netherlands, 2012.
19. Lumb, A.; Sharma, T.C.; Bibeault, J.F. A review of genesis and evolution of water quality index (WQI) and some future directions. *Water Qual. Expo. Health* **2011**, *3*, 11–24. [CrossRef]
20. Sutadian, A.D.; Muttil, N.; Yilmaz, A.G.; Perera, B.J.C. Development of river water quality indices—A review. *Environ. Monit. Assess.* **2016**, *188*, 58. [CrossRef]
21. Huang, J.; Zhang, Y.; Arhonditsis, G.B.; Gao, J.; Chen, Q.; Wu, N.; Dong, F.; Shi, W. How successful are the restoration efforts of China's lakes and reservoirs? *Environ. Int.* **2019**, *123*, 96–103. [CrossRef]
22. Vatanpour, N.; Malvandi, A.M.; Talouki, H.H.; Gattinoni, P.; Scesi, L. Impact of rapid urbanization on the surface water's quality: A long-term environmental and physicochemical investigation of Tajan river, Iran (2007–2017). *Environ. Sci. Pollut. Res.* **2020**, *27*, 8439–8450. [CrossRef]
23. Srivastava, P.K.; Mukherjee, S.; Gupta, M.; Singh, S.K. Characterizing monsoonal variation on water quality index of River Mahi in India using geographical information system. *Water Qual. Expo. Health* **2011**, *2*, 193–203. [CrossRef]
24. Verma, R.K.; Murthy, S.; Tiwary, R.K.; Verma, S. Development of simplified WQIs for assessment of spatial and temporal variations of surface water quality in upper Damodar river basin, eastern India. *Appl. Water Sci.* **2019**, *9*, 21. [CrossRef]
25. Şener, Ş.; Şener, E.; Davraz, A. Evaluation of water quality using water quality index (WQI) method and GIS in Aksu River (SW-Turkey). *Sci. Total Environ.* **2017**, *584*, 131–144. [CrossRef] [PubMed]
26. Tian, Y.; Jiang, Y.; Liu, Q.; Dong, M.; Xu, D.; Liu, Y.; Xu, X. Using a water quality index to assess the water quality of the upper and middle streams of the Luanhe River, northern China. *Sci. Total Environ.* **2019**, *667*, 142–151. [CrossRef]
27. Koçer, M.A.T.; Sevgili, H. Parameters selection for water quality index in the assessment of the environmental impacts of land-based trout farms. *Ecol. Indic.* **2014**, *36*, 672–681. [CrossRef]
28. Ma, Z.; Song, X.; Wan, R.; Gao, L. A modified water quality index for intensive shrimp ponds of Litopenaeus vannamei. *Ecol. Indic.* **2013**, *24*, 287–293. [CrossRef]
29. Wu, Y.; Chen, J. Investigating the effects of point source and nonpoint source pollution on the water quality of the East River (Dongjiang) in South China. *Ecol. Indic.* **2013**, *32*, 294–304. [CrossRef]
30. de Souza Pereira, M.A.; Cavalheri, P.S.; de Oliveira, M.Â.C.; Magalhães Filho, F.J.C. A multivariate statistical approach to the integration of different land-uses, seasons, and water quality as water resources management tool. *Environ. Monit. Assess.* **2019**, *191*, 549. [CrossRef]
31. Tripathi, M.; Singal, S.K. Use of Principal Component Analysis for parameter selection for development of a novel Water Quality Index: A case study of river Ganga India. *Ecol. Indic.* **2019**, *96*, 430–436. [CrossRef]
32. Tripathi, M.; Singal, S.K. Allocation of weights using factor analysis for development of a novel water quality index. *Ecotoxicol. Environ. Saf.* **2019**, *183*, 109510. [CrossRef]
33. Wu, Z.; Wang, X.; Chen, Y.; Cai, Y.; Deng, J. Assessing river water quality using water quality index in Lake Taihu Basin, China. *Sci. Total Environ.* **2018**, *612*, 914–922. [CrossRef]
34. Han, Q.; Tong, R.; Sun, W.; Zhao, Y.; Yu, J.; Wang, G.; Shrestha, S.; Jin, Y. Anthropogenic influences on the water quality of the Baiyangdian Lake in North China over the last decade. *Sci. Total Environ.* **2020**, *701*, 134929. [CrossRef]
35. Rodríguez-Romero, A.J.; Rico-Sánchez, A.E.; Mendoza-Martínez, E.; Gómez-Ruiz, A.; Sedeño-Díaz, J.E.; López-López, E. Impact of changes of land use on water quality, from tropical forest to anthropogenic occupation: A multivariate approach. *Water* **2018**, *10*, 1518. [CrossRef]

36. Cha, Y.; Cho, K.H.; Lee, H.; Kang, T.; Kim, J.H. The relative importance of water temperature and residence time in predicting cyanobacteria abundance in regulated rivers. *Water Res.* **2017**, *124*, 11–19. [CrossRef] [PubMed]
37. Gebler, D.; Wiegleb, G.; Szoszkiewicz, K. Integrating river hydromorphology and water quality into ecological status modelling by artificial neural networks. *Water Res.* **2018**, *139*, 395–405. [CrossRef] [PubMed]
38. Tong, S.T.; Chen, W. Modeling the relationship between land use and surface water quality. *J. Environ. Manag.* **2002**, *66*, 377–393. [CrossRef]
39. Whitehead, P.G.; Jin, L.; Bussi, G.; Voepel, H.E.; Darby, S.E.; Vasilopoulos, G.; Manley, R.; Rodda, C.; Hutton, C.; Hackney, C.; et al. Water quality modelling of the Mekong River basin: Climate change and socioeconomics drive flow and nutrient flux changes to the Mekong Delta. *Sci. Total Environ.* **2019**, *673*, 218–229. [CrossRef]
40. You, Q.; Fang, N.; Liu, L.; Yang, W.; Zhang, L.; Wang, Y. Effects of land use, topography, climate and socio-economic factors on geographical variation pattern of inland surface water quality in China. *PLoS ONE* **2019**, *14*, e0217840. [CrossRef]
41. Statistics Korea, 2017 Population and Housing Census of Korea. Available online: http://kostat.go.kr/ (accessed on 26 February 2020).
42. Kannel, P.R.; Lee, S.; Lee, Y.S.; Kanel, S.R.; Khan, S.P. Application of water quality indices and dissolved oxygen as indicators for river water classification and urban impact assessment. *Environ. Monit. Assess.* **2007**, *132*, 93–110. [CrossRef]
43. Pesce, S.F.; Wunderlin, D.A. Use of water quality indices to verify the impact of Córdoba City (Argentina) on Suquía River. *Water Res.* **2000**, *34*, 2915–2926. [CrossRef]
44. Sánchez, E.; Colmenarejo, M.F.; Vicente, J.; Rubio, A.; García, M.G.; Travieso, L.; Borja, R. Use of the water quality index and dissolved oxygen deficit as simple indicators of watersheds pollution. *Ecol. Indic.* **2007**, *7*, 315–328. [CrossRef]
45. QGIS Development Team. QGIS geographic information system. Open Source Geospatial Foundation Project; QGIS Development Team. 2018. Available online: http://qgis.osgeo.org (accessed on 4 January 2019).
46. Environmental Systems Research Institute. *ArcGIS Desktop: Release 10.3*; Environmental Systems Research Institute: Redlands, CA, USA, 2014.
47. Forina, M.; Armanino, C.; Raggio, V. Clustering with dendrograms on interpretation variables. *Anal. Chim. Acta* **2002**, *454*, 13–19. [CrossRef]
48. Shrestha, S.; Kazama, F. Assessment of surface water quality using multivariate statistical techniques: A case study of the Fuji river basin. *Jpn. Environ. Model. Softw.* **2007**, *22*, 464–475. [CrossRef]
49. Singh, K.P.; Malik, A.; Mohan, D.; Sinha, S. Multivariate statistical techniques for the evaluation of spatial and temporal variations in water quality of Gomti River (India)—A case study. *Water Res.* **2004**, *38*, 3980–3992. [CrossRef] [PubMed]
50. Jones, E.; Oliphant, T.; Peterson, P. SciPy: Open Source Scientific Tools for Python. 2001. Available online: https://www.scipy.org (accessed on 25 December 2019).
51. Van Rossum, G.; Drake, F.L., Jr. *Python Reference Manual*; Centrum voor Wiskunde en Informatica: Amsterdam, The Netherlands, 1995.
52. Dojlido, J.; Raniszewski, J.; Woyciechowska, J. Water quality index applied to rivers in the Vistula river basin in Poland. *Environ. Monit. Assess.* **1994**, *33*, 33–42. [CrossRef] [PubMed]
53. Kendall, M.G. *Rank Correlation Methods*; Griffin: Oxford, UK, 1948.
54. Mann, H.B. Nonparametric tests against trend. *Econometrica* **1945**, *13*, 245–259. [CrossRef]
55. Hirsch, R.M.; Slack, J.R.; Smith, R.A. Techniques of trend analysis for monthly water quality data. *Water Resour. Res.* **1982**, *18*, 107–121. [CrossRef]
56. Hussain, M.; Mahmud, I. pyMannKendall: A python package for non parametric Mann Kendall family of trend tests. *J. Open Source Softw.* **2019**, *4*, 1556. [CrossRef]
57. Kaiser, H.F. An index of factorial simplicity. *Psychometrika* **1974**, *39*, 31–36. [CrossRef]
58. Barlett, M.S. Properties of sufficiency and statistical tests. *Proc. R. Soc. Lond. A Math. Phys. Sci.* **1937**, *160*, 268–282.
59. Horn, J.L. A rationale and test for the number of factors in factor analysis. *Psychometrika* **1965**, *30*, 179–185. [CrossRef]
60. Thompson, B. *Exploratory and Confirmatory Factor Analysis: Understanding Concepts and Applications*; American Psychological Association: Washington, DC, USA, 2004.

61. Williams, B.; Onsman, A.; Brown, T. Exploratory factor analysis: A five-step guide for novices. *Australas. J. Paramed.* **2010**, *8*, 1–13. [CrossRef]
62. Revelle, W.R. Psych: Procedures for Personality and Psychological Research. 2017. Available online: https://CRAN.R-project.org/package=psych (accessed on 16 February 2019).
63. R Development Core Team. *R: A Language and Environment for Statistical Computing*; R Foundation for Statistical Computing: Vienna, Austria, 2019.
64. Liu, C.W.; Lin, K.H.; Kuo, Y.M. Application of factor analysis in the assessment of groundwater quality in a blackfoot disease area in Taiwan. *Sci. Total Environ.* **2003**, *313*, 77–89. [CrossRef]
65. Debels, P.; Figueroa, R.; Urrutia, R.; Barra, R.; Niell, X. Evaluation of water quality in the Chillán River (Central Chile) using physicochemical parameters and a modified water quality index. *Environ. Monit. Assess.* **2005**, *110*, 301–322. [CrossRef] [PubMed]
66. El Najjar, P.; Kassouf, A.; Probst, A.; Probst, J.L.; Ouaini, N.; Daou, C.; El Azzi, D. High-frequency monitoring of surface water quality at the outlet of the Ibrahim River (Lebanon): A multivariate assessment. *Ecol. Indic.* **2019**, *104*, 13–23. [CrossRef]
67. Liou, S.M.; Lo, S.L.; Wang, S.H. A generalized water quality index for Taiwan. *Environ. Monit. Assess.* **2004**, *96*, 35–52. [CrossRef] [PubMed]
68. Sun, W.; Xia, C.; Xu, M.; Guo, J.; Sun, G. Application of modified water quality indices as indicators to assess the spatial and temporal trends of water quality in the Dongjiang River. *Ecol. Indic.* **2016**, *66*, 306–312. [CrossRef]
69. Wang, J.; Fu, Z.; Qiao, H.; Liu, F. Assessment of eutrophication and water quality in the estuarine area of Lake Wuli, Lake Taihu, China. *Sci. Total Environ.* **2019**, *650*, 1392–1402. [CrossRef]
70. Nong, X.; Shao, D.; Zhong, H.; Liang, J. Evaluation of water quality in the South-to-North Water Diversion Project of China using the water quality index (WQI) method. *Water Res.* **2020**, *178*, 115781. [CrossRef]
71. Landwehr, J.M.; Deininger, R.A. A comparison of several water quality indexes. *J. Water Pollut. Control Fed.* **1976**, *48*, 954–958.
72. Korea Environment Institute. *River Management and Ecological Restoration in Response to Climate Change*; Korea Environment Institute: Sejong, Korea, 2012.
73. National Institute of Environmental Research. *A Study on the Improvement for TMDL System Enforcement—Analysis of the Pollutant Load Contribution and the Establishment of Monitoring Standards for Decentralized Wastewater Treatment System*; National Institute of Environmental Research: Incheon, Korea, 2018.
74. National Institute of Environmental Research. *Customized Policy Support for Nonpoint Pollution Management and Water Circulation Improvement (III)*; National Institute of Environmental Research: Incheon, Korea, 2018.
75. National Institute of Environmental Research. *Hydraulic and Hydrologic Scenario Modelling for Prevention and Outbreak Response of Algal Bloom (II)—Focused on Monitoring-Based Contaminant Transport*; National Institute of Environmental Research: Incheon, Korea, 2018.
76. Goonetilleke, A.; Thomas, E.; Ginn, S.; Gilbert, D. Understanding the role of land use in urban stormwater quality management. *J. Environ. Manag.* **2005**, *74*, 31–42. [CrossRef]
77. Lee, J.H.; Bang, K.W. Characterization of urban stormwater runoff. *Water Res.* **2000**, *34*, 1773–1780. [CrossRef]
78. Yang, Y.Y.; Toor, G.S. Sources and mechanisms of nitrate and orthophosphate transport in urban stormwater runoff from residential catchments. *Water Res.* **2017**, *112*, 176–184. [CrossRef] [PubMed]
79. Belabed, B.E.; Meddour, A.; Samraoui, B.; Chenchouni, H. Modeling seasonal and spatial contamination of surface waters and upper sediments with trace metal elements across industrialized urban areas of the Seybouse watershed in North Africa. *Environ. Monit. Assess.* **2017**, *189*, 265. [CrossRef]
80. Hasan, H.H.; Jamil, N.R.; Aini, N. Water quality index and sediment loading analysis in Pelus River, Perak, Malaysia. *Procedia Environ. Sci.* **2015**, *30*, 133–138. [CrossRef]
81. Lane, P.N.; Sheridan, G.J. Impact of an unsealed forest road stream crossing: Water quality and sediment sources. *Hydrol. Process.* **2002**, *16*, 2599–2612. [CrossRef]
82. O'Mullan, G.D.; Juhl, A.R.; Reichert, R.; Schneider, E.; Martinez, N. Patterns of sediment-associated fecal indicator bacteria in an urban estuary: Benthic-pelagic coupling and implications for shoreline water quality. *Sci. Total Environ.* **2019**, *656*, 1168–1177. [CrossRef]

83. Fairbairn, D.J.; Karpuzcu, M.E.; Arnold, W.A.; Barber, B.L.; Kaufenberg, E.F.; Koskinen, W.C.; Novak, P.J.; Rice, P.J.; Swackhamer, D.L. Sources and transport of contaminants of emerging concern: A two-year study of occurrence and spatiotemporal variation in a mixed land use watershed. *Sci. Total Environ.* **2016**, *551*, 605–613. [CrossRef]
84. Whitehead, P.G.; Wilby, R.L.; Rattarbee, R.W.; Kernan, M.; Wade, A.J. A review of the potential impacts of climate change on surface water quality. *Hydrol. Sci. J.* **2009**, *54*, 101–123. [CrossRef]
85. Xu, G.; Li, P.; Lu, K.; Tantai, Z.; Zhang, J.; Ren, Z.; Wang, X.; Yu, K.; Shi, P.; Cheng, Y. Seasonal changes in water quality and its main influencing factors in the Dan River basin. *Catena* **2019**, *173*, 131–140. [CrossRef]

Publisher's Note: MDPI stays neutral with regard to jurisdictional claims in published maps and institutional affiliations.

water

MDPI

Article

Integrated Framework for Detecting the Areas Prone to Flooding Generated by Flash-Floods in Small River Catchments

Romulus Costache [1,2,3,†], Alina Barbulescu [3,†] and Quoc Bao Pham [4,*]

1 Research Institute of the University of Bucharest, 90-92 Sos, Panduri, 5th District, 050663 Bucharest, Romania; romuluscostache2000@yahoo.com
2 National Institute of Hydrology and Water Management, București-Ploiești Road, 97E, 1st District, 013686 Bucharest, Romania
3 Department of Civil Engineering, Transilvania University of Brasov, 5, Turnului Str, 500152 Brasov, Romania; alinadumitriu@yahoo.com
4 Institute of Applied Technology, Thu Dau Mot University, Thu Dau Mot City 75000, Binh Duong Province, Vietnam
* Correspondence: phambaoquoc@tdmu.edu.vn
† These authors contributed equally to this work.

Abstract: In the present study, the susceptibility to flash-floods and flooding was studied across the Izvorul Dorului River basin in Romania. In the first phase, three ensemble models were used to determine the susceptibility to flash-floods. These models were generated by a combination of three statistical bivariate methods, namely frequency ratio (FR), weights of evidence (WOE), and statistical index (SI), with fuzzy analytical hierarchy process (FAHP). The result obtained from the application of the FAHP-WOE model had the best performance highlighted by an Area Under Curve—Receiver Operating Characteristics Curve (AUC-ROC) value of 0.837 for the training sample and another of 0.79 for the validation sample. Furthermore, the results offered by FAHP-WOE were weighted on the river network level using the flow accumulation method, through which the valleys with a medium, high, and very high torrential susceptibility were identified. Based on these valleys' locations, the susceptibility to floods was estimated. Thus, in the first stage, a buffer zone of 200 m was delimited around the identified valleys along which the floods could occur. Once the buffer zone was established, ten flood conditioning factors were used to determine the flood susceptibility through the analytical hierarchy process model. Approximately 25% of the total delimited area had a high and very high flood susceptibility.

Keywords: flooding; flash-floods; bivariate statistics; fuzzy multicriteria decision-making; small catchments; Romania

Citation: Costache, R.; Barbulescu, A.; Pham, Q.B. Integrated Framework for Detecting the Areas Prone to Flooding Generated by Flash-Floods in Small River Catchments. *Water* **2021**, *13*, 758. https://doi.org/10.3390/w13060758

Academic Editors: Marco Franchini and Salvador García-Ayllón Veintimilla

Received: 24 November 2020
Accepted: 9 March 2021
Published: 11 March 2021

1. Introduction

According to Hu et al. [1], a total number of 2.5 billion peoples were affected by flash-floods and floods between 1994 and 2013. In the same period, 0.16 billion fatalities occurred due to the same natural risk phenomena. Therefore, since flash-floods are extremely severe phenomena, they are also very dangerous for human life [2,3]. These phenomena appear most frequently in small river basins characterized by a high slope. Additionally, areas with smaller slopes favor the accumulation of the transported water and materials [4]. In this context, the identification of sections that favor the surface runoff occurrence, torrential valleys on which the flash-floods are propagated, and the flood susceptibility assessment in those regions, is one of the most important measures to combat the negative effects of these phenomena on water quality and human society. Additionally, the results provided by this type of analysis are very important in assessing a region's vulnerability and risk to flash floods. It should be noted that most of the procedures regarding the evaluation of flash-flood and flood risk assessment, which were adopted by the European countries includes the use of several traditional methods such as hydraulic and hydrological modeling. These techniques are time consuming and very expensive [5,6]. In this context, the need to find faster, more accurate, and cheaper techniques for determining the flood hazard has significantly increased.

In recent years, the scientific field of flash-food and flood susceptibility assessment has had a high dynamic due to the fast development of the techniques and software used to perform these analyses [7]. Thus, to assess the flood susceptibility, Geographic Information System (GIS) techniques, complex models of bivariate statistics, and machine learning are used [8]. The most used bivariate statistical techniques for assessing susceptibility to natural hazards are weights of evidence [9], frequency ratio [10], evidential belief function [11], certainty factor [12], statistical index [13], and index of entropy [14]. The most well-known machine learning models used in the study of susceptibility to floods are decision trees [15], multilayer perceptron [16], logistic regression [17], support vector machine [18], bagging [19], k-nearest neighbor [20], naïve Bayes [21], Decorate [22], Dagging [15], and adaptive neuro-fuzzy inference system [23]. Many researchers have assessed the risk of flash-floods and floods by using ensemble models resulting from the combination of several methods [15,21,24].

Nevertheless, in all the research papers where machine learning and bivariate statistics were used, the susceptibility was estimated separately for flash-floods and flooding. Up to now, there is no approach in which the susceptibility to these two phenomena, which are strongly related, can be estimated together. A first attempt to identify the torrential valleys, based on the flash-flood susceptibility, was done by Costache et al. [25]. In that study, the authors managed to detect the river valleys prone to flash-flood propagation using four hybrid models and the flow accumulation method. Nevertheless, the flooding susceptibility was not estimated along the torrential river valleys, this fact being a shortcoming that should be addressed.

In this context, we aimed to propose an integrated approach for estimating the surface runoff susceptibility and the susceptibility to floods. Thus, in the first stage, we follow the identification of areas susceptible to flash-floods by applying three overall models generated by combining frequency ratio, statistical index, and weights of evidence bivariate statistics models, on the one hand, and fuzzy analytical hierarchy process on the other hand. The models' performances were evaluated utilizing the ROC curve. The second stage of the study aims to identify the torrential valleys susceptible to the propagation of the upstream flash-floods by applying the flow accumulation method. Once the valleys with a medium, high, and very high potential for flash-flood propagation are identified, the flood susceptibility is calculated to determine the areas exposed to floods. Flood susceptibility is determined through the analytical hierarchy process stand-alone model.

It should be mentioned that this is the first time in the literature when the susceptibility of these two phenomena, flash-floods and flooding generated by them, were analyzed in an integrated way and in a spatial causal relationship. The previous studies carried out in Romania as well as in any part of the globe were focused on the estimation of flooding or flash-flood susceptibility without taking into account their strong spatial relationship.

2. Study Area

The present study focused on the Izvorul Dorului River basin located in the mountainous area of the central part of Romania. The surface of the study area is 33 km^2, which falls into the category of small-area basins that are frequently affected by rapid floods. The altitude inside the study zone varies from 763 m to 2202 m (Figure 1a). This high difference in altitude on a small area creates favorable premises for flash-flood genesis and their propagation from the upper to the lower part of the river basin. The river basin is characterized by an average high slope of 15.6°, which is another indicator of the high potential for flash-flood propagation along the valleys in the study area. According to the existing information and, as can be seen in Figure 1b, the afforestation degree of the river basin is around 50%. Additionally, from Figure 1b, one can remark that in the perimeter of the deforested surfaces located at the highest altitude also exists a very high potential for a rapid surface runoff on the slopes. This is another element indicating that the genesis of the flash-floods is related to the high-altitude region of the river basin from where they are propagated along the steep river valleys toward the lowland area. The lower part of the study area corresponds to the built space of Sinaia city, the most famous mountain tourist resort in Romania. This locality has been affected by floods multiple times, caused by Izvorul Dorului River and its tributaries. One of the most violent flash-floods took place in August 2010 when several dozens of buildings were affected as well as National Road 1, National Road 71, and the railroad between Bucharest and Brasov cities. Additionally, as a result of different strong floods, several landslides were activated and affected the houses from Sinaia.

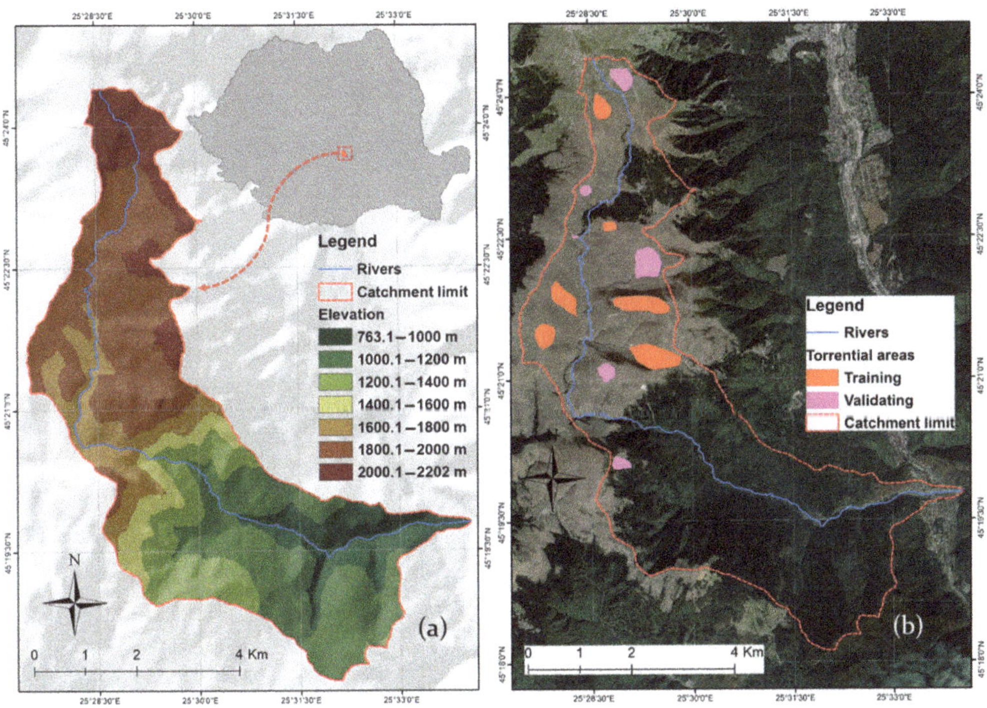

Figure 1. The study area. (**a**) Study area location in Romania; (**b**) torrential area location.

3. Methods

3.1. Background of the Models

3.1.1. Statistical Index

Proposed by van Westen [26], statistical index (SI) is a bivariate method widely used in natural risk susceptibility evaluation studies [13,27]. According to this model, the score of a predictor class can be computed by applying the natural algorithm to the ratio between the density of pixels associated with the phenomenon presence in the predictor class and the density of the same pixels across the study area [28]. Thus, a well-known formula to estimate the SI weight is the following:

$$W_{ij} = \ln\left(\frac{f_{ij}}{f}\right) = \ln\left(\frac{\frac{Npix(S_i)}{Npix(N_i)}}{\frac{\sum Npix(S_i)}{\sum Npix(N_i)}}\right), \tag{1}$$

where W_{ij} is the weight of the class/category i of predictor j; f_{ij} is the density of the phenomenon in class i of predictor j; f is the density of phenomenon in the study; *Npix(Si)* is the number of pixels associated with the phenomenon in class i; and *Npix(Ni)* is the sum of pixels of the same parameter class.

3.1.2. Frequency Ratio

Frequency ratio (*FR*) is a bivariate statistical model, widely applied to evaluate flood and landslide susceptibility mapping worldwide [9,10,13]. The relationship between food occurrences and conditioning parameters is used to analyze and calculate the frequency ratio. The mathematical expression of frequency ratio (i.e., the frequency ratio of class i of factor j) is given in Equation (2) [10]:

$$FR = \frac{\frac{Npix(1)}{Npix(2)}}{\frac{\sum Npix(3)}{\sum Npix(4)}} \tag{2}$$

where $Npix(1)$ is the total number of torrential points contained by a class/category of factor; $Npix(2)$ is the total number of pixels contained by each class/category; $\sum Npix(3)$ is the total number of torrential pixels within the study area; and $\sum Npix(4)$ is the total number of pixels within the study area.

After calculating the frequency ratio, each controlling factor summed up all the values to generate a map of flood vulnerability. If the frequency ratio is greater than 1, the conditioning factors strongly influence flooding, otherwise, there is a negative relationship between conditioning factors and flood occurrence.

3.1.3. Weights of Evidence

Weights of evidence (WOE) is a widely used statistical model for landslide, flood, and fire forest susceptibility assessment [29–31]. This method was first introduced for geological studies in 1992, then adopted for the analysis of different hazards (e.g., fire forest, flood, landslides) [27]. This method estimates the weights of evidence coefficients based on the relationship between each class of factors and the flood absence/presence. The positive weight (W^+) and the negative weight (W^-) are necessary for the computation. These weights reflect the presence and absence of areas affected by the flood, respectively, and can be computed using the following [29–31]:

$$W^+ = ln\frac{P\{B|S\}}{P\{B|\overline{S}\}} \tag{3}$$

$$W^- = ln\frac{P\{\overline{B}|S\}}{P\{\overline{B}|\overline{S}\}} \tag{4}$$

where B and $\overline{B}$ are the presence and absence of flood conditioning parameters, respectively; P is the probability; and S, and $\overline{S}$ are the presence and absence of flooding, respectively.

The output of the performed processes is used to implement Equations (3) and (4) in ArcGIS. Subsequently, the mathematical representation of these two equations are [29]:

$$W^+ = ln\frac{\frac{Npix_1}{Npix_1 + Npix_2}}{\frac{Npix_3}{Npix_3 + Npix_4}} \tag{5}$$

$$W^- = ln\frac{\frac{Npix_2}{Npix_1 + Npix_2}}{\frac{Npix_4}{Npix_3 + Npix_4}} \tag{6}$$

where W^+ and W^- are the positive and negative weights, respectively; $Npix_1$ and $Npix_2$ are the number of pixels with flood points inside *and* outside of the class, respectively;

and $Npix_3$ and $Npix_4$ are the number of pixels without flooding inside and outside of the class, respectively.

The final weights of evidence coefficients (Wf) assigned to each factor class can be obtained as follows [29]:

$$Wf = Wplus + Wmintotal - Wmin \tag{7}$$

where (Wf) is the final weight of evidence coefficients; *Wmintotal* is the total of all negative weights in a multiclass map; and *Wplus* and *Wmin* are the positive negative weights of a class factor, respectively.

3.1.4. Fuzzy Analytical Hierarchy Process

The analytical hierarchy process (AHP) is an algorithm used for flood, landslide, and fire forest susceptibility mapping [32–35]. Through a pairwise comparison matrix constructed based on expert knowledge, AHP was used to calculate the weights of relevant criterion map layers. Since AHP has several advantages such as its fuzzy extension, the fuzzy analytical hierarchy process (FAHP) was proposed and applied to solve the hierarchical fuzzy problems. It can be employed to increase the analysis quality, reducing the subjectivity in the estimation of weights criteria by a combination of the fuzzy set theory and the analytical hierarchy process [36]. The following steps show how to determine the weights of criteria in the FAHP.

The pairwise comparison matrices are constructed from flood conditioning factors (elevation, slope angle, stream density, curve number, rainfall, lithology, land use, soil texture, etc.). Linguistic terms are assigned to the pairwise comparison (Equation (8)) to establish the most important criteria [37]:

$$A' = \begin{bmatrix} 1' & a'_{12} & \cdots & a'_{1n} \\ a'_{21} & 1' & \cdots & a'_{1n} \\ \vdots & \vdots & \ddots & \vdots \\ a'_{n1} & a'_{n2} & \cdots & 1' \end{bmatrix} = \begin{bmatrix} 1' & a'_{12} & \cdots & a'_{1n} \\ 1/a'_{21} & 1' & \cdots & a'_{1n} \\ \vdots & \vdots & \ddots & \vdots \\ 1/a'_{n1} & 1/a'_{n2} & \cdots & 1' \end{bmatrix} \tag{8}$$

where a'_{ij} indicates a pair of criteria i and j.

The Buckley method [38] is utilized to calculate the fuzzy geometric mean and fuzzy weight of each criterion by:

$$r'_i = \left(a'_{i1} \otimes a'_{i2} \otimes \ldots \otimes a'_{in}\right)^{\frac{1}{n}}, \tag{9}$$

$$w'_i = r'_i \otimes \left(r'_1 \otimes \ldots \otimes r'_n\right)^{-1}, \tag{10}$$

where a'_{in} is the fuzzy comparison value between the pair criterion i and criterion n; and r'_1 is the geometric mean of the fuzzy comparison values for criterion i compared to each of the other criteria; w'_i is the fuzzy weighting of the ith criterion; and $w'_i = (lw_i,\ mw_i,\ uw_i)$, where lw_i, mw_i and uw_i are the values of the lower, middle, and upper, fuzzy weighting of the ith criterion, respectively [37,39].

The extent analysis algorithm was applied to determine the final values of the flood conditioning factor weights. The construction of a fuzzy triangular comparison matrix is the first step. This matrix is done by [40]:

$$A' = \left(a'_{ij}\right)_{nxn} = \begin{bmatrix} (1,1,1) & (l_{12},\ m_{12}, u_{12}) & \cdots & (l_{1n},\ m_{1n}, u_{1n}) \\ (l_{21},\ m_{21}, u_{21}) & (1,1,1) & \cdots & (l_{2n},\ m_{2n}, u_{2n}) \\ \vdots & \vdots & \ddots & \vdots \\ (l_{n1},\ m_{n1}, u_{n1}) & (l_{n2},\ m_{n2}, u_{n2}) & \cdots & (1,1,1) \end{bmatrix} \tag{11}$$

where $a'_{ij} = (l_{ij},\ m_{ij}, u_{ij})$ and $a_{ij}^{'-1} = (1/l_{ij},\ 1/m_{ij}, 1/u_{ij})$ for $i, j = 1, \ldots, n$ and $i \neq j$.

Next, we computed the priority vector of the triangular matrix. Then, the fuzzy arithmetic function was employed to sum up each row of the matrix A' in a first stage, as follows:

$$RS_i = \sum_{j=1}^{n} a'_{ij} = \left(\sum_{j=1}^{n} l_{ij}, \sum_{j=1}^{n} m_{ij}, \sum_{j=1}^{n} u_{ij} \right), \ i = 1, \ldots, n \quad (12)$$

Then, the value of the fuzzy synthetic extent in terms of the ith object is obtained through the normalization of the above relation, as follows [32]:

$$S'_i = \sum_{j}^{n} a'_{ij} \otimes \left[\sum_{k=1}^{n} \sum_{j=1}^{n} a'_{kj} \right]^{-1} = \left(\frac{\sum_{j=1}^{n} l_{ij}}{\sum_{k=1}^{n} \sum_{j=1}^{n} u_{kj}}, \frac{\sum_{j=1}^{n} m_{ij}}{\sum_{k=1}^{n} \sum_{j=1}^{n} m_{kj}}, \frac{\sum_{j=1}^{n} u_{ij}}{\sum_{k=1}^{n} \sum_{j=1}^{n} l_{kj}} \right), \ i = 1, \ldots, n. \quad (13)$$

The computation of the degree of possibility of $S'_i \geq S'_j$ represents the third step and is achieved through Equation (14):

$$V(S'_i \geq S'_j) = \begin{cases} 1, if\ m_i \geq m_j, \\ \frac{u_i - l_j}{(u_i - m_i) + (m_j - l_j)}, l_j \leq u_i, \ i, j = 1, \ \ldots, \ n; j \neq i \\ 0, \text{ otherwise} \end{cases} \quad (14)$$

where $S'_i = (l_i, m_i, u_i)$ and $S'_j = (l_j, m_j, u_j)$.

Considering that:

$$w'(a_i) = \min\{V(S'_i \geq S'_k)\}, \ k = 1, 2, \ldots, n; k \neq i \quad (15)$$

the weight vector values can be calculated by:

$$w'(a_i) = [w'(a_1), \ w'(a_2), \ \ldots, \ w'(a_n)]^T. \quad (16)$$

The weight vectors were obtained using the following equation after a normalization process:

$$w(a_i) = [w(a_1), \ w(a_2), \ \ldots, \ w(a_n)]^T \quad (17)$$

where w is a non-fuzzy number.

The present study was carried out by completing several methodological steps, as presented in Figure 5 and also briefly described below.

3.2. Data Used

3.2.1. Torrential Areas Inventory

Identifying the areas previously affected by a natural risk phenomenon is vital for detecting other zones where that phenomenon has a high probability of occurrence [41]. The appearance of any phenomenon will be favored in areas with characteristics similar to those where the phenomenon has already occurred [42]. For this reason, to estimate the susceptibility to the occurrence of rapid floods, torrential areas were inventoried and mapped. These areas were generated by the rapid surface runoff on the slopes. The modality of identification of such zones is presented in the studies [43]. Torrential areas are zones characterized by the unified presence of a torrential microform of relief such as ravines and gullies generated by surface runoff. Thus, through the satellite images made available through the Google Earth application (Figure 1), an area affected by intense torrential processes of about 170 hectares was vectorized, which is located in the upper part of the river basin where the absence of vegetation and the high slopes favor the apparition of such phenomena.

3.2.2. Flash-Flood and Flood Predictors

Whereas torrential zones represent an indicator of the rapid surface runoff on the slopes, certain geographical factors are the predictors of this phenomenon, or in other words, are the variables that generate and favor the surface runoff. Moreover, the genesis of floods generated by flash-floods also depends on the characteristics of geographical factors. Therefore, to identify as accurately as possible the surfaces favorable to the genesis of flash-floods and those susceptible to floods, twelve conditioning factors were taken into account. Eight morphometrical predictors were obtained by processing the digital elevation model, while the other four flood and flash-flood predictors were collected from the following vector databases: hydrological soil groups from the Digital Soil Map of Romania, 1:200,000; land use/cover from Corine Land Cover, 2018; lithology from the Digital Geological Map of Romania, 1:200,000; and distance from rivers was estimated with the help of the river network in an Environmental Systems Research Institute (ESRI) shapefile format. Below, the main characteristics of flood and flash-flood predictors are briefly presented.

The slope is the geographic factor that has the biggest influence on both the potential for rapid surface runoff and the flood potential [24]. Surfaces with steep slopes cause rapid water drainage, while the flat surfaces lead to the water accumulation process [44]. In our case study, the sloping relief had values between 0.1° and 54.1° (Figure 2a). This interval was divided into six classes according to the literature [43].

Land use/cover is another predictor that influences both flash-floods and floods [45]. Lands covered with pastures or without vegetation will favor the appearance of rapid runoff on the slopes, while areas covered with forests are characterized by a lower potential for runoff and flooding [21]. In the study area, the grassland and the forest shared equally almost all of the territory (Figure 2b). Additionally, the presence of the built space in the lower part of the Izvorul Dorului River basin was observed.

Hydrological soil group has a high influence on the flood. Thus, the flooding phenomenon will likely be over the areas with soils with high clay content such as those in hydrological group D, while water infiltration will be more pronounced on soils with a sandy texture [46–48]. Within the study area, the largest surface was occupied by hydrological soil group A (Figure 2c).

Convergence index (CI) is a predictor obtained from the DEM whose values show the concentration degree of the drainage network. CI values close to −100 indicate a high density of the river network whereas positive CI values are associated with the interfluvial surfaces. In the study area, the CI values are situated in the range from −86 to 84 (Figure 2d). These were divided into five classes according to the literature [43].

Profile curvature is a predictor whose negative values show the surfaces that favor the accelerated surface runoff, while the decelerated runoff manifests itself on the surfaces with positive values. The information from the literature was used to classify profile curvature values into the next classes: −2.3–−0.1; 0–0.1; 0.2–2.6 (Figure 3a).

The aspect factor obtained from the DEM is an indicator of the humidity potential that exists at the slope level [49]. In the case of the Izvorul Dorului basin, the southeast surfaces were the most extensive, these being followed by the southwest slopes (Figure 3b).

Topographic position index (TPI) is a predictor calculated from the DEM, which shows the relative position of a point in the research area in relation to the immediately neighboring regions [50]. The next TPI classes were established using the natural breaks method: −7.8–−1.8; −1.7–−0.5; −0.4–0.5; 0.6–1.9; 2–8.6 (Figure 3c). The following five classes of Topographic Wetness Index (TWI) were delimited using the natural break method: −4.4–4.7; 4.8–8.4; 8.5–11.8; 11.9–15; 15.1–23.1 (Figure 3d).

The elevation is a useful indicator for detecting the surfaces exposed to flooding processes that may occur as a result of flash-flood propagation from the upper part of river basins [7]. The lower relief zones have a higher sensitivity to flooding occurrence. For the study area, the range from 763.1 m to 2202 m was split into seven classes that generally succeeded at a difference of 200 m (Figure 4a).

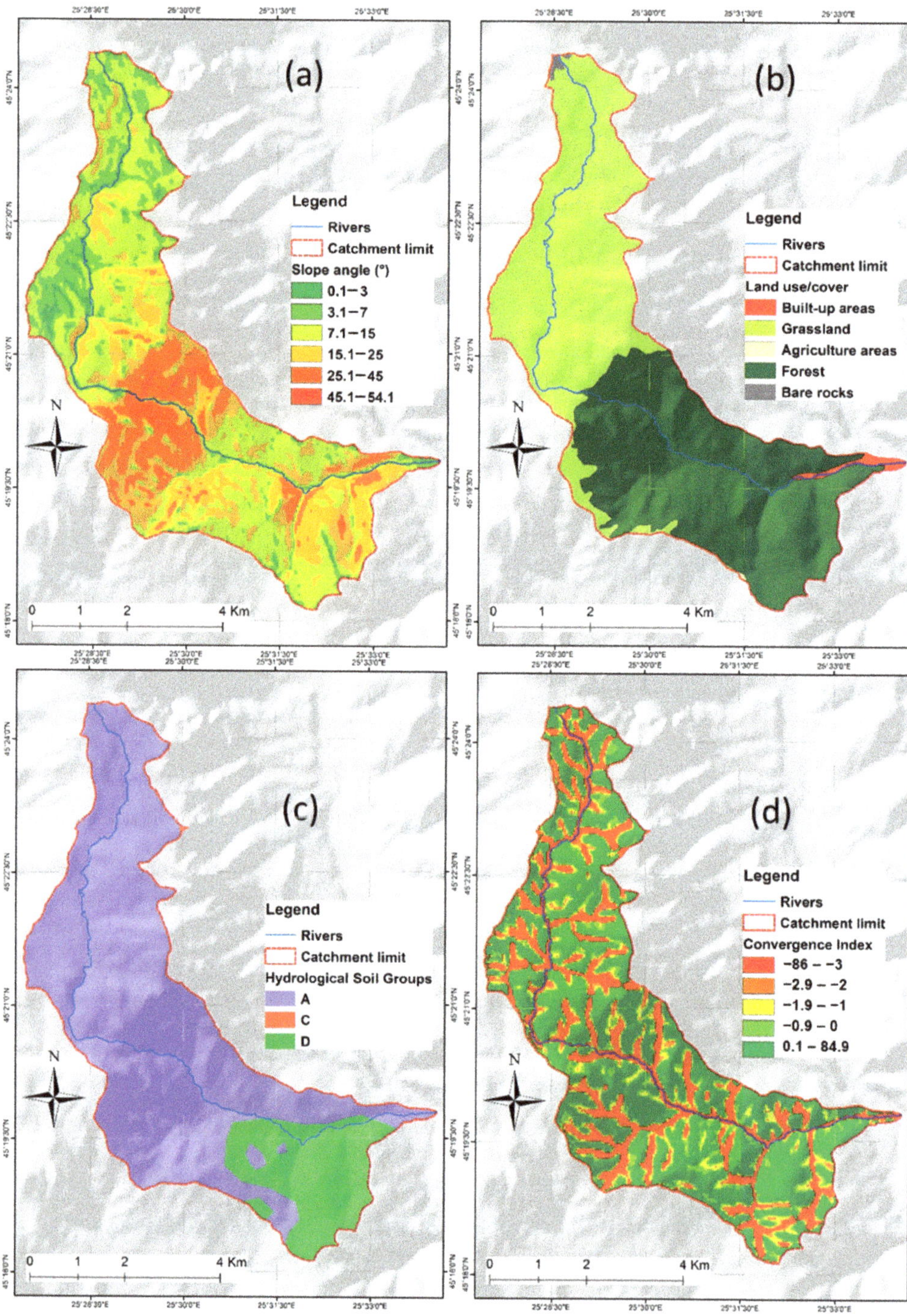

Figure 2. Flash-flood and flood predictors. (**a**) Slope; (**b**) Land use; (**c**) Hydrological Soil Group; (**d**) Convergence index.

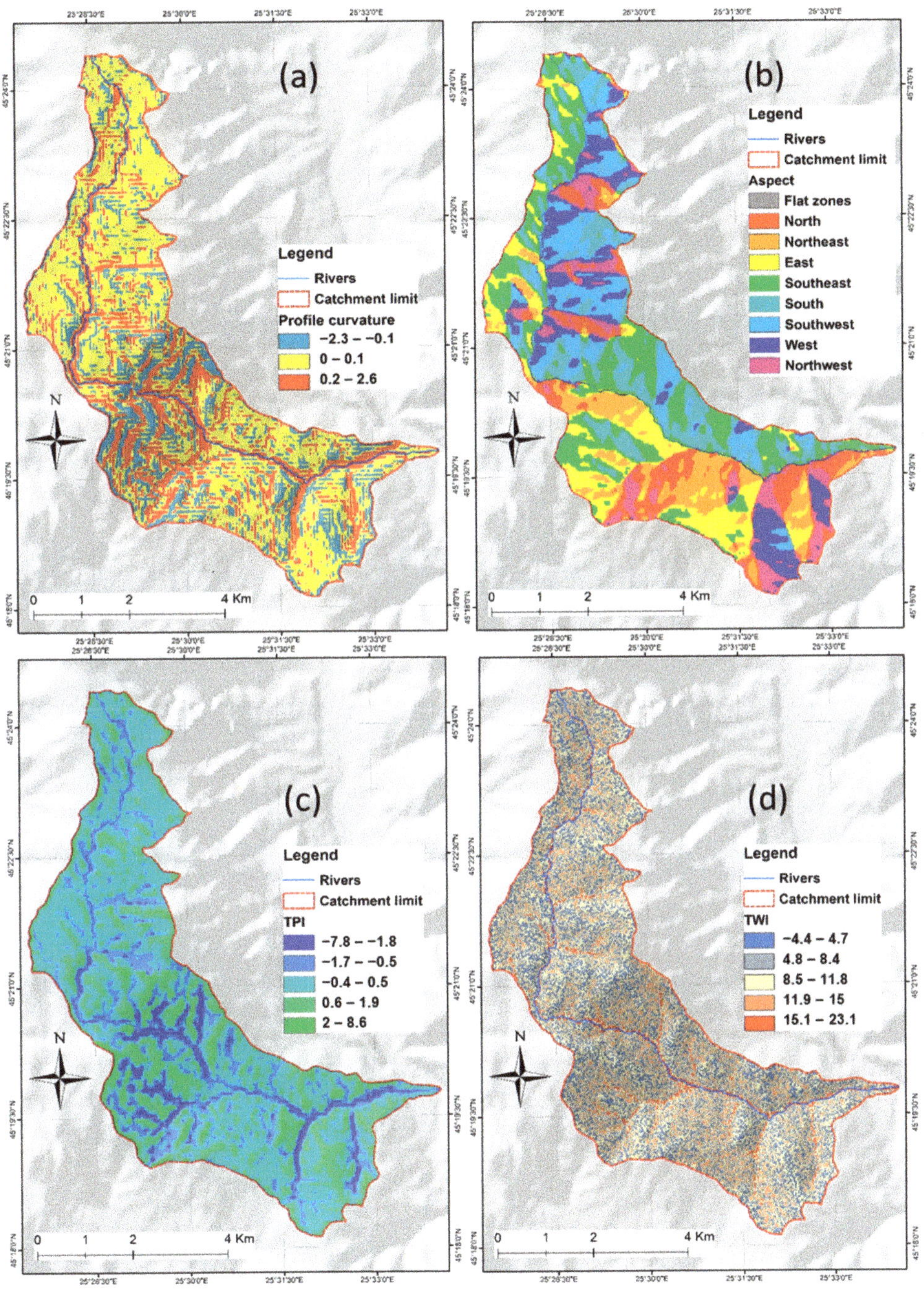

Figure 3. Flash-flood and flood predictors. (**a**) Profile curvature; (**b**) Aspect; (**c**) Topographic position index (TPI); (**d**) Topographic Wetness Index TWI.

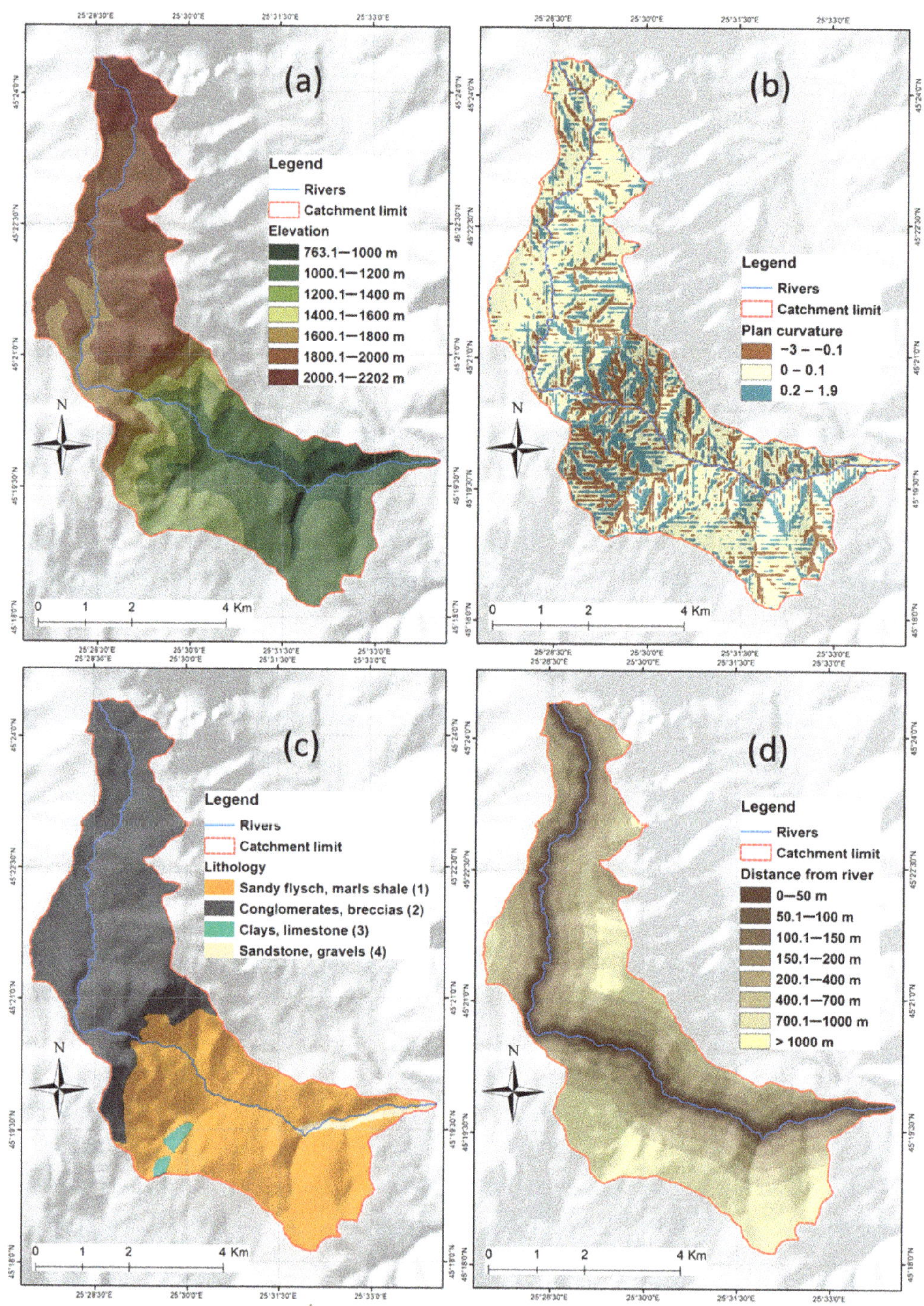

Figure 4. Flash-flood and flood predictors. (**a**). Elevation; (**b**) Plan curvature; (**c**) Lithology; (**d**) Distance from rive.

Plan curvature shows the difference between the surfaces on which the convergent and divergent runoff is manifested. Three classes were delimited for the plan curvature values (Figure 4b): −3–−0.1; 0–0.1; 0.2–1.9.

Lithology is a predictor that influences the infiltration capacity at the ground surface, so it should be considered in the studies concerning the flood and flash-flood potential. The conglomerates, breccias, sandy flysch, and marls shale are predominant in the study area (Figure 4c).

Distance from the river was generated using the Euclidean distance tool from ArcGIS 10.3 software. This is an important parameter that indicates the distance from different surfaces to the nearest watercourse. The surfaces in the vicinity of watercourses will be more prone to flash-floods and the floods generated by them. In the present study, the distance from the river predictor was classified into eight classes.

3.3. *Methodological Steps Implemented in the Present Study*

3.3.1. Step 1: Flash-Flood Database Preparation

The flash-flood database used in the present research consisted of 1965 torrential points collected from the delineated torrential surfaces and ten flash-flood conditioning factors. Building and processing the flash-flood database were done through ArcGIS 10.3 software. It should be noted that the torrential points were obtained by converting the torrential areas from a raster format, with a cell size of 30 m, to a point. Therefore, each point corresponds to a raster cell. According to the literature [51], the entire sample was divided into a training dataset (70%) and a validation dataset (30%). The training dataset was used to calculate the frequency ratio, weights of evidence, and statistical index coefficients, while the validation dataset was used to evaluate the accuracy of the results achieved.

3.3.2. Step 2: Computation of Flash-Flood Potential Index (FFPI)

The flash-flood potential index represents a qualitative indicator of the potential for torrential surface runoff, which exists at the slope level [52]. In the first stage, the frequency ratio, weights of evidence, and statistical index coefficients were determined by analyzing the spatial correlations between the torrential points included in the training sample and the ten flash-flood predictors. In this regard, the equations from Sections 3.1.1–3.1.3 were implemented in Excel and ArcGIS. The number of pixels used in the computation of the types of bivariate statistics coefficients was 1376. Furthermore, the second stage consisted of the computation of flash-flood predictors weights by the fuzzy analytical hierarchy process method. Finally, three variants of the flash-flood potential index were computed by the weighted sum between fuzzy analytical hierarchy process weights and the values of frequency ratio, weights of evidence, and statistical index coefficients.

3.3.3. Step 3: Evaluation of Results Accuracy Using Receiver Operating Characteristic (ROC) Curve

The results of FFPI were assessed using the receiver operating characteristic (ROC) curve. The ROC curve represents a graphical plot that highlights the ability of a binary model to classify a given dataset used in the modeling process into the presence or the absence of a specific phenomenon [53]. This is the most frequently used algorithm to validate the outcomes provided by a model for natural hazards susceptibility [42,49,54,55]. The ROC curves were constructed by comparing the existing torrential points with the flash-flood potential index results. Both the success rate, constructed with the training sample, and prediction rate constructed with the validation sample, will be used. The area under curve (AUC) will highlight the performance of each flash-flood potential index model.

3.3.4. Step 4: Computation the Flood Potential Index (FPI) Based on the Most Performant FFPI Result

To identify the valleys with a high torrential degree, the best performing flash-flood potential index that resulted was used in a *flow accumulation* procedure (Figure 5). Through the *flow accumulation* method, the flash-flood potential index values are weighted at the level of the river network within the study area. The weighted flash-flood potential index values are further classified into five categories: very low, low, medium, high, and very high. In the next stage, to select the river valleys along which the flood potential index will be calculated, the hydrographic network having assigned a medium, high, and very high flash-flood propagation susceptibility is selected. The flood potential index represents a qualitative indicator that highlights the degree to which a specific region can be affected by the flooding phenomenon [56]. The area on which the flood potential index will be computed was limited to a buffer zone of 200 m along with the selected river network. Eventually, the flood potential index values are obtained by involving the next ten flood conditioning factors in the analytical hierarchy process method: slope, land use, hydrological soil groups, convergence index, topographic position index, topographic wetness index, elevation, distance from the river, plan curvature, and lithology. The values of the FPI are then classified into five categories through which the areas prone to flooding generated by flash-floods will be detected.

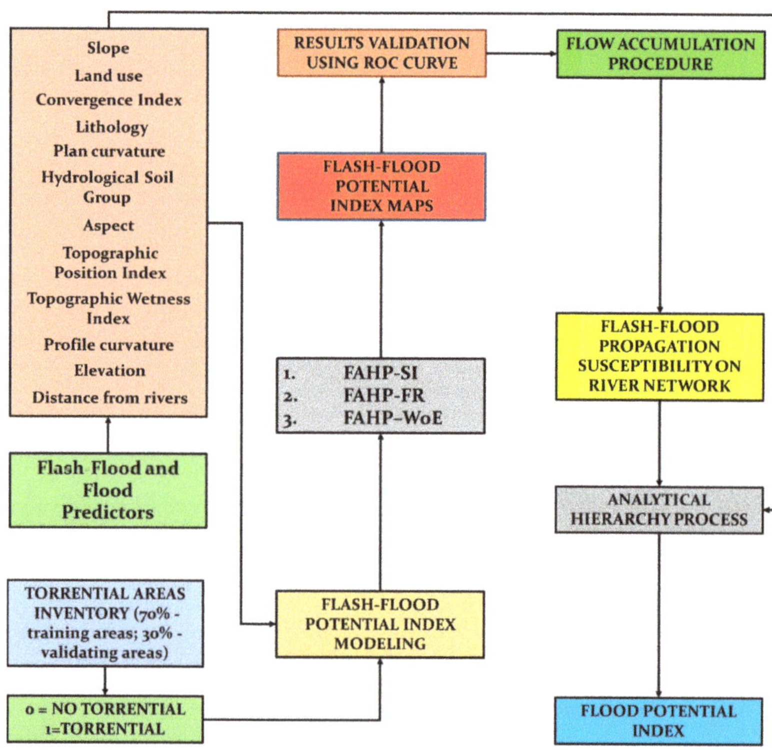

Figure 5. Flowchart of the present research.

4. Results

4.1. Bivariate Statistics Coefficients

The values of bivariate statistics coefficients highlight the spatial relationships between the location of torrential areas and the classes/categories of flash-flood predictors. According to Table 1, the lowest weights of evidence coefficients were assigned to hydrological soil group D (−15.04), lithological category of sandy flysch, marls shale (−10.3), lithological category of clays, limestone (−9.27), and hydrological soil group C (−8.91). The highest weights of evidence values were attributed to slope angles higher than 45° (3.34), grassland land use (1.52), lithological category of conglomerates, breccias (1.48), and slope angles between 15.1° and 25° (0.95). In terms of frequency ratio coefficients, the lowest values (0) were associated with agricultural zones, built-up areas, bare rocks, lithological categories of sandstone, gravels, clays, and limestone, zones with slope angles lower than 3°, and hydrological soil groups C and D. The highest frequency ratio values were assigned to zones with slope angles higher than 45° (13.5), grassland land use (2.07), lithological category of conglomerates, breccias (2.04), areas with slope angles between 15.1° and 25° (1.82), and convergence index class between −86 and −3 (1.78). In the case of SI coefficients, the lowest values were calculated for hydrological soil group D (−7.83), zones with slopes lower than 3° (−5.99), lithological category of sandstone and gravels (−5.25), lithological category of sandy flysch and marls shale (−5.12), and built-up areas (−4.94). The highest SI coefficients were obtained by zones with slope angles higher than 45° (2.6), grassland land use (0.73), lithological category of conglomerates, breccias (0.71), zones with slope angles between 15.1° and 25° (0.6), and convergence index class between −86 and −3 (0.58).

Table 1. Bivariate statistics of flash-floods conditioning factors classes.

Factor	Class	Class Pixels	Torrential Points	WOE	FR	SI
Slope	0–3°	1065	0	−6.05	0.00	−5.99
	3.1–7°	5008	230	0.26	1.22	0.20
	7.1–15°	15,098	299	−0.94	0.53	−0.64
	15.1–25°	9937	677	1.01	1.82	0.60
	25.1–45°	5407	93	−0.88	0.46	−0.78
	45.1–54°	152	77	3.34	13.50	2.60
Land use	Built-up areas	374	0	−6.78	0	−4.94
	Grassland	16,949	1316	1.52	2.07	0.73
	Agriculture areas	15	0	−3.55	0	−1.73
	Forest	19,218	60	−5.05	0.08	−2.49
	Bare rocks	111	0	−5.56	0	−3.73
Convergence index	−86–−3	7514	502	0.93	1.78	0.58
	−2.9–−2	2085	114	0.51	1.46	0.38
	−1.9–−1	2750	107	0.13	1.04	0.04
	−0.9–0	3939	101	−0.35	0.68	−0.38
	0.1–84.9	20,379	552	−0.56	0.72	−0.33
Lithology	Sandy flysch, marls shale	17,891	4	−10.30	0.01	−5.12
	Conglomerates, breccias	17,948	1372	1.48	2.04	0.71
	Clays, limestone	321	0	−9.27	0	−4.79
	Sandstone, gravels	507	0	−4.44	0	−5.25
Plan curvature	−3–−0.1	7202	345	0.29	1.28	0.244
	0–0.1	20,963	821	0.07	1.04	0.043
	0.2–1.9	8502	210	−0.57	0.66	−0.418
HSG	A	29,965	1376	0.42	1.22	0.20
	C	18	0	−8.91	0	−1.91
	D	6684	0	−15.04	0	−7.83
Aspect	Flat surfaces	93	2	−0.31	0.57	−0.56
	North	3306	97	−0.01	0.78	−0.25
	Northeast	4440	27	−1.70	0.16	−1.82
	East	5064	104	−0.43	0.55	−0.60
	Southeast	6455	181	−0.09	0.75	−0.29
	South	5225	334	0.95	1.70	0.53
	Southwest	5434	286	0.69	1.40	0.34
	West	3829	211	0.73	1.47	0.38
	Northwest	2821	134	0.53	1.27	0.24
TPI	−7.8–−1.8	2063	27	−1.21	0.35	−1.05
	−1.7–−0.5	8121	380	0.22	1.25	0.22
	−0.4–0.5	16,532	744	0.29	1.20	0.18
	0.6–1.9	8181	192	−0.68	0.63	−0.47
	2–8.6	1770	33	−0.83	0.50	−0.70
TWI	−4.4–4.7	7477	277	−0.03	0.99	−0.01
	4.8–8.4	9509	376	0.06	1.05	0.05
	8.5–11.8	9180	307	−0.17	0.89	−0.12
	11.9–15	9414	404	0.18	1.14	0.13
	15.1–23.1	1083	12	−1.28	0.30	−1.22
Profile curvature	−3–−0.1	7255	185	−0.65	0.68	−0.39
	0–0.1	21,678	957	0.30	1.18	0.16
	0.2–1.9	7734	234	−0.45	0.81	−0.22

4.2. Flash-Flood Potential Index Computation Using Fuzzy Analytical Hierarchy Process Based Ensembles

Following the methodological steps described in Section 3.1.4 the fuzzy analytical hierarchy process algorithm was applied to determine the weights of the flash-flood predictors. In the first step, the fuzzy analytical hierarchy process evaluation matrix was created based on expert judgment (Table 2). Furthermore, using the values included in the evaluation matrix, the synthesis values were calculated by using the formula:

$$\left[\sum_{k=1}^{n}\sum_{j=1}^{n} a'_{kj}\right]^{-1} = (88.48\ 130.16\ 182.17)^{-1} = (0.005\ 0.008\ 0.011) \quad (18)$$

Table 2. Fuzzy analytical hierarchy process evaluation matrix.

	1	**2**	**3**	**4**	**5**	**6**	**7**	**8**	**9**	**10**
Slope (1)										
l_1	1	1	2	1	1	2	3	3	2	1
m_1	1	2	3	2	2	3	4	4	3	2
u_1	1	3	4	3	3	4	5	5	4	3
Land use (2)										
l_2	0.33	1	1	1	1	1	2	1	1	1
m_2	0.5	1	2	1	2	2	3	2	2	1
u_2	1	1	3	1	3	3	4	3	3	1
Convergence index (3)										
l_3	0.25	0.33	1	0.33	0.33	1	1	1	1	0.33
m_3	0.33	0.5	1	0.5	0.5	1	2	2	1	0.5
u_3	0.5	1	1	1	1	1	3	3	1	1
Lithology (4)										
l_4	0.33	1	1	1	1	1	2	2	1	1
m_4	0.5	1	2	1	1	2	3	3	2	1
u_4	1	1	3	1	1	3	4	4	3	1
Plan curvature (5)										
l_5	0.33	1	1	1	1	1	2	2	1	1
m_5	0.5	1	2	1	1	2	3	3	2	1
u_5	1	1	3	1	1	3	4	4	3	1
HGS (6)										
l_6	0.25	0.33	1	0.33	0.33	1	1	1	1	0.33
m_6	0.33	0.5	1	0.5	0.5	1	2	2	1	0.5
u_6	0.5	1	1	1	1	1	3	3	1	1

Table 2. *Cont.*

	1	2	3	4	5	6	7	8	9	10
Aspect (7)										
l_7	0.2	0.25	0.33	0.25	0.25	0.33	1	1	1	0.33
m_7	0.25	0.33	0.5	0.33	0.33	0.5	1	1	1	0.5
u_7	0.33	0.5	1	0.5	0.5	1	1	1	1	1
TPI (8)										
l_8	0.2	0.25	0.33	0.25	0.25	0.33	1	1	1	0.33
m_8	0.25	0.33	0.5	0.33	0.33	0.5	1	1	1	0.5
u_8	0.33	0.5	1	0.5	0.5	1	1	1	1	1
TWI (9)										
l_9	0.25	0.33	1	0.33	0.33	1	1	1	1	0.33
m_9	0.33	0.5	1	0.5	0.5	1	2	2	1	0.5
u_9	0.5	1	1	1	1	1	3	3	1	1
Profile curvature (10)										
l_{10}	0.33	1	1	1	1	1	2	1	1	1
m_{10}	0.5	1	2	1	2	2	3	2	2	1
u_{10}	1	1	3	1	3	3	4	3	3	1

The synthesis values calculated above were used in the following step to calculate the fuzzy numbers for each flash-flood predictor. The fuzzy numbers were then compared using the degree of possibility procedure, which is exemplified in Table 3. Utilizing the results provided by the degree of possibility method, the weight vector values were calculated using the following relations:

$$w'(a_i) = \{1\ 0.71\ 0.32\ 0.68\ 0.68\ 0.32\ 0\ 0\ 0.32\ 0.71\}^T \quad (19)$$

$$w(a_i) = \{0.211\ 0.15\ 0.066\ 0.143\ 0.143\ 0.066\ 0\ 0\ 0.066\ 0.15\}^T \quad (20)$$

In the next step, employing the defuzzification procedure, the Triangular Fuzzy Numbers (TFN_s) were transformed into the crisp weights that will be attributed to each flash-flood predictor and multiplied with statistical index, frequency ratio, and weights of evidence values to obtain the flash-flood potential index.

Flash-flood potential index values were mapped using the map algebra capability from ArcGIS software. All three flash-flood potential indices were standardized between 0 and 1 and then reclassified into five classes using the natural break method. In the case of $FFPI_{FAHP\text{-}SI}$, very low values, situated from 0 to 0.25, were found in about 2.82% of the study area (Figure 6a). The values, ranging from 0.26 to 0.62, were mainly associated with the southern half and represent 28.9% of the entire river basin. The medium $FFPI_{FAHP\text{-}SI}$ class corresponded to approximately 13.86% of the Izvorul Dorului catchment. The high and very high potential were spread over a total of 54.43% of the entire catchment surface. The analysis of $FFPI_{FAHP\text{-}FR}$ revealed that the very low potential spanned 18.48% of the total study area and was present mainly in the southern half. The low flash-flood potential accounted for approximately 16.29% of the catchment surface, while the medium $FFPI_{FAHP\text{-}WOE}$, with values from 0.32 to 0.48, covered 26.66% of the study zone. A zone including 38.57% of the research area was characterized by a high and very high flash-flood potential (Figure 6b). Following the application of the FAHP-WOE ensemble, only 0.62% of the Izvorul Dorului catchment had a very low flash-flood potential (Figure 6c). The low flash-flood potential, with values between 0.26 and 0.44, covered around 10.62% of the entire territory, while the medium values quantified approximately 30.81% of the river basin. A percentage of 57.95% of the study area had high and very high $FFPI_{FAHP\text{-}WOE}$ values ranging from 0.64 to 1.

Table 3. The ordinate of the highest intersection point, the degree possibility for Triangular Fuzzy Numbers (TFNs), and the weights of the flash-flood predictors.

Slope = 1	Land Use = 2	CI = 3	Lithology = 4	Plan Curvature = 5
V(S1 ≥ S2) = 1	V(S2 ≥ S1) = 0.71	V(S3 ≥ S1) = 0.32	V(S4 ≥ S1) = 0.68	V(S5 ≥ S1) = 0.68
V(S1 ≥ S3) = 1	V(S2 ≥ S3) = 1	V(S3 ≥ S2) = 0.65	V(S4 ≥ S2) = 1	V(S5 ≥ S2) = 1
V(S1 ≥ S4) = 1	V(S2 ≥ S4) = 1	V(S3 ≥ S4) = 0.62	V(S4 ≥ S3) = 1	V(S5 ≥ S3) = 1
V(S1 ≥ S5) = 1	V(S2 ≥ S5) = 1	V(S3 ≥ S5) = 0.62	V(S4 ≥ S5) = 1	V(S5 ≥ S4) = 1
V(S1 ≥ S6) = 1	V(S2 ≥ S6) = 1	V(S3 ≥ S6) = 1	V(S4 ≥ S6) = 1	V(S5 ≥ S6) = 1
V(S1 ≥ S7) = 1	V(S2 ≥ S7) = 1	V(S3 ≥ S7) = 1	V(S4 ≥ S7) = 1	V(S5 ≥ S7) = 1
V(S1 ≥ S8) = 1	V(S2 ≥ S8) = 1	V(S3 ≥ S8) = 1	V(S4 ≥ S8) = 1	V(S5 ≥ S8) = 1
V(S1 ≥ S9) = 1	V(S2 ≥ S9) = 1	V(S3 ≥ S9) = 1	V(S4 ≥ S9) = 1	V(S5 ≥ S9) = 1
V(S1 ≥ S10) = 1	V(S2 ≥ S10) = 1	V(S3 ≥ S10) = 0.65	V(S4 ≥ S10) = 1	V(S5 ≥ S10) = 1
$\min\{V(S_1 \geq S_k)\} = 1$	$\min\{V(S_2 \geq S_k)\} = 0.71$	$\min\{V(S_3 \geq S_k)\} = 0.32$	$\min\{V(S_4 \geq S_k)\} = 0.68$	$\min\{V(S_5 \geq S_k)\} = 0.68$
Weight = 0.211	Weight = 0.15	Weight = 0.066	Weight = 0.143	Weight = 0.143
HSG = 6	**Aspect = 6**	**TPI = 7**	**TWI = 8**	**Profile Curvature = 10**
V(S6 ≥ S1) = 0.32	V(S7 ≥ S1) = 0	V(S8 ≥ S1) = 0	V(S9 ≥ S1) = 0.32	V(S10 ≥ S1) = 0.71
V(S6 ≥ S2) = 0.65	V(S7 ≥ S2) = 0.23	V(S8 ≥ S2) = 0.23	V(S9 ≥ S2) = 0.65	V(S10 ≥ S2) = 1
V(S6 ≥ S3) = 1	V(S7 ≥ S3) = 0.59	V(S8 ≥ S3) = 0.59	V(S9 ≥ S3) = 1	V(S10 ≥ S3) = 1
V(S6 ≥ S4) = 0.62	V(S7 ≥ S4) = 0.19	V(S8 ≥ S4) = 0.19	V(S9 ≥ S4) = 0.62	V(S10 ≥ S4) = 1
V(S6 ≥ S5) = 0.62	V(S7 ≥ S5) = 0.19	V(S8 ≥ S5) = 0.19	V(S9 ≥ S5) = 0.62	V(S10 ≥ S5) = 1
V(S6 ≥ S7) = 1	V(S7 ≥ S6) = 0.59	V(S8 ≥ S6) = 0.59	V(S9 ≥ S6) = 1	V(S10 ≥ S6) = 1
V(S6 ≥ S8) = 1	V(S7 ≥ S8) = 1	V(S8 ≥ S7) = 1	V(S9 ≥ S7) = 1	V(S10 ≥ S7) = 1
V(S6 ≥ S9) = 1	V(S7 ≥ S9) = 0.59	V(S8 ≥ S9) = 0.59	V(S9 ≥ S8) = 1	V(S10 ≥ S8) = 1
V(S6 ≥ S10) = 0.63	V(S7 ≥ S10) = 0.23	V(S8 ≥ S10) = 0.23	V(S9 ≥ S10) = 0.63	V(S10 ≥ S9) = 1
$\min\{V(S_6 \geq S_k)\} = 0.32$	$\min\{V(S_7 \geq S_k)\} = 0$	$\min\{V(S_8 \geq S_k)\} = 0$	$\min\{V(S_9 \geq S_k)\} = 0.32$	$\min\{V(S_{10} \geq S_k)\} = 0.71$
Weight = 0.066	Weight = 0	Weight = 0	Weight = 0.066	Weight = 0.15

4.3. Flash-Flood Potential Index Results Validation

Results validation is a mandatory step to establish the best ensemble model whose results will be used to identify the areas prone to flood generated by flash-floods. In this regard, the success rate and prediction rate were used. The success rate revealed that the highest performance was obtained by the results provided by FAHP-WOE (AUC = 0.837), followed by FAHP-SI (AUC = 0.833) and FAHP-FR (AUC = 0.723) (Figure 7a). The same hierarchy was also revealed by the construction of the prediction rate. Thus, the FAHP-WOE ranked first (AUC = 0.79), followed by FAHP-SI (AUC = 0.787) and FAHP-FR (AUC = 0.717) (Figure 7b). Therefore, following the results validation procedure, the $FFPI_{FAHP\text{-}WOE}$ was selected to be used in the next step of the analysis.

4.4. Flood Potential Index Computation

According to the description provided in Section 3.3.4, the flow accumulation method was applied to $FFPI_{FAHP\text{-}WOE}$ to evaluate the torrential degree of the river valleys across the study area. The results showed that a percentage of 21.59% of the total river valleys identified were characterized by a low and very low torrential degree and are, therefore, considered to be less favorable for flash-flood propagation (Figure 8a). For a 200 m buffer zone along with the other 78.41% of the river valleys, the flood potential index (FPI) was calculated. In this regard, the stand-alone analytical hierarchy process (AHP) multicriteria decision-making was used. It should be mentioned that through AHP, in the first stage, the weights of flash-flood predictors and classes/categories of flash-flood predictors were calculated. In terms of flash-flood predictors, the highest weight was detected for slope (0.224), followed by land use (0.137), elevation (0.137), distance from river (0.137), lithology (0.085), plan curvature (0.081), hydrological soil groups (0.064), convergence index (0.055), TPI (0.046), and TWI (0.031) (Table 4). The analysis of the weights attributed to the classes/categories of flash-flood predictors revealed that the highest value was obtained for hydrological soil group D (0.66), followed by the plan curvature class between −3 and −0.1 (0.539), the TPI class between −7.8 and −1.8 (0.439), the TWI class between −4.4 and 4.7 (0.433), the conglomerates and breccias lithological categories (0.423), the convergence index class between −86 and −3 (0.42), and the slope angle class lower than 3° (0.379).

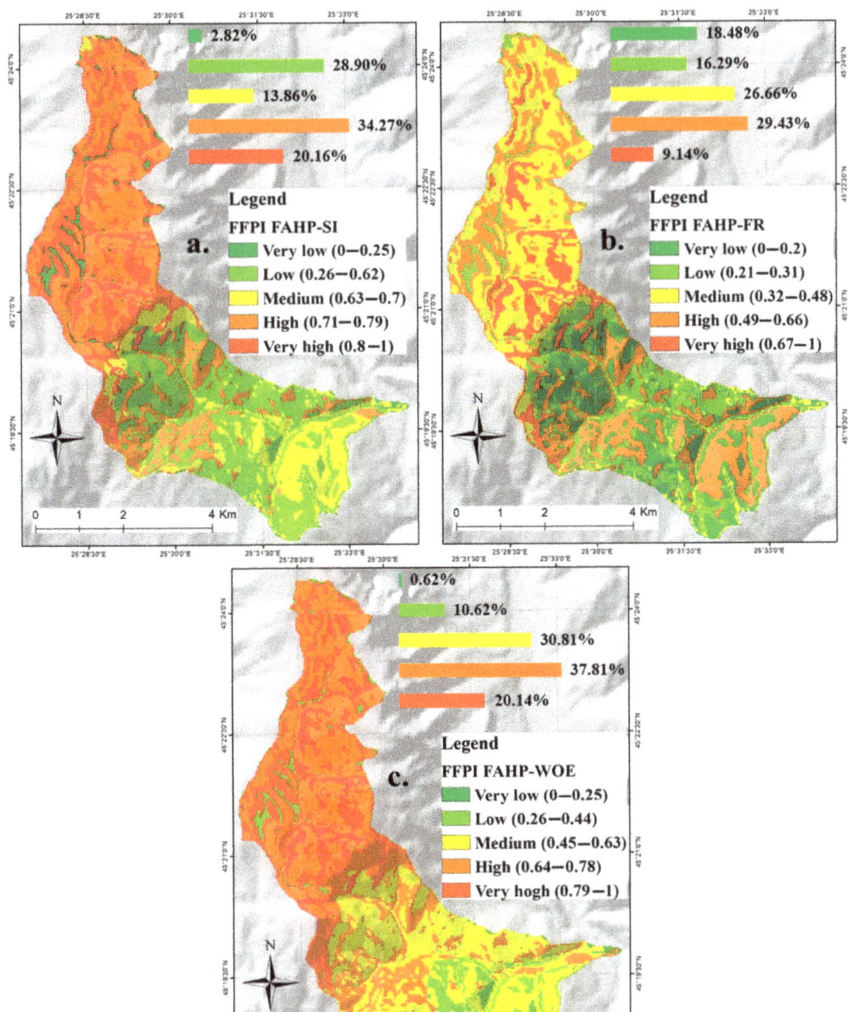

Figure 6. Flash-flood potential index (FFPI) values. (**a**) Fuzzy Analytical Hierarchy Process—Statistical Index (FAHP-SI); (**b**) Fuzzy Analytical Hierarchy Process—Frequency Ratio (FAHP-FR); (**c**) Fuzzy Analytical Hierarchy Process—Weights of Evidence (FAHP-WOE).

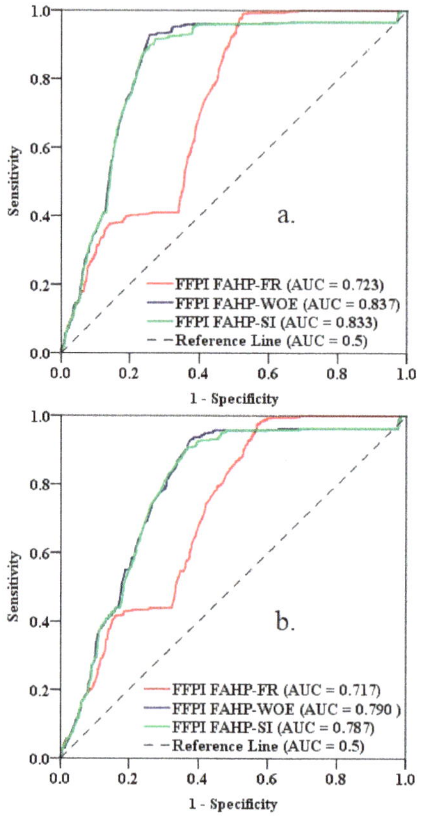

Figure 7. Receiver operating characteristic (ROC) Curves. (**a**) Success rate; (**b**) Prediction rate.

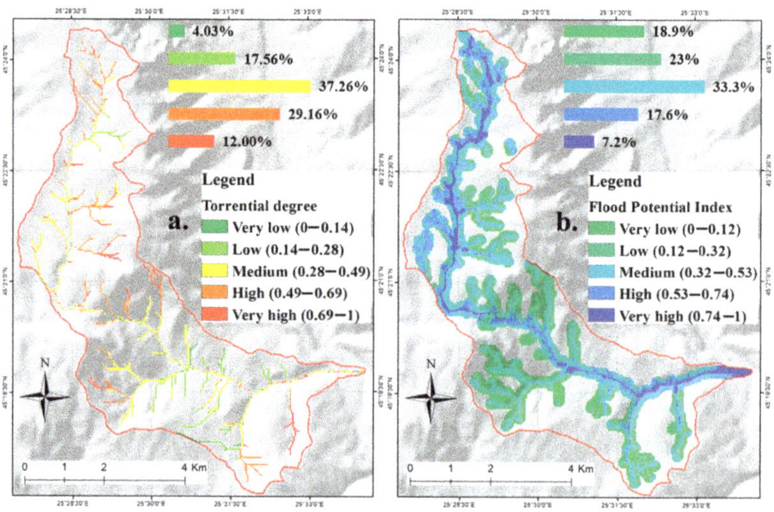

Figure 8. (**a**) River valleys torrential degree of the river; (**b**) Flood potential index classes.

Table 4. Pair-wise comparison matrix and normalized weights for each factor and class/category.

Factor and Classes/Categories Flood Predictors	Pair-Wise Comparison Matrix [1]	[2]	[3]	[4]	[5]	[6]	[7]	[8]	[9]	[10]	AHP Weights
[1] Slope	1										0.224
[2] TPI	1/4	1									0.046
[3] TWI	1/5	1/2	1								0.031
[4] Land use	1/2	3	4	1							0.137
[5] Lithology	1/3	2	3	1/2	1						0.085
[6] Elevation	1/2	3	4	1	2	1					0.137
[7] Distance from river	1/2	3	4	1	2	1	1				0.137
[8] Plan curvature	1/3	2	2	1/2	1	1/2	1/2	1			0.081
[9] CI	1/4	1	2	1/3	1/2	1/3	1/3	1/2	1		0.055
[10] HSG	1/3	2	3	1/2	1/2	1/2	1/2	1/2	1/2	1	0.064
Classes in each factor											
Slope angle											
[1] 0–3°	1										0.379
[2] 3.1–7°	1/2	1									0.249
[3] 7.1–15°	1/3	1/2	1								0.160
[4] 15.1–25°	1/4	1/3	1/2	1							0.102
[5] 25.1–45°	1/5	1/4	1/3	1/2	1						0.065
[6] 45.1–54°	1/6	1/5	1/4	1/3	1/2	1					0.043
TPI											
[1] −7.8–−1.8	1										0.439
[2] −1.7–−0.5	1/2	1									0.255
[3] −0.4–0.5	1/3	1/2	1								0.156
[4] 0.6–1.9	1/5	1/3	1/2	1							0.092
[5] 2–8.6	1/6	1/4	1/3	1/2	1						0.058
TWI											
[1] −4.4–4.7	1										0.433
[2] 4.8–8.4	1/2	1									0.251
[3] 8.5–11.8	1/3	1/2	1								0.164
[4] 11.9–15	1/5	1/3	1/2	1							0.100
[5] 15.1–23.1	1/6	1/4	1/3	1/3	1						0.052
Land use											
[1] Built-up areas	1										0.328
[2] Grassland	1/2	1									0.189
[3] Agriculture areas	1/3	1/2	1								0.120
[4] Forest	1/8	1/6	1/5	1							0.034
[5] Bare rocks	1	2	3	8	1						0.328
Lithology											
[1] Sandy flysch, marls shale	1										0.227
[2] Conglomerates, breccias	2	1									0.423
[3] Clays, limestone	1/2	1/3	1								0.123
[4] Sandstone, gravels	1	1/2	2	1							0.227
Plan curvature											
[1] −3–−0.1	1										0.539
[2] 0–0.1	1/2	1									0.297
[3] 0.2–1.9	1/3	1/2	1								0.164
Elevation											
[1] 763.1–1000 m	1										0.350
[2] 1000.1–1200 m	1/2	1									0.237
[3] 1200.1–1400 m	1/3	1/2	1								0.159
[4] 1400.1–1600 m	1/4	1/3	1/2	1							0.107
[5] 1600.1–1800 m	1/5	1/4	1/3	1/2	1						0.071
[6] 1800.1–2000 m	1/6	1/5	1/4	1/3	1/2	1					0.049
[7] 2000.1–2202 m	1/8	1/7	1/6	1/5	1/4	1/3	1				0.026
Distance from river											
[1] 0–50 m	1										0.327
[2] 50.1–100 m	1/2	1									0.227
[3] 100.1–150 m	1/3	1/2	1								0.157
[4] 150.1–200 m	1/4	1/3	1/2	1							0.108
[5] 200.1–400 m	1/5	1/4	1/3	1/2	1						0.073
[6] 400.1–700 m	1/6	1/5	1/4	1/3	1/2	1					0.050
[7] 700.1–1000 m	1/7	1/6	1/5	1/4	1/3	1/2	1				0.034
[8] >1000 m	1/8	1/7	1/6	1/5	1/4	1/3	1/2	1			0.024
Convergence index											
[1] −86–−3	1										0.420
[2] −2.9–−2	1/2	1									0.299
[3] −1.9–−1	1/3	1/3	1								0.141
[4] −0.9–0	1/4	1/4	1/2	1							0.088
[5] 0.1–84.9	1/7	1/5	1/3	1/2	1						0.052
HSG											
[1] A	1										0.117
[2] C	3	1									0.224
[3] D	4	5	1								0.660

The consistency of judgments was evaluated using the consistency ratio (CR) values. The results from Table 5 show that the CR values were less than 0.1, indicating that all the comparisons within the matrices were consistent. Table 5 also contains the values of some parameters involved in the CR computation.

Table 5. Properties of comparison matrices in the previous table.

Factors	N	λ_{max}	CI	RI	CR
All	10	10.32	0.036	1.49	0.024
Slope	6	6.123	0.025	1.24	0.020
TPI	5	5.046	0.012	1.12	0.010
TWI	5	5.121	0.030	1.12	0.030
Land use	5	5.063	0.016	1.12	0.010
Lithology	4	4.010	0.003	0.90	0.004
Elevation	7	7.248	0.041	1.32	0.030
Distance from river	8	8.292	0.042	1.41	0.030
Plan curvature	3	3.009	0.005	0.58	0.010
CI	5	5.087	0.022	1.12	0.020
HSG	3	3.203	0.102	0.58	0.018

To derive the flood potential index across the study area, the AHP weights, together with the raster dataset associated with the flood predictors, were used in map algebra of ArcGIS software. The normalized values of FPI were classified into five classes using the natural break method. The very low class, between 0 and 0.12, covered about 18.9% of the total area and was mainly spread along the valleys located in the southern part of the catchment. Another 23% of the delimited zone was characterized by a low flood potential. The medium FPI values (between 0.32 and 0.53) were associated with about 33.3% of the delimited perimeter. The high and very high potential was spread around 24.8% and was associated with FPI values higher than 0.53 (Figure 8b).

5. Discussion

In the last ten years, the study of the susceptibility to hydrological risk phenomena has developed significantly as a result of the combined application of geospatial analysis techniques with statistical models or algorithms from artificial intelligence [49]. It is well known that small river basins located in mountainous areas favor the occurrence of flash-floods and their propagation toward the areas located in the lower zones of the basins. The mountainous area of Romania is not an exception and is often affected by severe flash-flood events that generate property damage and loss of life. In this context, the present study aimed to identify the areas exposed to floods generated by flash-floods within the Izvorul Dorului River basin located in the Romanian Carpathians, which could produce pollution such as the transport of polycyclic aromatic hydrocarbons resulting from different sources [57].

The present study included a first part in which the potential for rapid water runoff on the slopes was determined, the second part referred to the identification of valleys with high torrential potential, followed by the evaluation of flood susceptibility existing along these valleys. The potential for rapid surface runoff, expressed through the FFPI, was calculated by applying three ensemble models resulting from the combination of three statistical bivariate methods and the fuzzy AHP model.

The decision to apply three ensemble models was taken after a careful review of the literature according to which hybrid models have higher performance than stand-alone ones [15]. The models applied for the calculation of the FFPI showed that the hydrographic basin of the Izvorul Dorului River has a high and very high potential for a rapid surface runoff with a percentage between 38% and 58% of its surface. It was also highlighted that in particular, the upper and middle basin is characterized by these values of FFPI. Since the genesis of rapid water runoff on the slopes is finally reflected in the flooded areas along the rivers, it was decided to continue the study with the identification of valleys with a high potential for flash-flood propagation, along with the identification of the floodplains. In this regard, the three FFPI models were evaluated, and the result provided by FAHP-WOE, characterized by an AUC-ROC curve of 0.837 in the case of training data and 0.79 in the case of test data, was identified as the most accurate. Using the methodology proposed by Costache et al. [25], the valleys in the study area were identified and classified according to the degree of torrentiality. Valleys with a small and very small propagation potential were eliminated from the analysis, with only those characterized by a medium, high, and very high potential being used. The AHP model was further used to calculate the flood potential index along the torrential valleys and at the same time to determine the potential for flooding generated by the flash-flood propagation. It is worth mentioning that following the flash-flood genesis (which is facilitated by the torrential areas highlighted in Figure 1) and their propagation, the areas located along the torrential valleys are the most affected regions because the water flow from the slopes will be concentrated on the main river network. Therefore, it is very important to indicate the surfaces that are finally affected by these complex phenomena. This resulted in 24% of the delimited surface having a high and very high potential for flooding.

In a previous study, Costache et al. [58] estimated only the flooding susceptibility along the large river valleys within the Trotuș River basin from Romania, unlike the present study which analyzed the following three elements in close spatial and causal connection: (i) flash-flood potential at the slopes level; (ii) river valleys torrential degree; and (iii) flood potential along the river basins within this small catchment. In addition to this difference regarding the methodological approaches, the present study also differed from that conducted by Costache et al. [58] by the methods proposed for determining the susceptibility to the analyzed hydrological hazards. Thus, in the present study, three ensemble models of the fuzzy analytical hierarchy process with bivariate statistical methods for the estimation of flash-flood potential at the slopes level were applied, while in the previous study, three other ensemble models of the adaptive neuro-fuzzy inference system (ANIFS) were applied to determine the flooding potential at the large river valley level. Moreover, the flow accumulation procedure was applied in the present research in order to identify the torrential valleys. Another example where the fuzzy multicriteria decision making analysis was proposed to estimate the flood susceptibility was the study carried out by Azareh et al. [59]. In that research, which focused on the Haraz watershed in Iran, the authors used a combination between DEMATEL, analytical network process, and fuzzy logic in order to estimate the flood susceptibility. Like in the present case, the performance of the applied model was very good, which was revealed by an AUC-ROC curve between 0.8 and 0.9. Nevertheless, the main difference between the present study and the research work developed by Azareh et al. [59] is given by the fact that the mentioned study only included the evaluation of the terrain surface potential along the river valley, to produce the flooding phenomenon and did not also include an evaluation of the slopes for surface runoff genesis.

6. Conclusions

The assessment of flash-floods and flood susceptibility is an actual scientific topic due to the high potential of the studies to propose solutions for reducing the economic damage and diminishing the number of victims. The new approach developed in the present research is useful because it provides a complete overview regarding the susceptibility of the entire phenomenon composed of rapid surface runoff on the slopes, the propagation of flash-floods generated by the surface runoff, and the potential for flooding along torrential valleys. The water quality in the floodplains will be lower because the flash-flood waves will be accompanied by the massive transport of materials from the slopes and inside the forest vegetation. Furthermore, the decision-makers will have a clearer image regarding the places they must implement measures to reduce the water runoff on the slopes, to arrange the torrential valleys, and to protect the areas exposed to floods.

Author Contributions: Conceptualization, R.C., Q.B.P. and A.B.; Data curation, R.C. and Q.B.P. Methodology, R.C. and A.B.; Writing—original draft R.C., Q.B.P. and A.B.; Writing—review and editing, R.C., Q.B.P. and A.B. All authors have read and agreed to the published version of the manuscript.

Funding: This work was supported by a grant of the Romanian Ministry of Education and Research, CNCS-UEFISCDI, project number PN-III-P1-1.1-PD-2019-0424, within PNCDI III.

Institutional Review Board Statement: Not applicable.

Informed Consent Statement: Not applicable

Conflicts of Interest: The authors declare no conflict of interest.

References

1. Hu, P.; Zhang, Q.; Shi, P.; Chen, B.; Fang, J. Flood-Induced Mortality across the Globe: Spatiotemporal Pattern and Influencing Factors. *Sci. Total Environ.* **2018**, *643*, 171–182. [CrossRef]
2. Lee, B.-J.; Kim, S. Gridded Flash Flood Risk Index Coupling Statistical Approaches and TOPLATS Land Surface Model for Mountainous Areas. *Water* **2019**, *11*, 504. [CrossRef]
3. Sattar, A.; Bonakdari, H.; Gharabaghi, B.; Radecki-Pawlik, A. Hydraulic Modeling and Evaluation Equations for the Incipient Motion of Sandbags for Levee Breach Closure Operations. *Water* **2019**, *11*, 279. [CrossRef]
4. Sattar, A.M.A.; Gharabaghi, B. Gene Expression Programming in Open Channel Hydraulics. *Open Channel Hydraul. River Hydraul. Struct. Fluv. Geomorphol. Eng. Geomorphol. Phys. Geogr.* **2017**, *196*, 196–211.
5. Bakuła, K.; StĘpnik, M.; Kurczyński, Z. Influence of Elevation Data Source on 2D Hydraulic Modelling. *Acta Geophys.* **2016**, *64*, 1176–1192. [CrossRef]
6. Bonakdari, H.; Moradi, F.; Ebtehaj, I.; Gharabaghi, B.; Sattar, A.A.; Azimi, A.H.; Radecki-Pawlik, A. A Non-Tuned Machine Learning Technique for Abutment Scour Depth in Clear Water Condition. *Water* **2020**, *12*, 301. [CrossRef]
7. Nhu, V.-H.; Thi Ngo, P.-T.; Pham, T.D.; Dou, J.; Song, X.; Hoang, N.-D.; Tran, D.A.; Cao, D.P.; Aydilek, İ.B.; Amiri, M. A New Hybrid Firefly–PSO Optimized Random Subspace Tree Intelligence for Torrential Rainfall-Induced Flash Flood Susceptible Mapping. *Remote Sens.* **2020**, *12*, 2688. [CrossRef]
8. Khosravi, K.; Shahabi, H.; Pham, B.T.; Adamowski, J.; Shirzadi, A.; Pradhan, B.; Dou, J.; Ly, H.-B.; Gróf, G.; Ho, H.L. A Comparative Assessment of Flood Susceptibility Modeling Using Multi-Criteria Decision-Making Analysis and Machine Learning Methods. *J. Hydrol.* **2019**, *573*, 311–323. [CrossRef]
9. Khosravi, K.; Nohani, E.; Maroufinia, E.; Pourghasemi, H.R. A GIS-Based Flood Susceptibility Assessment and Its Mapping in Iran: A Comparison between Frequency Ratio and Weights-of-Evidence Bivariate Statistical Models with Multi-Criteria Decision-Making Technique. *Nat. Hazards* **2016**, *83*, 947–987. [CrossRef]
10. Cao, C.; Xu, P.; Wang, Y.; Chen, J.; Zheng, L.; Niu, C. Flash Flood Hazard Susceptibility Mapping Using Frequency Ratio and Statistical Index Methods in Coalmine Subsidence Areas. *Sustainability* **2016**, *8*, 948. [CrossRef]
11. Althuwaynee, O.F.; Pradhan, B.; Park, H.-J.; Lee, J.H. A Novel Ensemble Bivariate Statistical Evidential Belief Function with Knowledge-Based Analytical Hierarchy Process and Multivariate Statistical Logistic Regression for Landslide Susceptibility Mapping. *Catena* **2014**, *114*, 21–36. [CrossRef]
12. Chen, Z.; Liang, S.; Ke, Y.; Yang, Z.; Zhao, H. Landslide Susceptibility Assessment Using Evidential Belief Function, Certainty Factor and Frequency Ratio Model at Baxie River Basin, NW China. *Geocarto Int.* **2019**, *34*, 348–367. [CrossRef]
13. Chen, W.; Li, W.; Hou, E.; Bai, H.; Chai, H.; Wang, D.; Cui, X.; Wang, Q. Application of Frequency Ratio, Statistical Index, and Index of Entropy Models and Their Comparison in Landslide Susceptibility Mapping for the Baozhong Region of Baoji, China. *Arab. J. Geosci.* **2015**, *8*, 1829–1841. [CrossRef]
14. Devkota, K.C.; Regmi, A.D.; Pourghasemi, H.R.; Yoshida, K.; Pradhan, B.; Ryu, I.C.; Dhital, M.R.; Althuwaynee, O.F. Landslide Susceptibility Mapping Using Certainty Factor, Index of Entropy and Logistic Regression Models in GIS and Their Comparison at Mugling–Narayanghat Road Section in Nepal Himalaya. *Nat. Hazards* **2013**, *65*, 135–165. [CrossRef]
15. Yariyan, P.; Janizadeh, S.; Phong, T.V.; Nguyen, H.D.; Costache, R.; Le, H.V.; Pham, B.T.; Pradhan, B.; Tiefenbacher, J.P. Improvement of Best First Decision Trees Using Bagging and Dagging Ensembles for Flood-Risk Mapping. *Water Resour. Manag.* **2020**. [CrossRef]
16. Pham, B.T.; Bui, D.T.; Prakash, I.; Dholakia, M. Hybrid Integration of Multilayer Perceptron Neural Networks and Machine Learning Ensembles for Landslide Susceptibility Assessment at Himalayan Area (India) Using GIS. *Catena* **2017**, *149*, 52–63. [CrossRef]
17. Debella-Gilo, M.; Etzelmüller, B. Spatial Prediction of Soil Classes Using Digital Terrain Analysis and Multinomial Logistic Regression Modeling Integrated in GIS: Examples from Vestfold County, Norway. *Catena* **2009**, *77*, 8–18. [CrossRef]
18. Kalantar, B.; Pradhan, B.; Naghibi, S.A.; Motevalli, A.; Mansor, S. Assessment of the Effects of Training Data Selection on the Landslide Susceptibility Mapping: A Comparison between Support Vector Machine (SVM), Logistic Regression (LR) and Artificial Neural Networks (ANN). *Geomat. Nat. Hazards Risk* **2018**, *9*, 49–69. [CrossRef]

19. Chen, W.; Shahabi, H.; Zhang, S.; Khosravi, K.; Shirzadi, A.; Chapi, K.; Pham, B.T.; Zhang, T.; Zhang, L.; Chai, H. Landslide Susceptibility Modeling Based on Gis and Novel Bagging-Based Kernel Logistic Regression. *Appl. Sci.* **2018**, *8*, 2540. [CrossRef]
20. Mansuy, N.; Thiffault, E.; Paré, D.; Bernier, P.; Guindon, L.; Villemaire, P.; Poirier, V.; Beaudoin, A. Digital Mapping of Soil Properties in Canadian Managed Forests at 250 m of Resolution Using the K-Nearest Neighbor Method. *Geoderma* **2014**, *235*, 59–73. [CrossRef]
21. Hosseini, F.S.; Choubin, B.; Mosavi, A.; Nabipour, N.; Shamshirband, S.; Darabi, H.; Haghighi, A.T. Flash-Flood Hazard Assessment Using Ensembles and Bayesian-Based Machine Learning Models: Application of the Simulated Annealing Feature Selection Method. *Sci. Total Environ.* **2020**, *711*, 135161. [CrossRef]
22. Zhang, C.-X.; Wang, G.-W.; Zhang, J.-S. An Empirical Bias–Variance Analysis of DECORATE Ensemble Method at Different Training Sample Sizes. *J. Appl. Stat.* **2012**, *39*, 829–850. [CrossRef]
23. Ahmadlou, M.; Karimi, M.; Alizadeh, S.; Shirzadi, A.; Parvinnejhad, D.; Shahabi, H.; Panahi, M. Flood Susceptibility Assessment Using Integration of Adaptive Network-Based Fuzzy Inference System (ANFIS) and Biogeography-Based Optimization (BBO) and BAT Algorithms (BA). *Geocarto Int.* **2019**, *34*, 1252–1272. [CrossRef]
24. Antonetti, M.; Horat, C.; Sideris, I.V.; Zappa, M. Ensemble Flood Forecasting Considering Dominant Runoff Processes–Part 1: Set-up and Application to Nested Basins (Emme, Switzerland). *Nat. Hazards Earth Syst. Sci.* **2019**, *19*, 19–40. [CrossRef]
25. Costache, R.; Hong, H.; Wang, Y. Identification of Torrential Valleys Using GIS and a Novel Hybrid Integration of Artificial Intelligence, Machine Learning and Bivariate Statistics. *Catena* **2019**, *183*, 104179. [CrossRef]
26. van Westen, C. *Statistical Landslide Hazards Analysis, ILWIS 2.1 for Windows Application Guide*; ITC Publication: Enschede, The Netherlands, 1997.
27. Khosravi, K.; Pourghasemi, H.R.; Chapi, K.; Bahri, M. Flash Flood Susceptibility Analysis and Its Mapping Using Different Bivariate Models in Iran: A Comparison between Shannon's Entropy, Statistical Index, and Weighting Factor Models. *Environ. Monit. Assess.* **2016**, *188*, 656. [CrossRef] [PubMed]
28. Rautela, P.; Lakhera, R.C. Landslide Risk Analysis between Giri and Tons Rivers in Himachal Himalaya (India). *Int. J. Appl. Earth Obs. Geoinf.* **2000**, *2*, 153–160. [CrossRef]
29. Ali, S.A.; Parvin, F.; Pham, Q.B.; Vojtek, M.; Vojteková, J.; Costache, R.; Linh, N.T.T.; Nguyen, H.Q.; Ahmad, A.; Ghorbani, M.A. GIS-Based Comparative Assessment of Flood Susceptibility Mapping Using Hybrid Multi-Criteria Decision-Making Approach, Naïve Bayes Tree, Bivariate Statistics and Logistic Regression: A Case of Topl'a Basin, Slovakia. *Ecol. Indic.* **2020**, *117*, 106620. [CrossRef]
30. Chen, W.; Li, H.; Hou, E.; Wang, S.; Wang, G.; Panahi, M.; Li, T.; Peng, T.; Guo, C.; Niu, C. GIS-Based Groundwater Potential Analysis Using Novel Ensemble Weights-of-Evidence with Logistic Regression and Functional Tree Models. *Sci. Total Environ.* **2018**, *634*, 853–867. [CrossRef]
31. Lee, S.; Kim, Y.-S.; Oh, H.-J. Application of a Weights-of-Evidence Method and GIS to Regional Groundwater Productivity Potential Mapping. *J. Environ. Manag.* **2012**, *96*, 91–105. [CrossRef]
32. Roodposhti, M.S.; Rahimi, S.; Beglou, M.J. PROMETHEE II and Fuzzy AHP: An Enhanced GIS-Based Landslide Susceptibility Mapping. *Nat. Hazards* **2014**, *73*, 77–95. [CrossRef]
33. Eskandari, S. A New Approach for Forest Fire Risk Modeling Using Fuzzy AHP and GIS in Hyrcanian Forests of Iran. *Arab. J. Geosci.* **2017**, *10*, 190. [CrossRef]
34. Hategekimana, Y.; Yu, L.; Nie, Y.; Zhu, J.; Liu, F.; Guo, F. Integration of Multi-Parametric Fuzzy Analytic Hierarchy Process and GIS along the UNESCO World Heritage: A Flood Hazard Index, Mombasa County, Kenya. *Nat. Hazards* **2018**, *92*, 1137–1153. [CrossRef]
35. Kahraman, C.; Cebeci, U.; Ulukan, Z. Multi-Criteria Supplier Selection Using Fuzzy AHP. *Logist. Inf. Manag.* **2003**, *16*, 382–394. [CrossRef]
36. Costache, R.; Popa, M.C.; Bui, D.T.; Diaconu, D.C.; Ciubotaru, N.; Minea, G.; Pham, Q.B. Spatial Predicting of Flood Potential Areas Using Novel Hybridizations of Fuzzy Decision-Making, Bivariate Statistics, and Machine Learning. *J. Hydrol.* **2020**, *585*, 124808. [CrossRef]
37. Feizizadeh, B.; Shadman Roodposhti, M.; Jankowski, P.; Blaschke, T. A GIS-Based Extended Fuzzy Multi-Criteria Evaluation for Landslide Susceptibility Mapping. *Comput. Geosci.* **2014**, *73*, 208–221. [CrossRef] [PubMed]
38. Buckley, J.J. Fuzzy Hierarchical Analysis. *Fuzzy Sets Syst.* **1985**, *17*, 233–247. [CrossRef]
39. Sun, C.-C. A Performance Evaluation Model by Integrating Fuzzy AHP and Fuzzy TOPSIS Methods. *Expert Syst. Appl.* **2010**, *37*, 7745–7754. [CrossRef]
40. Wang, T.-C.; Chen, Y.-H. Applying Fuzzy Linguistic Preference Relations to the Improvement of Consistency of Fuzzy AHP. *Inf. Sci.* **2008**, *178*, 3755–3765. [CrossRef]
41. Zaharia, L.; Costache, R.; Prăvălie, R.; Ioana-Toroimac, G. Mapping Flood and Flooding Potential Indices: A Methodological Approach to Identifying Areas Susceptible to Flood and Flooding Risk. Case Study: The Prahova Catchment (Romania). *Front. Earth Sci.* **2017**, *11*, 229–247. [CrossRef]
42. Gioia, A.; Totaro, V.; Bonelli, R.; Esposito, A.A.; Balacco, G.; Iacobellis, V. *Flood Susceptibility Evaluation on Ephemeral Streams of Southern Italy: A Case Study of Lama Balice*; Springer: Berlin/Heidelberg, Germany, 2018; pp. 334–348.
43. Costache, R. Flash-Flood Potential Index Mapping Using Weights of Evidence, Decision Trees Models and Their Novel Hybrid Integration. *Stoch. Environ. Res. Risk Assess.* **2019**, *33*, 1375–1402. [CrossRef]

44. Hapciuc, O.-E.; Romanescu, G.; Minea, I.; Iosub, M.; Enea, A.; Sandu, I. Flood Susceptibility Analysis of the Cultural Heritage in the Sucevita Catchment (Romania). *Int. J. Conserv. Sci.* **2016**, *7*, 501–510.
45. Zhao, G.; Pang, B.; Xu, Z.; Peng, D.; Xu, L. Assessment of Urban Flood Susceptibility Using Semi-Supervised Machine Learning Model. *Sci. Total Environ.* **2019**, *659*, 940–949. [CrossRef]
46. Costache, R.; Bui, D.T. Identification of Areas Prone to Flash-Flood Phenomena Using Multiple-Criteria Decision-Making, Bivariate Statistics, Machine Learning and Their Ensembles. *Sci. Total Environ.* **2020**, *712*, 136492. [CrossRef]
47. Gessesse, B.; Bewket, W.; Bräuning, A. Model-based Characterization and Monitoring of Runoff and Soil Erosion in Response to Land Use/Land Cover Changes in the Modjo Watershed, Ethiopia. *Land Degrad. Dev.* **2015**, *26*, 711–724. [CrossRef]
48. Bărbulescu, A.; Maftei, C.; Dumitriu, C.S. The Modelling of the Climateric Process That Participates at the Sizing of an Irrigation System. *Bull. Appl. Comput. Math.* **2002**, *2048*, 11–20.
49. Chapi, K.; Singh, V.P.; Shirzadi, A.; Shahabi, H.; Bui, D.T.; Pham, B.T.; Khosravi, K. A Novel Hybrid Artificial Intelligence Approach for Flood Susceptibility Assessment. *Environ. Model. Softw.* **2017**, *95*, 229–245. [CrossRef]
50. Skentos, A. Topographic Position Index Based Landform Analysis of Messaria (Ikaria Island, Greece). *Acta Geobalcanica* **2018**, *4*, 7–15. [CrossRef]
51. Arabameri, A.; Rezaei, K.; Cerdà, A.; Conoscenti, C.; Kalantari, Z. A Comparison of Statistical Methods and Multi-Criteria Decision Making to Map Flood Hazard Susceptibility in Northern Iran. *Sci. Total Environ.* **2019**, *660*, 443–458. [CrossRef]
52. Brewster, J. *Development of the Flash Flood Potential Index (FFPI) for Central NY & Northeast PA*; WFO: Binghamton, NY, USA, 2010; pp. 2–4.
53. Campbell, G.; Ratnaparkhi, M.V. An Application of Lomax Distributions in Receiver Operating Characteristic (ROC) Curve Analysis. *Commun. Stat. Theory Methods* **1993**, *22*, 1681–1687. [CrossRef]
54. Hong, H.; Tsangaratos, P.; Ilia, I.; Liu, J.; Zhu, A.-X.; Chen, W. Application of Fuzzy Weight of Evidence and Data Mining Techniques in Construction of Flood Susceptibility Map of Poyang County, China. *Sci. Total Environ.* **2018**, *625*, 575–588. [CrossRef]
55. Samela, C.; Troy, T.J.; Manfreda, S. Geomorphic Classifiers for Flood-Prone Areas Delineation for Data-Scarce Environments. *Adv. Water Resour.* **2017**, *102*, 13–28. [CrossRef]
56. Popa, M.C.; Peptenatu, D.; Drăghici, C.C.; Diaconu, D.C. Flood Hazard Mapping Using the Flood and Flash-Flood Potential Index in the Buzău River Catchment, Romania. *Water* **2019**, *11*, 2116. [CrossRef]
57. Krein, A.; Schorer, M. Road Runoff Pollution by Polycyclic Aromatic Hydrocarbons and Its Contribution to River Sediments. *Water Res.* **2000**, *34*, 4110–4115. [CrossRef]
58. Costache, R.; Țîncu, R.; Elkhrachy, I.; Pham, Q.B.; Popa, M.C.; Diaconu, D.C.; Avand, M.; Costache, I.; Arabameri, A.; Bui, D.T. New Neural Fuzzy-Based Machine Learning Ensemble for Enhancing the Prediction Accuracy of Flood Susceptibility Mapping. *Hydrol. Sci. J.* **2020**, *65*, 2816–2837. [CrossRef]
59. Azareh, A.; Rafiei Sardooi, E.; Choubin, B.; Barkhori, S.; Shahdadi, A.; Adamowski, J.; Shamshirband, S. Incorporating Multi-Criteria Decision-Making and Fuzzy-Value Functions for Flood Susceptibility Assessment. *Geocarto Int.* **2019**, 1–21. [CrossRef]

water

MDPI

Article

Computing the Beta Parameter in IDW Interpolation by Using a Genetic Algorithm

Alina Bărbulescu [1], Cristina Şerban [2,*] and Marina-Larisa Indrecan [2]

[1] Department of Civil Engineering, Transilvania University of Braşov, 5 Turnului Street, 900152 Braşov, Romania; alina.barbulescu@unitbv.ro

[2] Department of Mathematics and Computer Science, Ovidius University of Constanta, 124 Mamaia Av., 900527 Constanta, Romania; maryna_larysa@yahoo.com

* Correspondence: cgherghina@gmail.com

Abstract: This article proposes a new approach for determining the optimal parameter (β) in the Inverse Distance Weighted Method (IDW) for spatial interpolation of hydrological data series. This is based on a genetic algorithm (GA) and finds a unique β for the entire study region, while the classical one determines different βs for different interpolated series. The algorithm is proposed in four scenarios crossover/mutation: single-point/uniform, single-point/swap, two-point/uniform, and two-point swap. Its performances are evaluated on data series collected for 41 years at ten observation sites, in terms of mean absolute error (MAE) and mean standard error (MSE). The smallest errors are obtained in the two-point swap scenario. Comparisons of the results with those of the ordinary kriging (KG), classical IDW (with β = 2 and the optimum beta found by our algorithm), and the Optimized IDW with Particle Swarm Optimization (OIDW) for each study data series show that the present approach better performs in 70% (80%) cases.

Keywords: genetic algorithm (GA); IDW; spatial interpolation

Citation: Bărbulescu, A.; Şerban, C.; Indrecan, M.-L. Computing the Beta Parameter in IDW Interpolation by Using a Genetic Algorithm. *Water* **2021**, *13*, 863. https://doi.org/10.3390/w13060863

Academic Editors: Giuseppe Pezzinga and Elias Dimitriou

Received: 14 January 2021
Accepted: 20 March 2021
Published: 22 March 2021

1. Introduction

Evaluating and predicting the effects of atmospheric factors dynamics, like precipitation and temperature, are of major importance for human activity, especially for zones with arid or rainy climates. Since water scarcity impacts billions of people worldwide, it is important to assess the water resources availability at ungauged locations [1]. Spatial interpolation methods are utilized for estimating the values of environmental variables using data recorded at neighbor locations. The most utilized approaches are classified as deterministic, geostatistical, and combined (or hybrid) [2,3]. The Inverse Distance Weighting (IDW) is a deterministic (mechanical) technique. The attribute values of any pair of points are related to each other, their similarity being inversely proportional to the distance between the two locations [4,5].

Since IDW does not involve advanced computational knowledge, researchers widely utilized it for spatial interpolation problems. Different authors presented comparable IDW performances with other spatial interpolation methods [6–11]. In [6,7], it is shown that IDW provided better or comparative results as ordinary kriging (OK) in the spatial interpolation of precipitation in Taiwan and Norfolk Island. Ly et al. [8] reported that OK and IDW provided the smallest root mean squared error in a study concerning the daily rainfall at the catchment scale in Belgium. Dong et al. [9] found that Ordinary CoKriging (OCK) performed better than OK and IDW when interpolating daily rainfall in a river basin from China. IDW, Thiessen Polygons Method (TPM), and kriging have been evaluated against the Most Probable Precipitation Method (MPPM) on annual, monthly, seasonal, and annual monthly maximum precipitation series from ten stations of 41 data [10]. IDW over performed TPM and OK, but underperformed MPPM. Chen et al. [11] proposed an improved regression-based scheme (PCRR) that was superior to IDW and multiple linear regression (MLR) interpolation methods on data from the mesoscale catchment of the Fuhe River.

Even if the classical IDW (with the value of the parameter $\beta = 2$) was successfully employed for a long period for spatial interpolation problems, being easy to use, improving its performances was targeted by scientists. For example, Lu and Wong [4] proposed the weights' modification depending on the neighboring locations' distribution density around the unsampled place. Golkhatmi et al. [12] introduced altitude as a new variable in the IDW interpolation (keeping $\beta = 2$) and reported good results in the case study.

Another direction is finding the best β. This is an optimization problem by itself, targeted by many scientists [13–19]. For example, Noori et al. [13] employed IDW for estimating the distribution of precipitation in Iran, the value of the parameter (β) being recursively searched in the interval (1, 5], increasing its value each time. However, this grid-search procedure is time-consuming for small step sizes [5]. To avoid this drawback, Mei et al. [14] designed and implementing parallel adaptive inverse distance weighting (AIDW) interpolation algorithms by using the graphics processing unit (GPU) for accelerating the parameter finding. Gholipour et al. [15] propose a hybridization of IDW with a harmony search, which improves the convergence rate and reduces the search time.

In the same idea, hybrid methods have been proposed. Zhang et al. [16] combined Support Vector Machines (SVM) with IDW obtaining the SVM residual IDW, obtaining superior results by comparison to IDW and OK for the spatial interpolation of the multi-year average annual precipitation in the Three Gorges Region basin. Nourani et al. [17] used a two-stage framework for spatial interpolation of precipitation, employing, in the first stage, three artificial intelligence models that generate the input for the second stage, where they utilize IDW for spatial interpolation. Bărbulescu et al. [18] proposed a Particle Swarm Optimization approach (called OIDW) for finding a single β in IDW interpolation of maximum annual precipitation from the Dobrogea region (Romania). Chang et al. [19] applied a genetic algorithm (GA) to find the optimal distances between the gauged stations to minimize the estimation errors in IDW. Still, based on our knowledge, no attempt to optimize the choice of β parameter of IDW using a GA has been made so far.

On the other hand, GAs are widely used for solving real-life problems. For example, Ratnam et al. [20] improved seasonal air temperature forecasts using a genetic algorithm. Nasseri et al. [21] presented an optimized scenario for rainfall forecasting using a genetic algorithm coupled with an artificial neural network using rainfall hyetograph of recording rain gauges in the Upper Parramatta catchment (Sydney, Australia). Using the ability of GAs to search complex decision spaces, Sen and Ôztopal [22] utilized such an algorithm for optimizing the classification of rainy and non-rainy day occurrences using atmospheric data (temperature, humidity, dew point, vertical velocity). Heat conduction and control problems have also been solved by utilizing GAs [23,24].

In this context, this article proposes a new approach that optimizes the finding of the beta parameter of IDW. This is based on a genetic algorithm and finds a unique β for the entire study region, while the classical one determines different βs for different interpolated series. The algorithm is proposed in four scenarios crossover/mutation: single-point/uniform, single-point/swap, two-point/uniform, and two-point swap. Comparisons of its performances with those of the classical IDW (with $\beta = 2$ and the optimal beta found in our algorithm), ordinary kriging, and two versions of the optimized IDW by using Particle Swarm Optimization (OIDW) are also provided.

2. Methodology and Data Series

2.1. IDW Interpolation

The study problem is estimating a variable's values at ungagged locations employing the same variable's known values, registered at the neighboring observation sites [18]. In terms of mathematics, one can formulate the problem as follows. Given a set of spatial data of a variable z at n observation sites, $s_1, \ldots, s_n$ determine the same variable's values at the study site, s_0.

The IDW interpolation formula is:

$$\widehat{z(s_0)} = \sum_{i=1}^{n} \frac{1/d(s_0, s_i)^{\beta}}{\sum_{i=1}^{n}\left(1/d(s_0, s_i)^{\beta}\right)} z(s_i),\ \beta > 1 \quad (1)$$

where $\widehat{z(s_0)}$ is the value computed for the site s_0, $z(s_i)$ is the value recorded at the site s_i, $d(s_0, s_i)$ is the distance between s_0 and s_i, and β is a parameter whose value is either given or determined by different optimization methods. In the original algorithm, $\beta = 2$ [25,26].

The interpolation quality depends on β which is generally determined after running a grid search. The time spent for finding the parameters is inverse proportional with the step of the grid.

2.2. Genetic Algorithms

A genetic algorithm (GA) is a metaheuristic method inspired by natural selection laws that try to find optimal solutions to complex problems to which deterministic approaches usually cannot find a good result. The genetic operators, selection, crossover, and mutation establish a balance between the exploration and exploitation of the search space [27,28]. Exploration means that the algorithm searches for new solutions in new regions, while exploitation refers to making refinement to existing solutions to improve their quality. A function called fitness measures the quality of the solutions, which are represented by chromosomes.

A GA starts with a population of some random chromosomes and (by applying the principle of 'survival of the fittest') produces multiple generations by selecting in each one the fittest individuals for breeding. The mutation is then applied to increase the population diversity. Along with the generations, better individuals, i.e., better approximations to the solution, are obtained. The process continues until the fittest individual (the optimal solution) is found or the maximum number of generations is reached.

Using a genetic algorithm to solve a problem means finding the representation of the problem's solutions (encoding of the chromosomes), the fitness function, and the genetic operators. A chromosome is a feasible solution to the problem. In our case, a chromosome represents a real value of the parameter $a \le \beta \le b$. Thus, we apply a value encoding and get a binary string with the length l, calculated using the following formula (the default encoding of real values to binary strings):

$$2^{l-1} < (b - a) * 10^{z} < 2^{l} \quad (2)$$

where z represents the given number of β's decimals. In this study, $l = 9$ bits.

The decimal value, *val*, of the binary chromosome representation, is computed by (3). We get the real value of a chromosome (β) by applying (4).

$$val = (\beta - a) \times (2^{l} - 1)/(b - a) \quad (3)$$

or

$$\beta = a + val \times (b - a)/(2^{l} - 1) \quad (4)$$

The fitness function controls the possibility of individuals' reproduction. The better chromosome is (i.e., the better fitness is), the more likely it is to be selected for breeding the next generation. Since our goal is to minimize the error between the results obtained by the spatial interpolation and those recorded at the meteorological stations, the fitness function will record the mean standard error (MSE) between the known data and those computed by IDW. A GA performs best when a feasible solution maximizes the fitness function. Hence, we apply one of the most commonly adopted fitness mapping (inversion scaling), which does not alter the minimum location, but converts a minimization problem to an equivalent maximization one.

We use the mean standard error and mean absolute deviation (MAE) to evaluate the GA's performance. The lowest the MAE or MSE is, the better the algorithm performs.

Genetic operators are used for producing new generations of individuals with more diverse properties. There are three operators, selection, crossover, and mutation, which can set and, most of the time, find a good ratio between exploration and exploitation of the search space.

A selection operator determines the best individuals' regions that will exchange information to create a new generation. In this paper, the roulette wheel selection method [27] is used.

A crossover operator combines two or more parents to generate one or two offspring. It implements the idea that a swap of information between good individuals will generate an even better one. In this paper, the single-point crossover and the two-point crossover are used to create new offspring [29].

A mutation operator randomly modifies chromosomes with a given probability, *pm*, called mutation rate, leading to an increased population's structural diversity. Thus, a mutation operator facilitates the recovery of genetic material lost during the selection step and exploring new solutions. Here we used the uniform mutation [27] (one gene is randomly chosen and its value is modified) and swap mutation [27] (two positions on the chromosome are randomly selected, and their values are interchanged).

One may configure several control parameters in a genetic algorithm to achieve a balance between exploration and exploitation. If the population size is large, the search space is more explored than when the population size is small [28]. However, the runtime of the algorithm would increase. If crossover and mutation rates are high, the search will explore much of the solutions space, but there is a high chance of missing good solutions, the GA acting more like a random search. If crossover and mutation rates are low, the search space remains unexplored, and in this case, the GA resembles the hill-climbing algorithms. Therefore, we investigated the influence of the population size, crossover rate, mutation rate, and stop condition on the GA results. We performed each test ten times and averaged the results to increase their precision (as suggested in the literature). We implemented two crossover operators and two mutation operators to find the ones which are best suited for our problem. We also ran several tests for each pair of operators to see the relationship between the control parameters and the fitness value. Details are presented in the following sections.

2.3. New Approach for Estimating Beta

The genetic algorithm we implemented is presented in the following.

Input: The distances between stations and the precipitation series recorded at these stations.
Output: The optimal parameter value of β

Begin

1. **Generate** a random population of n individuals represented as binary strings of length $l = 9$
2. **Compute** the fitness function
    a. **Select** some chromosomes for crossover operation (the number of selected individuals is defined by the crossover rate)
    b. **Apply** one of the crossover operators described in 2.2 to generate two new offspring
    c. **Copy** the remaining chromosomes (that were not recombined) to the next generation
3. **Select** a few chromosomes for the mutation (the number of selected individuals is defined by the mutation rate)
4. If the number of generations is reached, **then** stop, **else** go to step 2

End

For a better understanding, a flowchart of the procedure of determination of the beta parameter is presented in Figure 1.

In order to find the best parameters settings for our problem, we fine tune our algorithm, which means creating several GA variants to test and find the best one, by slightly changing of GA parameters (population size, number of generations, crossover and mutation rates). We change only one parameter at a time, and try out several evolutionary literature-based test values. For example, for the crossover rate, the most used values in applications are in interval [0.6,1), whereas the mutation rate should be less than 10%. We run these GA variants on our problem, accept that parameter value at which the GA performs best, and continue to the next GA parameter, and so forth, to the last one. More precisely, to select the best population size, stop condition, and crossover rate the following steps are done.

Step 1. We start with predefined values for the stop condition (10 generations), crossover rate (0.75), and mutation rate (0.015). Then, we vary the population size and compute the values of the fitness function. For each pair of operators (single-point/uniform, single-point/swap, two-point/uniform, two-point/swap), we chose the optimal population size to be the lowest value from which population growth does not significantly influence the modification of the fitness value.

Step 2. With the population value determined at Step 1, the crossover rate, and the mutation rate kept at the same values as in Step 1, we run the algorithm to determine the best number of generations.

Step 3. With the number of individuals determined at Step1, the number of generations determined at Step 2, and the mutation kept at the same value as in Step 1, we run the algorithm using different crossover rates, to determine the best crossover rate.

Step 4. To find the best mutation rate, we set the best parameters from the previous steps and run the algorithm with different mutation rates.

Step 5. The algorithm is run in each scenario with the new parameters determined in the previous steps.

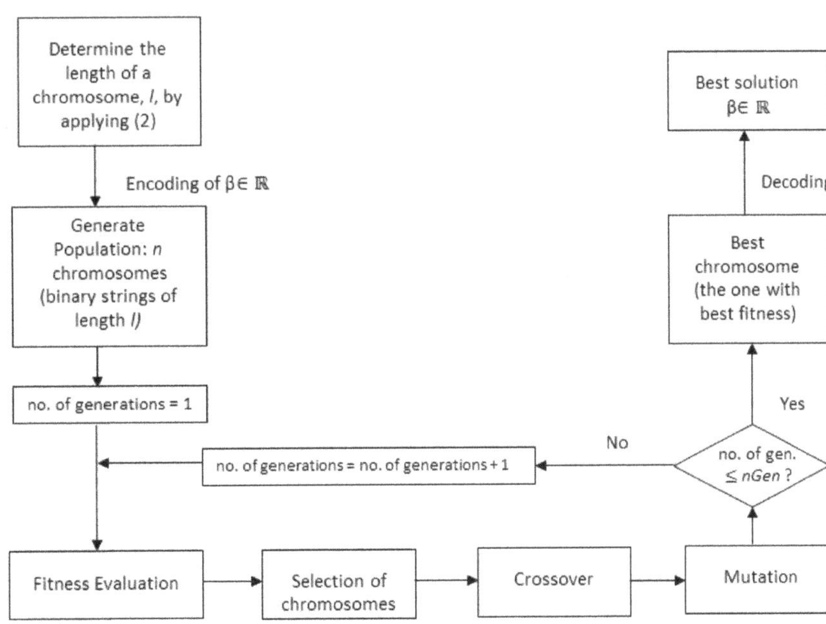

Figure 1. The procedure flow chart. *nGen* represents the maximum number of generations.

Although the complexity of GAs has a probabilistic convergence time [30], the settings of our genetic algorithm are not complex, and, based on the experimental results that show that it converges in a short time, we may state that the best convergence time is logarithmic, $O(log(n))$, whereas the worst is linear, $O(n)$. In the Results and Discussion, we present the recorded execution time (in seconds), which shows that the algorithms stop in short time for each test we did.

2.4. Data Series

Dobrogea is a region covering a surface between the Romanian Littoral of the Black Sea, the lower Danube River, and the Danube Delta, situated in the southeast part of Romania and characterized by long droughts periods. Records show the absence of precipitation for 4–6 months per year after 1961, which affects agricultural activities. Researchers analyzed precipitation and temperature evolution in this zone, especially after 2010, to mitigate the drought effects [31–33].

The data we are working with is formed by the maximum annual precipitation series recorded during a period of 41 years at 10 main meteorological stations from the Dobrogea region (Figure 2).

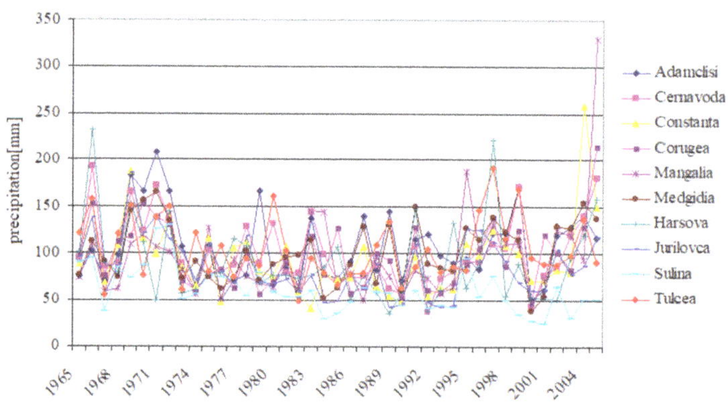

Figure 2. Maximum annual precipitation series.

3. Results and Discussion

Firstly, we ran several tests to find the settings of control parameters that are most likely to produce the best results. We started with predefined values from the literature [27,28] for the stop condition (10 generations), crossover rate (0.75), and mutation rate (0.015). Then, we varied the population size (from 10 to 80, with a step of 5) and computed the fitness function's corresponding values, run time, and β. Table 1 shows the relationship between the fitness value and population size.

For each pair of operators (single-point/uniform, single-point/swap, two-point/uniform, two-point/swap), we chose the optimal population size to be the lowest value from which population growth does not significantly influence the modification of the fitness value. This is 45, 40, 30, and 35 individuals, respectively, and the fitness function value is 0.0317. The corresponding β values obtained in the four scenarios are 1.256, 1.372, 1.308, and 1.336, respectively. These results are highlighted in Table 1.

Table 1. The impact of population size on the GA accuracy. The best results are highlighted.

Scenario	Single Point/Uniform			Single Point/Swap			Two-Point/Uniform			Two-Point/Swap		
Pop. Size	Fitness	Time (s)	β	Fitness	Time (s)	β	Fitness	Time (s)	β	Fitness	Time (s)	β
10	0.0311	0.300	2.528	0.0312	0.3156	2.316	0.0313	0.3500	2.186	0.0316	0.3625	1.580
15	0.0313	0.4625	1.500	0.0315	0.4688	1.612	0.0313	0.4719	2.154	0.0316	0.4781	1.568
20	0.0314	0.6062	1.862	0.0315	0.6031	1.742	0.0316	0.6375	1.514	0.0309	0.675	2.886
25	0.0315	0.8031	1.766	0.0315	0.8125	1.346	0.0315	0.7906	1.772	0.0313	0.7969	2.092
30	0.0316	0.9375	1.594	0.0315	0.9094	1.708	0.0317	0.9563	1.308	0.0315	0.9531	1.758
35	0.0316	1.0656	1.538	0.0316	1.0625	1.542	0.0317	1.1125	1.382	0.0317	1.100	1.336
40	0.0316	1.2813	1.494	0.0317	1.2188	1.372	0.0317	1.2813	1.298	0.0316	1.2437	1.620
45	0.0317	1.4312	1.256	0.0316	1.3687	1.382	0.0317	1.4094	1.208	0.0317	1.4781	1.212
50	0.0317	1.6781	1.480	0.0317	1.5781	1.128	0.0317	1.5375	1.308	0.0317	1.5594	1.274
55	0.0317	1.6906	1.324	0.0316	1.8375	1.552	0.0317	1.6938	1.132	0.0316	1.700	1.720
60	0.0317	1.8156	1.690	0.0317	1.850	1.258	0.0317	1.8750	1.122	0.0317	1.8469	1.192
65	0.0317	1.9625	1.212	0.0316	1.9781	1.510	0.0317	2.0156	1.294	0.0317	2.0938	1.246
70	0.0317	2.1781	1.248	0.0317	2.2062	1.056	0.0317	2.2250	1.200	0.0317	2.1656	1.210
75	0.0317	2.2437	1.306	0.0317	2.3062	1.126	0.0317	2.3406	1.262	0.0317	2.3687	1.282
80	0.0317	2.4406	1.162	0.0316	2.4469	1.382	0.0317	2.4500	1.320	0.0317	2.5844	1.306

Since we used a predefined number of generations as the stop condition, in the second stage, we had to determine its optimal value. To find it, for each pair of operators, we ran tests with several values of the number of generations, the population size previously estimated (45, 40, 30, 35, respectively—Table 1), keeping the mutation rate set to 0.015, and the crossover rate set to 0.75. Table 2 and Figure 3 show that the fitness value does not improve after a certain number of generations (which is the optimal number of generations).

Table 2. The impact of the number of generations on the GA performance. The best results are highlighted.

Scenario	Single Point/Uniform					Single Point/Swap				
Crossover Rate	Pop. Size	No. Gen	Fitness	Time (s)	β	Pop. Size	No. Gen	Fitness	Time (s)	β
0.6	45	9	0.0316	1.2469	1.476	40	5	0.0317	0.6219	1.242
0.65	45	9	0.0316	1.2344	1.372	40	5	0.0317	0.6094	1.170
0.7	45	9	0.0316	1.2469	1.26	40	5	0.0317	0.6156	1.310
0.75	45	9	0.0316	1.2875	1.550	40	5	0.0317	0.6094	1.174
0.8	45	9	0.0317	1.2500	1.228	40	5	0.0317	0.6094	1.344
0.85	45	9	0.0317	1.2563	1.064	40	5	0.0317	0.6000	1.148
0.9	45	9	0.0317	1.2313	1.556	40	5	0.0316	0.6125	1.408
0.95	45	9	0.0317	1.2563	1.386	40	5	0.0316	0.6094	1.192
Scenario	**Two-Point/Uniform**					**Two-Point/Swap**				
Cross Rate	**Pop. Size**	**No. Gen**	**Fitness**	**Time (s)**	**β**	**Pop Size**	**No. Gen**	**Fitness**	**Time (s)**	**β**
0.6	30	5	0.0316	0.4594	1.424	35	5	0.0316	0.5625	1.538
0.65	30	5	0.0315	0.4750	1.964	35	5	0.0316	0.5594	1.578
0.7	30	5	0.0316	0.4719	1.564	35	5	0.0317	0.5344	1.362
0.75	30	5	0.0317	0.4594	1.122	35	5	0.0317	0.5313	1.252
0.8	30	5	0.0317	0.4562	1.532	35	5	0.0317	0.5375	1.208
0.85	30	5	0.0317	0.4688	1.400	35	5	0.0317	0.5313	1.348
0.9	30	5	0.0317	0.4688	1.326	35	5	0.0317	0.5469	1.124
0.95	30	5	0.0316	0.4531	1.406	35	5	0.0317	0.5469	1.274

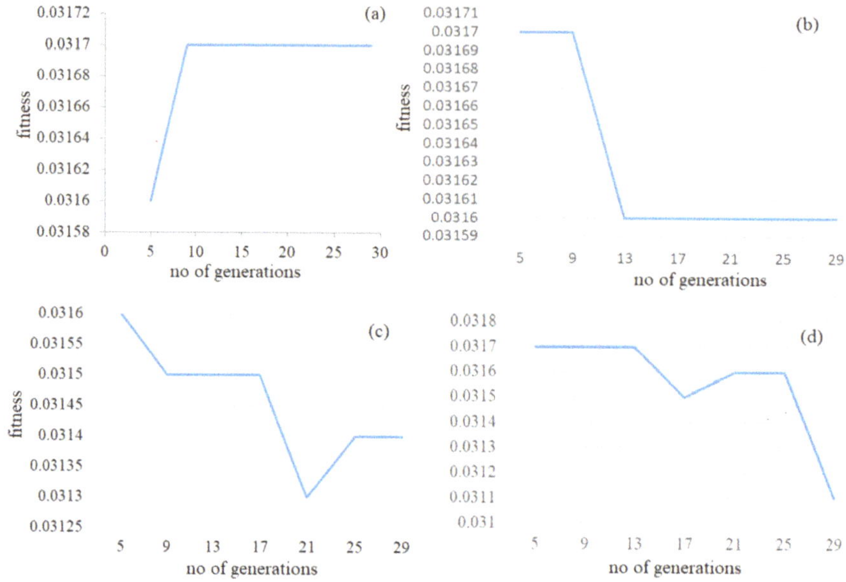

Figure 3. The impact of the number of generations on fitness value in (**a**) single-point/uniform, (**b**) single-point/swap, (**c**) two point/uniform, (**d**) two point /swap scenarios.

Based on the highest fitness value, the optimum number of generations determined was nine for the single-point/uniform mutation scenario, whereas, for the other scenarios, the number of generations was five. The corresponding β-values obtained are 1.228, 1.242, 1.122, and 1.362, respectively (highlighted in Table 2, together with the fitness and beta values). Since in our algorithm, β is represented by a chromosome, and several genetic operators have been used, different chromosomes (β) could produce the same fitness value.

The next step is the determination of the optimal crossover rate. For this aim, we ran the algorithm in each scenario with different values of the crossover rate (from 0.6 – value suggested in the literature to 0.95, with a step of 0.05), the population size and number of generations previously determined (45, 40, 30, and 35 individuals, respectively; 9, 5, 5, 5 generations, respectively), and the mutation rate kept to 0.015. We chose the optimal crossover rate for each pair of operators to be the value that gives the best (highest) fitness.

From Table 3 it results that the best crossover rates are 0.8 when using a single-point/uniform scenario, 0.6 for single-point/swap, 0.75 for two-point/uniform, and 0.7 for two-point/swap. These values correspond to the highlighted sequences of values in Table 3.

The last step was the determination of the best mutation rate. Therefore, we analyzed the impact the mutation rate has on the GA's results. We considered the population size, the number of generations, and the crossover rates we established in previous stages, and we performed new tests aiming at detecting the value of the mutation rate. For example, for single-point uniform mutation, we took the population size = 45, the number of generations = 9, the crossover rate = 0.80, and ran the tests for a mutation rate from 0.02 to 0.1, with a step size of 0.01. For each pair of operators, we search the optimal mutation rate for which the fitness value evolves to a maximum along with the generations. Table 4 contains the values of the fitness function obtained after running the algorithm in the four scenarios, with different mutation rates. For example, in Table 4a we present the values of the fitness function obtained for each generation (from 1 to 9) and the mutation rates from 0.02 to 0.1, in the single-point crossover/uniform mutation scenario. The highest fitness value is obtained after nine generations in the single-point crossover/uniform mutation, with a mutation rate of 0.06 (the sixth column–the highlighted values).

Table 3. The impact of crossover rate on the GA accuracy. The best results are highlighted.

Scenario	Single-Point/Uniform					Single-Point/Swap				
Crossover Rate	Pop. Size	No. Gen	Fitness	Time (s)	β	Pop. Size	No. Gen	Fitness	Time (s)	β
0.6	45	9	0.0316	1.2469	1.476	40	5	0.0317	0.6219	1.242
0.65	45	9	0.0316	1.2344	1.372	40	5	0.0317	0.6094	1.170
0.7	45	9	0.0316	1.2469	1.26	40	5	0.0317	0.6156	1.310
0.75	45	9	0.0316	1.2875	1.550	40	5	0.0317	0.6094	1.174
0.8	45	9	0.0317	1.2500	1.228	40	5	0.0317	0.6094	1.344
0.85	45	9	0.0317	1.2563	1.064	40	5	0.0317	0.6000	1.148
0.9	45	9	0.0317	1.2313	1.556	40	5	0.0316	0.6125	1.408
0.95	45	9	0.0317	1.2563	1.386	40	5	0.0316	0.6094	1.192
	Two-Point/Uniform					Two-Point/Swap				
Cross Rate	Pop. Size	No. Gen	Fitness	Time (s)	β	Pop Size	No. Gen	Fitness	Time (s)	β
0.6	30	5	0.0316	0.4594	1.424	35	5	0.0316	0.5625	1.538
0.65	30	5	0.0315	0.4750	1.964	35	5	0.0316	0.5594	1.578
0.7	30	5	0.0316	0.4719	1.564	35	5	0.0317	0.5344	1.362
0.75	30	5	0.0317	0.4594	1.122	35	5	0.0317	0.5313	1.252
0.8	30	5	0.0317	0.4562	1.532	35	5	0.0317	0.5375	1.208
0.85	30	5	0.0317	0.4688	1.400	35	5	0.0317	0.5313	1.348
0.9	30	5	0.0317	0.4688	1.326	35	5	0.0317	0.5469	1.124
0.95	30	5	0.0316	0.4531	1.406	35	5	0.0317	0.5469	1.274

Table 4. The impact of mutation rate on the GA accuracy. The best results are highlighted.

a. Single-Point Crossover/Uniform Mutation

	Mutation Rate								
Gener.	0.02	0.03	0.04	0.05	0.06	0.07	0.08	0.09	0.1
1	0.0307	0.0309	0.0309	0.0309	0.0308	0.0309	0.0310	0.0309	0.0309
2	0.0309	0.0310	0.0310	0.0310	0.0310	0.0309	0.0310	0.0310	0.0309
3	0.0309	0.0310	0.0310	0.0310	0.0310	0.0310	0.0311	0.0312	0.0309
4	0.0310	0.0311	0.0312	0.0310	0.0310	0.0311	0.0312	0.0312	0.0309
5	0.0310	0.0313	0.0312	0.0311	0.0312	0.0312	0.0312	0.0313	0.0309
6	0.0311	0.0313	0.0312	0.0311	0.0312	0.0312	0.0312	0.0313	0.0309
7	0.0311	0.0313	0.0313	0.0311	0.0312	0.0312	0.0312	0.0313	0.0309
8	0.0312	0.0313	0.0313	0.0311	0.0313	0.0312	0.0311	0.0312	0.0310
9	0.0313	0.0313	0.0313	0.0311	0.0314	0.0313	0.0311	0.0312	0.0312

b. Single-Point Crossover/Swap Mutation

	Mutation Rate								
Gener.	0.02	0.03	0.04	0.05	0.06	0.07	0.08	0.09	0.1
1	0.0308	0.0309	0.031	0.0309	0.0309	0.0309	0.0309	0.0308	0.0308
2	0.0308	0.0311	0.031	0.0309	0.0308	0.0309	0.0311	0.0311	0.0310
3	0.0308	0.0312	0.0312	0.0309	0.0308	0.031	0.0311	0.0311	0.0310
4	0.0309	0.0312	0.0312	0.031	0.0310	0.031	0.0312	0.0312	0.0311
5	0.0308	0.0313	0.0313	0.031	0.0310	0.0311	0.0315	0.0312	0.0311

c. Two-Point Crossover/Uniform Mutation

	Mutation Rate								
Gener.	0.02	0.03	0.04	0.05	0.06	0.07	0.08	0.09	0.1
1	0.0308	0.0308	0.0309	0.0308	0.0309	0.0311	0.0311	0.0308	0.0308
2	0.0309	0.0309	0.031	0.0309	0.0309	0.031	0.0313	0.0309	0.0309
3	0.0311	0.0309	0.0311	0.0311	0.0309	0.0311	0.0313	0.0311	0.0309
4	0.0311	0.0308	0.0312	0.0309	0.031	0.031	0.031	0.0311	0.0308
5	0.0311	0.0308	0.0313	0.0312	0.031	0.0312	0.0312	0.0311	0.0308

Table 4. *Cont.*

d. Two-Point Crossover/Swap Mutation									
Gener.	0.02	0.03	0.04	0.05	0.06	0.07	0.08	0.09	0.1
1	0.0308	0.0309	0.0311	0.0308	0.0308	0.031	0.0309	0.0309	0.0309
2	0.0309	0.031	0.0312	0.031	0.0311	0.031	0.031	0.031	0.0308
3	0.0308	0.031	0.0311	0.0312	0.0309	0.031	0.0311	0.031	0.0309
4	0.031	0.0311	0.0312	0.0313	0.0308	0.0309	0.0311	0.0312	0.0310
5	0.0311	0.031	0.0312	0.0313	0.0309	0.031	0.0311	0.0313	0.0309

In the single-point crossover/swap mutation (Table 4b), the highest fitness value is 0.0315, obtained after 5 generations, with a mutation rate of 0.08 (the eighth column of Table 4b). For the two-point crossover/uniform mutation and two-point crossover/swap mutation (Table 4c,d), the best mutation rates are 0.04 and 0.05, respectively, and the corresponding value of the fitness function is 0.0313 (contained in the highlighted columns—the fourth and the fifth, respectively).

After setting the optimal parameters, determined in the previous stages, we finally ran the algorithm to determine the optimum beta parameters. Table 5 summarizes the parameters used to implement the proposed genetic algorithm (columns 2–5), the fitness function obtained after running the algorithm with these parameters (column 6), the execution time (column 7), and the value obtained for the IDW's parameter (last column).

Remark that the values of β are different when using different scenarios, even if the fitness value is the same. This is due to the specifics of the individuals' selection and operations in GAs.

The lowest execution time (0.6188) is obtained when using the single-point/swap scenario and the highest one (10.5875 s) when using a two-point/swap procedure. Even if in the two-point/uniform case, the population size and the number of generations are the smallest, the execution time is high (the second-highest).

Table 5. The control parameters settings for the GA.

Crossover/Mutation	Pop. Size	No. of Gen.	Crossover Rate	Mutation Rate	Fitness	Time (s)	β
Single-Point/Uniform	45	9	0.8	0.06	0.0317	1.2437	1.318
Single-Point/Swap	40	5	0.6	0.08	0.0317	0.6188	1.124
Two-Point/Uniform	30	5	0.75	0.04	0.0317	7.5687	1.064
Two-Point/Swap	35	5	0.7	0.05	0.0317	10.5875	1.042

Table 6 contains the MSE and MAE for each station and the average (the last row of the table) computed after running the algorithm in each scenario. Comparing the MSEs in the two-point/swap and single-point/uniform (single-point/swap, and two-point/uniform) scenario, they are smaller in 70% (70%, 70%) cases, so our algorithm, in two-point/swap scenario, performs better in 70% cases compared to the other three scenarios. The MSEs' averages (31.5874, 31.5306, 31.5188, 31.5153) are comparable, the smallest being obtained in the two-point/swap scenario, followed by the third one.

Comparing the MAEs in the two-point/swap and single-point/uniform (single-point/swap, and two-point/uniform) scenario, they are smaller in 80% (80%, 80%) cases, so our algorithm, in two-point/swap scenario, performs better in 80% cases compared to the other three scenarios. The MAEs' averages (23.6352, 23.5542, 23.5308, 23.5228) are comparable, the smallest being obtained in the two-point/swap scenario, followed by the third one.

The corresponding values computed for beta in the best two cases are 1.042 (in two-point/swap) and 1.064 (in two-point/uniform).

Table 6. MSEs and MAEs in GA. The best results are highlighted.

Station	Single-Point/Uniform		Single-Point/Swap		Two-Point/Uniform		Two-Point/Swap	
	MSE	MAE	MSE	MAE	MSE	MAE	MSE	MAE
Adamclisi	36.2300	27.0536	36.1639	27.047	36.1372	27.0440	36.1264	27.0436
Cernavoda	22.9755	16.4861	22.7221	16.2628	22.6682	16.2290	22.6529	16.2254
Constanta	28.8453	17.0195	28.9273	17.2437	28.9601	17.3086	28.9731	17.3311
Corugea	21.0101	15.8130	21.0132	15.7836	21.0419	15.7770	21.0561	15.7744
Harsova	39.4587	26.2462	39.8004	26.2647	39.9151	26.2710	39.9581	26.2731
Jurilovca	24.5367	20.3591	24.4805	20.2806	24.4658	20.2606	24.4607	20.2534
Mangalia	36.4302	27.1703	36.3815	27.0490	36.3661	27.0040	36.3605	26.9864
Medgidia	26.8786	21.8044	26.5124	21.5704	26.3874	21.4822	26.3396	21.4471
Tulcea	33.5917	26.2378	33.5510	26.1170	33.5348	26.0791	33.5283	26.0667
Sulina	45.9171	38.1624	45.7538	37.9231	45.7115	37.8524	45.6972	37.8272
Average	31.5874	23.6352	31.5306	23.5542	31.5188	23.5308	**31.5153**	**23.5228**

For comparison reasons, we performed the classical IDW, with β = 2 (the value used in most applications) and β = 1.042. The MAE and MSE values computed for each station are presented in Table 7.

Comparing the MSEs in the two-point/swap algorithm (Table 6, column 8) with those from the IDW with β = 2 (Table 7 column 2), they are smaller in 70% cases (the first four stations, the sixth, eighth, and ninth), so our algorithm performs better in 70% cases. Comparing the MSEs in the two-point/swap algorithm with those from the IDW with β = 1.042 (Table 7 column 4), they are smaller in 60% cases (the second, third, fourth, sixth, eighth, and ninth stations), so our algorithm better performs in 60% cases.

In terms of the average MSEs, that in the two-point/swap approach is smaller than those of the IDW (β = 2), IDW (β = 1.042), and slightly higher than in KG (Table 7, the last column). Still, our approach is preferable against KG since it is difficult to determine the kriging parameters, requiring special knowledge of geostatistics.

Table 7. MSE and MAE in the classical IDW for β = 2 and β = 1.042. MSE in ordinary kriging (KG).

Station	IDW (β = 2)		IDW (β = 1.042)		KG *	OIDW **	OIDW *
	MSE	MAE	MSE	MAE	MSE	MSE	MSE
Adamclisi	36.32	27.03	35.73	26.96	32.73	32.4184	28.5774
Cernavoda	23.78	17.33	28.03	15.88	23.40	22.9189	22.6820
Constanta	37.29	27.80	35.83	27.09	30.22	30.3249	30.1713
Corugea	35.67	27.69	34.34	26.52	22.48	22.4062	22.3283
Hârșova	37.95	28.06	35.51	27.16	35.03	35.6281	35.2566
Jurilovca	34.89	27.89	34.35	27.31	23.04	24.1018	23.9813
Mangalia	35.65	26.86	35.03	26.81	42.73	42.0429	41.9517
Medgidia	35.89	27.41	34.89	26.82	22.58	22.3809	22.3073
Tulcea	38.55	29.84	36.15	28.19	43.54	44.0204	43.4118
Sulina	36.99	29.04	35.62	28.03	34.47	33.0844	33.0501
Average	35.30	26.90	34.55	26.08	31.02	30.93	30.37

* Results from [18], Table 2; ** Results from [18], Table 1.

The MAEs in the two-point/swap algorithm are smaller than those from the IDW with β = 2, in 80% cases (all, but the first and sixth station), and comparable for the first station.

The MAEs in the two-point/swap algorithm are smaller than those from the IDW (β = 1.042), they in 60% cases (all, but the first and sixth station), and comparable for the first station.

The average MAE in the two-point/swap approach (23.5228) is significantly smaller than those in the IDW with β = 2 (26.90), and IDW with β = 1.042 (26.08).

From the computational viewpoint, the highest computational time in our experiment was 10.5875 (in the two-point scenario), while in the grid search to estimate beta with 3 decimals takes 60 seconds for each series, so, a total of 60*10 stations = 600 seconds, which is 56.67 times higher than in our approach.

The last two columns of in Table 7 contains the MSE in the optimized IDW, denoted by OIDW [18], in two scenarios, as described in [18]—with different beta, found using a Particle Swarm Optimization (PSO) approach (column 7), or with a single best beta found by the same approach.

In term of MSE, our GA algorithm (Table 7, column 8) performs better that OIDW (Table 7, columns 7 and 8) in 80% of cases. Therefore, we can say that a significant improvement of the interpolation performances are obtained, that may reflect in the water management policy.

4. Conclusions

In this article, we presented a new approach to finding the beta parameter in IDW, using a GA implemented in four scenarios. The settings of this GA were optimized for finding the best fitness function and, by consequence, the best parameter beta, for all the study sites, not only for some of them.

It is shown that the algorithm proposed here performs better (in all scenarios) than the classical one (with $\beta = 2$ and $\beta = 1.042$) in terms of average MSE and MAE. When compared the MSEs and MAEs for the individual stations, the following results are obtained:

- In IDW with $\beta = 2$, MSE is smaller only for Hârșova, Mangalia, and Sulina, compared to the GA with a two-point swap.
- In IDW with $\beta = 2$, MAE is smaller only for Adamclisi and Mangalia, compared to the GA with a two-point swap.
- In IDW with $\beta = 1.042$, MSE is smaller than in GA (with two-point/swap) only for Adamclisi, Mangalia, and Sulina.
- In IDW with $\beta = 1.042$, MAE is smaller than in GA with a two-point swap only for Adamclisi, Cernavoda, and Mangalia.

The algorithm performs faster than the classical IDW, for which the running time on the same problem is 60s for each interpolated data series (so 600s for all ten series). It is easy to be implemented and used and can be applied to similar problems only by changing the input data.

Compared with other artificial intelligence methods used for finding beta (OIDW) our approach shows superior performances in 80% of cases.

Another advantage is that our algorithm provides a single beta for all the stations, optimizing the interpolation.

The results obtained in all four GA's scenarios are comparable. Since the execution time is the highest in the best scenario (Table 5), the other alternatives can be successfully used for the spatial interpolation when the number of series or the number of records per station is very high.

Author Contributions: Conceptualization, A.B. and C.Ș.; methodology, A.B. and C.Ș.; software, C. Ș. and M.-L.I.; validation, A.B. and C.Ș.; writing—review and editing, A.B., C.Ș. and M.-L.I., visualization, C.Ș. and M.-L.I.; supervision, A.B. All authors have read and agreed to the published version of the manuscript.

Funding: This research received no external funding.

Conflicts of Interest: The authors declare no conflict of interest.

References

1. Bărbulescu, A.; Popescu-Bodorin, N. History-based long-term predictability of regional monthly fuzzy data. *Stoch. Environ. Res. Risk A* **2019**, *33*, 1435–1441. [CrossRef]
2. Li, J.; Heap, A.D. Spatial interpolation methods applied in the environmental sciences: A review. *Environ. Modell. Softw.* **2014**, *53*, 173–189. [CrossRef]
3. Li, J.; Heap, A.D.; Potter, A.; Daniell, J.J. Application of machine learning methods to spatial interpolation of environmental variable. *Environ. Modell. Softw.* **2011**, *26*, 1647–1659. [CrossRef]

4. Lu, G.Y.; Wong, D.W. An adaptive inverse-distance weighting spatial interpolation technique. *Comput. Geosci.* **2008**, *34*, 1044–1055. [CrossRef]
5. Maleika, W. Inverse distance weighting method optimization in the process of digital terrain model creation based on data collected from a multibeam echosounder. *Appl. Geomat.* **2020**, *12*, 397–407. [CrossRef]
6. Hsieh, H.H.; Cheng, S.J.; Liou, J.Y.; Chou, S.C.; Siao, B.R. Characterization of spatially distributed summer daily rainfall. *J. Chin. Agr. Eng.* **2006**, *52*, 47–55.
7. Dirks, K.N.; Hay, J.E.; Stow, C.D.; Harris, D. High resolution studies of rainfall on Norfolk island part II: Interpolation of rainfall data. *J. Hydrol.* **1998**, *208*, 187–193. [CrossRef]
8. Ly, S.; Charles, C.; Degré, A. Geostatistical interpolation of daily rainfall at catchment scale: The use of several variogram models in the Ourthe and Ambleve catchments, Belgium. *Hydrol. Earth Sys. Sci.* **2011**, *15*, 2259–2274. [CrossRef]
9. Dong, X.H.; Bo, H.J.; Deng, X.; Su, J.; Wang, X. Rainfall spatial interpolation methods and their applications to Qingjiang river basin. *J. China Three Gorges Univ.* **2009**, *31*, 6–10.
10. Bărbulescu, A. A new method for estimation the regional precipitation. *Water Resour. Manag.* **2016**, *30*, 33–42. [CrossRef]
11. Chen, T.; Ren, L.; Yuan, F.; Yang, X.; Jiang, S.; Tang, T.; Zhang, L. Comparison of Spatial Interpolation Schemes for Rainfall Data and Application in Hydrological Modeling. *Water* **2017**, *9*, 342. [CrossRef]
12. Golkhatmi, N.S.; Sanaeinejad, S.H.; Ghahraman, B.; Pazhand, H.R. Extended modified inverse distance method for interpolation rainfall. *Int. J. Eng. Invent.* **2012**, *1*, 57–65.
13. Noori, M.J.; Hassan, H.; Mustafa, Y.T. Spatial estimation of rainfall distribution and its classification in Duhok governorate using GIS. *J. Water. Resour. Protect.* **2014**, *6*, 75–82. [CrossRef]
14. Mei, G.; Xu, L.; Xu, N. Accelerating Adaptive IDW Interpolation Algorithm on a Single GPU. *R. Soc. Open Sci.* **2017**, *4*, 170436. [CrossRef] [PubMed]
15. Gholipour, Y.; Shahbazi, M.M.; Behnia, A.R.A.S.H. An improved version of Inverse Distance Weighting metamodel assisted Harmony Search algorithm for truss design optimization. *Latin Am. J. Solids Struct.* **2013**, *10*, 283–300. [CrossRef]
16. Zhang, X.; Liu, G.; Wang, H.; Li, X. Application of a hybrid interpolation method based on support vector machine in the precipitation spatial interpolation of basins. *Water* **2017**, *9*, 760. [CrossRef]
17. Nourani, V.; Behfar, N.; Uzelaltinbulat, S.; Sadikoglu, F. Spatiotemporal precipitation modeling by artificial intelligence-based ensemble approach. *Environ. Earth Sci.* **2020**, *79*, 6. [CrossRef]
18. Bărbulescu, A.; Băutu, A.; Băutu, E. Particle Swarm Optimization for the Inverse Distance Weighting Distance method. *Appl. Sci.* **2020**, *10*, 2054. [CrossRef]
19. Chang, C.L.; Lo, S.L.; Yu, S.-L. The parameter optimization in the inverse distance method by genetic algorithm for estimating precipitation. *Environ. Monit. Assess.* **2006**, *117*, 145–155. [CrossRef]
20. Ratnam, J.V.; Dijkstra, H.A.; Doi, T.; Morioka, Y.; Nonaka, M.; Behera, S.K. Improving seasonal forecasts of air temperature using a genetic algorithm. *Sci. Rep.* **2019**, *9*, 12781. [CrossRef]
21. Nasseri, M.; Asghari, K.; Abedini, M.J. Optimized scenario for rainfall forecasting using genetic algorithm coupled with artificial neural network. *Expert Syst. Appl.* **2008**, *35*, 1415–1421. [CrossRef]
22. Sen, Z.; Ôztopal, A. Genetic algorithms for the classification and prediction of precipitation occurrence. *Hydrol. Sci. J.* **2001**, *46*, 255–267. [CrossRef]
23. Kadri, M.B.; Khan, W.A. Application of Genetic Algorithms in Nonlinear Heat Conduction Problems. *Sci. World J.* **2014**, *2014*, 451274. [CrossRef]
24. Reddy, C.S.; Balaji, K. A Genetic Algorithm (GA)-PID Controller for Temperature Control in Shell and Tube Heat Exchanger. *IOP Conf. Ser. Mater. Sci. Eng.* **2020**, *925*, 012020. [CrossRef]
25. Shepard, D. A Two-Dimensional Interpolation for Irregularly Spaced Data Function. In Proceedings of the 1968 ACM National Conference, New York, NY, USA, 27–29 August 1968; Blue, R.B., Rosenberg, A.M., Eds.; Association for Computing Machinery: New York, NY, USA, 1968; pp. 517–523.
26. Lafitte, P. *Traité d'Informatique Geologique*; Masson: Paris, France, 1972.
27. Sivanandam, S.N.; Deepa, S.N. *Introduction to Genetic Algorithms*; Springer: Berlin/Heidelberg, Germany, 2008.
28. Haupt, R.L.; Haupt, S.E. *Practical Genetic Algorithms*; John Wiley & Sons Inc.: New York, NY, USA, 1998.
29. Umbarkar, A.J.; Sheth, P.D. Crossover Operators in Genetic Algorithms: A Review. *ICTACT J. Soft. Comp.* **2015**, *6*, 1083–1092.
30. Oliveto, P.S.; Witt, C. Improved time complexity analysis of the simple genetic algorithm. *Theor. Comput. Sci.* **2015**, *605*, 21–41. [CrossRef]
31. Bărbulescu, A. Models for temperature evolution in Constanța area (Romania). *Rom. J. Phys.* **2016**, *61*, 676–686.
32. Bărbulescu, A.; Deguenon, J. About the variations of precipitation and temperature evolution in the Romanian Black Sea Littoral. *Rom. Rep. Phys.* **2015**, *67*, 625–637.
33. Bărbulescu, A.; Maftei, C.; Dumitriu, C.S. The modelling of the climateric process that participates at the sizing of an irrigation system. *Bull. Appl. Comput. Math.* **2002**, *2048*, 11–20.

MDPI
St. Alban-Anlage 66
4052 Basel
Switzerland
Tel. +41 61 683 77 34
Fax +41 61 302 89 18
www.mdpi.com

Water Editorial Office
E-mail: water@mdpi.com
www.mdpi.com/journal/water

www.ingramcontent.com/pod-product-compliance
Lightning Source LLC
Chambersburg PA
CBHW040805250926
51850CB00007BA/1089
9783036563596